STUDENT'S SOLUTIONS MANUAL FOR

Introductory Algebra

FIFTH EDITION

Keedy/Bittinger

STUDENT'S SOLUTIONS MANUAL FOR

Introductory Algebra

FIFTH EDITION

Keedy/Bittinger

Judith A. Beecher
Indiana University-
Purdue University at Indianapolis

ADDISON-WESLEY PUBLISHING COMPANY
Reading, Massachusetts • Menlo Park, California • Don Mills, Ontario
Wokingham, England • Amsterdam • Sydney • Singapore • Tokyo
Madrid • Bogotá • Santiago • San Juan

Reproduced by Addison-Wesley from camera-ready copy supplied by the author.

ISBN 0-201-15275-4
CDEFGHIJ-AL-89

TABLE OF CONTENTS

Special thanks are extended to Julie Stephenson for her excellent typing. The combination of her mathematical knowledge and her exceptional typing skills made the author's work much easier. Her dedication to perfection is greatly appreciated.

Exercise Set 1.1

1. We substitute 14 for x.

$x - 4 = 14 - 4 = 10$

Chris is 10 when Lowell is 14.

We substitute 29 for x.

$x - 4 = 29 - 4 = 25$

Chris is 25 when Lowell is 29.

We substitute 52 for x.

$x - 4 = 52 - 4 = 48$

Chris is 48 when Lowell is 52.

3. We substitute 16 for b and 9 for h.

$A = \frac{1}{2}\cdot b\cdot h$

$A = \frac{1}{2}\cdot 16\cdot 9 = 72$

The area of the triangle is 72 sq. yd.

5. We substitute 8 for y.

$7y = 7\cdot 8 = 56$

7. We substitute 9 for x and 3 for y.

$\frac{x}{y} = \frac{9}{3} = 3$

9. We substitute 20 for m and 4 for n.

$\frac{m - n}{8} = \frac{20 - 4}{8} = \frac{16}{8} = 2$

11. We substitute 8 for z and 2 for y.

$\frac{5z}{y} = \frac{5\cdot 8}{2} = \frac{40}{2} = 20$

13. 6 more than m

$6 + m$, or $m + 6$

15. 9 less than c

$c - 9$

17. 6 greater than q

$6 + q$, or $q + 6$

19. b more than a

$b + a$, or $a + b$

21. x less than y

$y - x$

23. 98% of x

$98\%\cdot x$, or $98\%x$

25. The sum of r and s

$r + s$, or $s + r$

27. Twice x

$2x$

29. 5 times t

$5t$

31. The difference of 3 and b

This algebraic expression has two interpretations.

$3 - b$ or $b - 3$

33. Smaller number y, Larger number $y + 6$

35. Larger number x, Smaller number $x - 4$

37. A number x plus three times y

$x + 3y$

39. The perimeter is $\ell + \ell + w + w$, or $2\ell + 2w$.

41. Substitute 1 for y and 4 for x.

$\frac{256y}{32x} = \frac{256\cdot 1}{32\cdot 4} = \frac{256}{128} = 2$

Exercise Set 1.2

1. $1\cdot 21$, $3\cdot 7$

3. $1\cdot 144$, $2\cdot 72$, $3\cdot 48$, $4\cdot 36$, $6\cdot 24$, $8\cdot 18$, $9\cdot 16$, $12\cdot 12$, $2\cdot 2\cdot 3\cdot 3\cdot 4$ There are many others.

5. $14 = 2\cdot 7$

7. $33 = 3\cdot 11$

9. $9 = 3\cdot 3$

11. $49 = 7\cdot 7$

13. We begin by factoring 18 any way we can and continue factoring until each factor is prime.

$18 = 2\cdot 9 = 2\cdot 3\cdot 3$ or $18 = 3\cdot 6 = 3\cdot 2\cdot 3$

15. We begin by factoring 40 any way we can and continue factoring until each factor is prime.

$40 = 4\cdot 10 = 2\cdot 2\cdot 2\cdot 5$ or
$40 = 5\cdot 8 = 5\cdot 2\cdot 4 = 5\cdot 2\cdot 2\cdot 2$ or
$40 = 2\cdot 20 = 2\cdot 2\cdot 10 = 2\cdot 2\cdot 2\cdot 5$

17. $90 = 9\cdot 10 = 3\cdot 3\cdot 2\cdot 5$ or
$90 = 3\cdot 30 = 3\cdot 3\cdot 10 = 3\cdot 3\cdot 2\cdot 5$ or
$90 = 2\cdot 45 = 2\cdot 5\cdot 9 = 2\cdot 5\cdot 3\cdot 3$

19. $210 = 10\cdot 21 = 2\cdot 5\cdot 3\cdot 7$, or $2\cdot 3\cdot 5\cdot 7$

21. We cannot divide 91 by 2, 3, or 5 without remainders, so we try dividing by the next prime, 7. $91 = 7\cdot 13$

23. $119 = 7\cdot 17$

25. The multiples of 3 are:

3, 6, 9, 12, 15, 18, 21, 24, 27, 30, 33, 36, 39, 42, 45, 48, 51, 54, 57, 60, 63, ...

The multiples of 7 are:

7, 14, 21, 28, 35, 42, 49, 56, 63, ...

The first three common multiples are 21, 42, and 63.

27. The multiples of 12 are:

12, 24, 36, 48, 60, 72, 84, 96, 108, 120, 132, 144, ...

The multiples of 16 are:

16, 32, 48, 64, 80, 96, 112, 128, 144, ...

The first three common multiples are 48, 96, and 144.

29. $12 = 2\cdot2\cdot3$
$18 = 2\cdot3\cdot3$
$\text{LCM} = 2\cdot2\cdot3\cdot3$, or 36

31. $45 = 3\cdot3\cdot5$
$72 = 2\cdot2\cdot2\cdot3\cdot3$
$\text{LCM} = 2\cdot2\cdot2\cdot3\cdot3\cdot5$, or 360

33. $30 = 2\cdot3\cdot5$
$50 = 2\cdot5\cdot5$
$\text{LCM} = 2\cdot3\cdot5\cdot5$, or 150

35. $30 = 2\cdot3\cdot5$
$40 = 2\cdot2\cdot2\cdot5$
$\text{LCM} = 2\cdot2\cdot2\cdot3\cdot5$, or 120

37. $18 = 2\cdot3\cdot3$
$24 = 2\cdot2\cdot2\cdot3$
$\text{LCM} = 2\cdot2\cdot2\cdot3\cdot3$, or 72

39. $35 = 5\cdot7$
$45 = 3\cdot3\cdot5$
$\text{LCM} = 3\cdot3\cdot5\cdot7$, or 315

41. 2, 3, and 5 are all prime. If two or more numbers have no common prime factor, the LCM is the product of the numbers.

$\text{LCM} = 2\cdot3\cdot5$, or 30

43. $24 = 2\cdot2\cdot2\cdot3$
$36 = 2\cdot2\cdot3\cdot3$
$12 = 2\cdot2\cdot3$
$\text{LCM} = 2\cdot2\cdot2\cdot3\cdot3$, or 72

45. 5 is prime
$12 = 2\cdot2\cdot3$
$15 = 3\cdot5$
$\text{LCM} = 2\cdot2\cdot3\cdot5$, or 60

47. $6 = 2\cdot3$
$12 = 2\cdot2\cdot3$
$18 = 2\cdot3\cdot3$
$\text{LCM} = 2\cdot2\cdot3\cdot3$, or 36

49. Substitute 7 for x and 42 for y.

$xy = 7\cdot42 = 294$

51. Jupiter: $12 = 2\cdot2\cdot3$
Saturn: $30 = 2\cdot3\cdot5$

The LCM of 12 and 30 is $2\cdot2\cdot3\cdot5$, or 60. Jupiter and Saturn will appear in the same direction every 60 years.

53. Jupiter: $12 = 2\cdot2\cdot3$
Saturn: $30 = 2\cdot3\cdot5$
Uranus: $84 = 2\cdot2\cdot3\cdot7$

The LCM of 12, 30, and 84 is $2\cdot2\cdot3\cdot5\cdot7$, 420. Jupiter, Saturn, and Uranus will appear in the same direction every 420 years.

Exercise Set 1.3

1.-7. Answers may vary.

1. $\frac{4}{3} = \frac{4}{3}\cdot1 = \frac{4}{3}\cdot\frac{5}{5} = \frac{20}{15}$ $\quad$ $\frac{4}{3} = \frac{4}{3}\cdot1 = \frac{4}{3}\cdot\frac{20}{20} = \frac{80}{60}$

$\frac{4}{3} = \frac{4}{3}\cdot1 = \frac{4}{3}\cdot\frac{12}{12} = \frac{48}{36}$ $\quad$ $\frac{4}{3} = \frac{4}{3}\cdot1 = \frac{4}{3}\cdot\frac{25}{25} = \frac{100}{75}$

3. $\frac{6}{11} = \frac{6}{11}\cdot1 = \frac{6}{11}\cdot\frac{3}{3} = \frac{18}{33}$ $\quad$ $\frac{6}{11} = \frac{6}{11}\cdot1 = \frac{6}{11}\cdot\frac{10}{10} = \frac{60}{110}$

$\frac{6}{11} = \frac{6}{11}\cdot1 = \frac{6}{11}\cdot\frac{7}{7} = \frac{42}{77}$ $\quad$ $\frac{6}{11} = \frac{6}{11}\cdot1 = \frac{6}{11}\cdot\frac{18}{18} = \frac{108}{198}$

5. $\frac{2}{11} = \frac{2}{11}\cdot1 = \frac{2}{11}\cdot\frac{4}{4} = \frac{8}{44}$

$\frac{2}{11} = \frac{2}{11}\cdot1 = \frac{2}{11}\cdot\frac{11}{11} = \frac{22}{121}$

$\frac{2}{11} = \frac{2}{11}\cdot1 = \frac{2}{11}\cdot\frac{40}{40} = \frac{80}{440}$

$\frac{2}{11} = \frac{2}{11}\cdot1 = \frac{2}{11}\cdot\frac{1000}{1000} = \frac{2000}{11{,}000}$

7. $5 = 5\cdot1 = \frac{5}{1}\cdot\frac{2}{2} = \frac{10}{2}$

$5 = 5\cdot1 = \frac{5}{1}\cdot\frac{6}{6} = \frac{30}{6}$

$5 = 5\cdot1 = \frac{5}{1}\cdot\frac{13}{13} = \frac{65}{13}$

$5 = 5\cdot1 = \frac{5}{1}\cdot\frac{30}{30} = \frac{150}{30}$

9. $\frac{8}{6} = \frac{4\cdot2}{3\cdot2} = \frac{4}{3}\cdot\frac{2}{2} = \frac{4}{3}$

11. $\frac{17}{34} = \frac{1\cdot17}{2\cdot17} = \frac{1}{2}\cdot\frac{17}{17} = \frac{1}{2}$

13. $\frac{100}{50} = \frac{50\cdot2}{50\cdot1} = \frac{50}{50}\cdot\frac{2}{1} = 2$

15. $\frac{250}{75} = \frac{10\cdot25}{3\cdot25} = \frac{10}{3}\cdot\frac{25}{25} = \frac{10}{3}$

17. $\frac{1}{4}\cdot\frac{1}{2} = \frac{1\cdot1}{4\cdot2} = \frac{1}{8}$

19. $\frac{17}{2}\cdot\frac{3}{4} = \frac{17\cdot3}{2\cdot4} = \frac{51}{8}$

21. $\frac{1}{2} + \frac{1}{2} = \frac{1+1}{2} = \frac{2}{2} = 1$

23. The LCM of the denominators is 18.

$\frac{4}{9} + \frac{13}{18} = \frac{4}{9}\cdot\frac{2}{2} + \frac{13}{18} = \frac{8}{18} + \frac{13}{18} = \frac{21}{18} = \frac{3\cdot7}{3\cdot6} = \frac{3}{3}\cdot\frac{7}{6} = \frac{7}{6}$

25. The LCM of the denominators is 30.

$\frac{3}{10} + \frac{8}{15} = \frac{3}{10}\cdot\frac{3}{3} + \frac{8}{15}\cdot\frac{2}{2} = \frac{9}{30} + \frac{16}{30} = \frac{25}{30} = \frac{5\cdot5}{5\cdot6} = \frac{5}{6}$

27. $\frac{5}{4} - \frac{3}{4} = \frac{5-3}{4} = \frac{2}{4} = \frac{1\cdot2}{2\cdot2} = \frac{1}{2}\cdot\frac{2}{2} = \frac{1}{2}$

29. The LCM of the denominators is 18.

$\frac{13}{18} - \frac{4}{9} = \frac{13}{18} - \frac{4}{9}\cdot\frac{2}{2} = \frac{13}{18} - \frac{8}{18} = \frac{5}{18}$

31. The LCM of the denominators is 60.

$\frac{11}{12} - \frac{2}{5} = \frac{11}{12}\cdot\frac{5}{5} - \frac{2}{5}\cdot\frac{12}{12} = \frac{55}{60} - \frac{24}{60} = \frac{31}{60}$

33. The sum of p and q

$p + q$, or $q + p$

35. $48 = 2\cdot24$
$= 2\cdot2\cdot12$
$= 2\cdot2\cdot2\cdot6$
$= 2\cdot2\cdot2\cdot2\cdot3$

37. $\frac{3x}{4x} = \frac{3}{4}\cdot\frac{x}{x} = \frac{3}{4}$

39. $\frac{4\cdot9\cdot16}{2\cdot8\cdot15} = \frac{2\cdot2\cdot3\cdot3\cdot2\cdot2\cdot2\cdot2}{2\cdot2\cdot2\cdot2\cdot3\cdot5} = \frac{2\cdot2\cdot2\cdot2\cdot3}{2\cdot2\cdot2\cdot2\cdot3}\cdot\frac{3\cdot2\cdot2}{5} = \frac{12}{5}$

Exercise Set 1.4

1. $29.1 = \frac{29.1}{1} \cdot \frac{10}{10} = \frac{291}{10}$

3. $4.67 = \frac{4.67}{1} \cdot \frac{100}{100} = \frac{467}{100}$

5. $3.62 = \frac{3.62}{1} \cdot \frac{100}{100} = \frac{362}{100}$

7. $18.789 = \frac{18.789}{1} \cdot \frac{1000}{1000} = \frac{18,789}{1000}$

9.
```
     0.5
  2 ⟌1.0
     1 0
       0
```
$\frac{1}{2} = 0.5$

11.
```
     0.6
  5 ⟌3.0
     3 0
       0
```
$\frac{3}{5} = 0.6$

13.
```
     0.222...
  9 ⟌2.000
     1 8
       20
       18
        20
        18
         2
```
$\frac{2}{9} = 0.222\ldots,\ 0.\overline{2}$

15.
```
     0.125
  8 ⟌1.000
       8
       20
       16
        40
        40
         0
```
$\frac{1}{8} = 0.125$

17.
```
      0.4545...
  11 ⟌5.0000
      4 4
        60
        55
         50
         44
          60
          55
           5
```
$\frac{5}{11} = 0.454545\ldots$, or $0.\overline{45}$

19.
```
      0.08333...
  12 ⟌1.00000
        96
         40
         36
          40
          36
           40
           36
            4
```
$\frac{1}{12} = 0.08333\ldots$, or $0.08\overline{3}$

21. 5^2 means $5 \cdot 5$

23. m^3 means $m \cdot m \cdot m$

25. x^3 means that x is used as a factor three times, xxx

27. y^4 means that y is used as factor four times, yyyy

29. $y^0 = 1$

31. $p^1 = p$

33. $M^1 = M$

35. $(\frac{a}{b})^0 = 1$

37. Substitute 3 for m.
$m^3 = 3^3 = 3 \cdot 3 \cdot 3 = 27$

39. Substitute 19 for p.
$p^1 = 19^1 = 19$

41. Substitute 4 for x.
$x^4 = 4^4 = 4 \cdot 4 \cdot 4 \cdot 4 = 256$

43. Substitute 5 for n.
$n^0 = 5^0 = 1$

45. Substitute 2 for z.
$z^5 = 2^5 = 2 \cdot 2 \cdot 2 \cdot 2 \cdot 2 = 32$

47. Substitute 24 for s.
$A = s^2 = (24m)^2 = 24m \cdot 24m = 576\ m^2$

49. Substitute 2 for y.
$(2y)^4 = (2 \cdot 2)^4 = 4^4 = 4 \cdot 4 \cdot 4 \cdot 4 = 256$
$2y^4 = 2 \cdot 2^4 = 2 \cdot 16 = 32$

51. x minus y
$x - y$

53. $18 = \underline{2 \cdot 3 \cdot 3}$
$27 = \underline{3 \cdot 3 \cdot 3}$
$54 = \underline{2 \cdot 3 \cdot 3 \cdot 3}$
LCM $= 2 \cdot 3 \cdot 3 \cdot 3$, or 54

55. Substitute 2 for x, 1 for y, and 3 for z
$x^3y^2z = 2^3 \cdot 1^2 \cdot 3 = 8 \cdot 1 \cdot 3 = 24$

57. $xxxyyyy = x^3y^4$ There are three factors of x and four factors of y.

Exercise Set 1.5

1. $(10 + 4) + 8 = 14 + 8 = 22$

3. $(10 \cdot 7) + 19 = 70 + 19 = 89$

5. $8 + 4 + 5 + 2 + 6 + 15 + 1$
$= (8 + 2) + (4 + 6) + (5 + 15) + 1$
$= 10 + 10 + 20 + 1$
$= 41$

7. $14 + 3 + 12 + 7 + 8 + 6 + 9$
$= (14 + 6) + (3 + 7) + (12 + 8) + 9$
$= 20 + 10 + 20 + 9$
$= 59$

9. 67 + 3 = 3 + 67 illustrates the commutative law of addition. The order has been changed.

11. 6 + (9 + 5) = (6 + 9) + 5 illustrates the associative law of addition. The grouping has been changed.

13. $(6 + 7)\cdot 4 = 13\cdot 4 = 52$

$(6 + 7)\cdot 4 = 6\cdot 4 + 7\cdot 4 = 24 + 28 = 52$

15. $\underline{9}x + \underline{9}y = \underline{9}(x + y)$

17. $\underline{\frac{1}{2}} a + \underline{\frac{1}{2}} b = \underline{\frac{1}{2}}(a + b)$

19. $\underline{1.5}x + \underline{1.5}z = \underline{1.5}(x + z)$

21. $\underline{4}x + \underline{4}y + \underline{4}z = \underline{4}(x + y + z)$

23. $\underline{\frac{4}{7}} a + \underline{\frac{4}{7}} b + \underline{\frac{4}{7}} c + \underline{\frac{4}{7}} d = \underline{\frac{4}{7}}(a + b + c + d)$

25. $9x + 9y = 9(x + y)$

Substitute 5 for x and 10 for y.

$9x + 9y = 9\cdot 5 + 9\cdot 10 = 45 + 90 = 135$

$9(x + y) = 9(5 + 10) = 9\cdot 15 = 135$

27. $10x + 10y = 10(x + y)$

Substitute 5 for x and 10 for y.

$10x + 10y = 10\cdot 5 + 10\cdot 10 = 50 + 100 = 150$

$10(x + y) = 10(5 + 10) = 10\cdot 15 = 150$

29. $5a + 5b = 5(a + b)$

Substitute 0 for a and 9 for b.

$5a + 5b = 5\cdot 0 + 5\cdot 9 = 0 + 45 = 45$

$5(a + b) = 5(0 + 9) = 5\cdot 9 = 45$

31. $20a + 20b = 20(a + b)$

Substitute 0 for a and 9 for b.

$20a + 20b = 20\cdot 0 + 20\cdot 9 = 0 + 180 = 180$

$20(a + b) = 20(0 + 9) = 20\cdot 9 = 180$

33. $P = 2\ell + 2w = 2(\ell + w)$

We substitute 40 for ℓ and 20 for w.

$P = 2\ell + 2w = 2\cdot 40 + 2\cdot 20 = 80 + 40 = 120$ ft

$P = 2(\ell + w) = 2(40 + 20) = 2\cdot 60 = 120$ ft

35. Substitute 2 for x.

$2x^3 = 2\cdot 2^3 = 2\cdot 8 = 16$

$(2x)^3 = (2\cdot 2)^3 = 4^3 = 64$

37. $96 = 2\cdot 48$

$= 2\cdot 2\cdot 24$

$= 2\cdot 2\cdot 2\cdot 12$

$= 2\cdot 2\cdot 2\cdot 2\cdot 6$

$= 2\cdot 2\cdot 2\cdot 2\cdot 2\cdot 3$

39. a) $3\cdot 4 + 2 = 12 + 2 = 14$

b) $3(4 + 2) = 3\cdot 6 = 18$

c) $3\cdot 4 + 3\cdot 2 = 12 + 6 = 18$

d) $3(2 + 4) = 3\cdot 6 = 18$

Expressions b), c), and d) represent the same number.

41. $5(x + y) = 5x + 5y = 5y + 5x$

Thus, a), b), and c) are equivalent.

Exercise set 1.6

1. $3(x + 1) = 3\cdot x + 3\cdot 1 = 3x + 3$

3. $4(1 + y) = 4\cdot 1 + 4\cdot y = 4 + 4y$

5. $9(4t + 3z) = 9\cdot 4t + 9\cdot 3z = 36t + 27z$

7. $7(x + 4 + 6y) = 7\cdot x + 7\cdot 4 + 7\cdot 6y$

$= 7x + 28 + 42y$

9. $5(3x + 9 + 7y) = 5\cdot 3x + 5\cdot 9 + 5\cdot 7y$

$= 15x + 45 + 35y$

11. $2x + 4 = 2\cdot x + 2\cdot 2 = 2(x + 2)$

13. $6x + 24 = 6\cdot x + 6\cdot 4 = 6(x + 4)$

15. $9x + 3y = 3\cdot 3x + 3\cdot y = 3(3x + y)$

17. $14x + 21y = 7\cdot 2x + 7\cdot 3y = 7(2x + 3y)$

19. $5 + 10x + 15y = 5\cdot 1 + 5\cdot 2x + 5\cdot 3y$

$= 5(1 + 2x + 3y)$

21. $8a + 16b + 64 = 8\cdot a + 8\cdot 2b + 8\cdot 8$

$= 8(a + 2b + 8)$

23. $3x + 18y + 15z = 3\cdot x + 3\cdot 6y + 3\cdot 5z$

$= 3(x + 6y + 5z)$

25. $2x + 3 + 3x + 9 = 2x + 3x + 3 + 9$

$= (2 + 3)x + (3 + 9)$

$= 5x + 12$

27. $10a + a = 10a + 1a$

$= (10 + 1)a$

$= 11a$

29. $2x + 9z + 6x = 2x + 6x + 9z$

$= (2 + 6)x + 9z$

$= 8x + 9z$

31. $41a + 90c + 60c + 2a = 41a + 2a + 90c + 60c$
$= (41 + 2)a + (90 + 60)c$
$= 43a + 150c$

33. $x + 0.09x + 0.2t + t = 1x + 0.09x + 0.2t + 1t$
$= (1 + 0.09)x + (0.2 + 1)t$
$= 1.09x + 1.2t$

35. $8u + 3t + 10u + 6u + 2t$
$= (8 + 10 + 6)u + (3 + 2)t$
$= 24u + 5t$

37. $23 + 5t + 7y + t + y + 27$
$= 23 + 5t + 7y + 1t + 1y + 27$
$= (5 + 1)t + (7 + 1)y + (23 + 27)$
$= 6t + 8y + 50$

39. $\frac{1}{2}b + \frac{2}{3} + \frac{1}{2}b + \frac{2}{3}$
$= (\frac{1}{2} + \frac{1}{2})b + (\frac{2}{3} + \frac{2}{3})$
$= b + \frac{4}{3}$

41. $2y + \frac{1}{4}y + y$
$= 2y + \frac{1}{4}y + 1y$
$= (2 + \frac{1}{4} + 1)y$
$= \frac{13}{4}y \qquad (3\frac{1}{4} = \frac{13}{4})$

43. $\frac{15}{16} \cdot \frac{8}{9} = \frac{120}{144} = \frac{12 \cdot 10}{12 \cdot 12} = \frac{10}{12} = \frac{2 \cdot 5}{2 \cdot 6} = \frac{5}{6}$

45. $3x + 10y + 2x + 5y + 25$
$= (3 + 2)x + (10 + 5)y + 25$
$= 5x + 15y + 25$ (Collecting like terms)
$= 5 \cdot x + 5 \cdot 3y + 5 \cdot 5$
$= 5(x + 3y + 5)$ (Factoring)

Exercise Set 1.7

1. The reciprocal of $\frac{3}{4}$ is $\frac{4}{3}$ because $\frac{3}{4} \cdot \frac{4}{3} = \frac{12}{12} = 1$.

3. The reciprocal of $\frac{1}{8}$ is 8 because $\frac{1}{8} \cdot 8 = \frac{8}{8} = 1$.

5. The reciprocal of 1 is 1 because $1 \cdot 1 = 1$.

7. $\frac{7}{6} \div \frac{3}{5} = \frac{7}{6} \cdot \frac{5}{3} = \frac{35}{18}$

9. $\frac{8}{9} \div \frac{4}{15} = \frac{8}{9} \cdot \frac{15}{4} = \frac{120}{36} = \frac{12 \cdot 10}{12 \cdot 3} = \frac{10}{3}$

11. $\frac{1}{4} \div \frac{1}{2} = \frac{1}{4} \cdot \frac{2}{1} = \frac{2}{4} = \frac{1}{2}$

13. $\dfrac{\frac{13}{12}}{\frac{39}{5}} = \frac{13}{12} \cdot \frac{5}{39} = \frac{13 \cdot 5}{12 \cdot 3 \cdot 13} = \frac{5}{36}$

15. $100 \div \frac{1}{5} = 100 \cdot \frac{5}{1} = 500$

17. $\frac{3}{4} \div 10 = \frac{3}{4} \cdot \frac{1}{10} = \frac{3}{40}$

19. $\frac{5}{4}$, or $1\frac{1}{4}$, is $\frac{1}{4}$ of the way from 1 to 2.

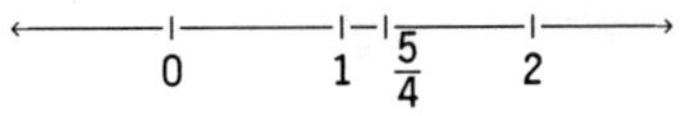

21. $\frac{15}{16}$ is $\frac{15}{16}$ of the way from 0 to 1.

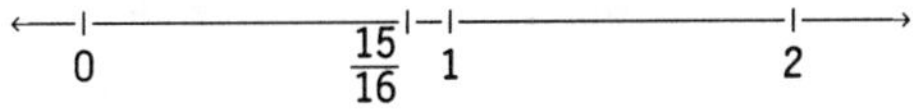

23. $\frac{1}{2} = \frac{1}{2} \cdot \frac{2}{2} = \frac{2}{4}$
Since the numerators (and the denominators) are the same, $\frac{1}{2} = \frac{2}{4}$.

25. $\frac{11}{15} = \frac{11}{15} \cdot \frac{8}{8} = \frac{88}{120}$ $\qquad \frac{13}{24} = \frac{13}{24} \cdot \frac{5}{5} = \frac{65}{120}$
Since $88 > 65$, it follows that $\frac{88}{120} > \frac{65}{120}$, so $\frac{11}{15} > \frac{13}{24}$.

27. $\frac{13}{8} = \frac{13}{8} \cdot \frac{5}{5} = \frac{65}{40}$ $\qquad \frac{8}{5} = \frac{8}{5} \cdot \frac{8}{8} = \frac{64}{40}$
Since $65 > 64$, it follows that $\frac{65}{40} > \frac{64}{40}$, so $\frac{13}{8} > \frac{8}{5}$.

29. $\frac{4}{5} = \frac{4}{5} \cdot \frac{2}{2} = \frac{8}{10}$
Since the numerators (and the denominators) are the same, $\frac{4}{5} = \frac{8}{10}$.

31. $\frac{7}{22} = \frac{7}{22} \cdot \frac{3}{3} = \frac{21}{66}$ $\qquad \frac{1}{3} = \frac{1}{3} \cdot \frac{22}{22} = \frac{22}{66}$
Since $21 < 22$, it follows that $\frac{21}{66} < \frac{22}{66}$, so $\frac{7}{22} < \frac{1}{3}$.

33. $A = \frac{1}{2}bh$

We substitute 48 cm for b and 17 cm for h.

$A = \frac{1}{2} \cdot 48\text{ cm} \cdot 17\text{ cm} = 408\text{ cm}^2$.

35. $45 = 3\cdot3\cdot5$
$55 = 5\cdot11$
$75 = 3\cdot5\cdot5$
$\text{LCM} = 3\cdot3\cdot5\cdot5\cdot11$, or 2475

37. $0.3125 = \frac{3125}{10,000}$

The reciprocal of $\frac{3125}{10,000}$ is $\frac{10,000}{3125}$, or 3.2.

39. $\frac{1439}{2007} = \frac{1439}{2007} \cdot \frac{2876}{2876} = \frac{4,138,564}{5,772,132}$

$\frac{2359}{2876} = \frac{2359}{2876} \cdot \frac{2007}{2007} = \frac{4,734,513}{5,772,132}$

Since $4,138,564 < 4,734,513$, it follows that

$\frac{4,138,564}{5,772,132} < \frac{4,734,513}{5,772,132}$, so $\frac{1439}{2007} < \frac{2359}{2876}$.

Exercise Set 1.8

1. $x + 8 = 10$

If we replace x by 2, we get a true equation $2 + 8 = 10$. No other replacement makes the equation true, so the only solution is 2.

3. $x - 4 = 5$

If we replace x by 9, we get a true equation $9 - 4 = 5$. No other replacement makes the equation true, so the only solution is 9.

5. $5x = 25$

If we replace x by 5, we get a true equation $5\cdot5 = 25$. No other replacement makes the equation true, so the only solution is 5.

7. $5x + 7 = 107$

If we replace x by 20, we get a true equation $5\cdot20 + 7 = 107$. No other replacement makes the equation true, so the only solution is 20.

9. $7x - 1 = 48$

If we replace x by 7, we get a true equation $7\cdot7 - 1 = 48$. No other replacement makes the equation true, so the only solution is 7.

11.
$$x + 17 = 22$$
$$x + 17 - 17 = 22 - 17 \quad \text{(Subtracting 17)}$$
$$x + 0 = 5$$
$$x = 5$$

Check:

$x + 17 = 22$	
$5 + 17$	22
22	

13.
$$x + 56 = 75$$
$$x + 56 - 56 = 75 - 56 \quad \text{(Subtracting 56)}$$
$$x + 0 = 19$$
$$x = 19$$

15.
$$x + 2.78 = 8.44$$
$$x + 2.78 - 2.78 = 8.44 - 2.78 \quad \text{(Subtracting 2.78)}$$
$$x + 0 = 5.66$$
$$x = 5.66$$

Check:

$x + 2.78 = 8.44$	
$5.66 + 2.78$	8.44
8.44	

17.
$$x + 5064 = 7882$$
$$x + 5064 - 5064 = 7882 - 5064 \quad \text{(Subtracting 5064)}$$
$$x + 0 = 2818$$
$$x = 2818$$

19.
$$x + \frac{1}{4} = \frac{2}{3}$$
$$x + \frac{1}{4} - \frac{1}{4} = \frac{2}{3} - \frac{1}{4} \quad \text{(Subtracting } \tfrac{1}{4}\text{)}$$
$$x + 0 = \frac{8}{12} - \frac{3}{12}$$
$$x = \frac{5}{12}$$

Check:

$x + \frac{1}{4} = \frac{2}{3}$	
$\frac{5}{12} + \frac{1}{4}$	$\frac{2}{3}$
$\frac{5}{12} + \frac{3}{12}$	
$\frac{8}{12}$	
$\frac{2}{3}$	

21.
$$x + \frac{2}{3} = \frac{5}{6}$$
$$x + \frac{2}{3} - \frac{2}{3} = \frac{5}{6} - \frac{2}{3} \quad \text{(Subtracting } \tfrac{2}{3}\text{)}$$
$$x + 0 = \frac{5}{6} - \frac{4}{6}$$
$$x = \frac{1}{6}$$

23.
$$6x = 24$$
$$\frac{1}{6} \cdot 6x = \frac{1}{6} \cdot 24 \quad \text{(Dividing by 6, or multiplying by } \tfrac{1}{6}\text{)}$$
$$1x = \frac{24}{6}$$
$$x = 4$$

Check:

$6x = 24$	
$6\cdot4$	24
24	

25. $4x = 5$

$\frac{1}{4} \cdot 4x = \frac{1}{4} \cdot 5$ (Dividing by 4, or multiplying by $\frac{1}{4}$)

$1x = \frac{5}{4}$

$x = \frac{5}{4}$

27. $10y = 2.4$

$\frac{1}{10} \times 10y = \frac{1}{10} \times 2.4$ (Dividing by 10, or multiplying by $\frac{1}{10}$)

$1y = \frac{2.4}{10}$

$y = 0.24$

Check:

$10y = 2.4$	
$10 \cdot 0.24$	2.4
2.4	

29. $2.9y = 8.99$

$\frac{1}{2.9} \times 2.9y = \frac{1}{2.9} \times 8.99$ (Dividing by 2.9, or multiplying by $\frac{1}{2.9}$)

$1y = \frac{8.99}{2.9}$

$y = 3.1$

31. $6.2x = 52.7$

$\frac{1}{6.2} \times 6.2x = \frac{1}{6.2} \times 52.7$ (Dividing by 6.2, or multiplying by $\frac{1}{6.2}$)

$1x = \frac{52.7}{6.2}$

$x = 8.5$

33. $\frac{3}{4} x = 35$

$\frac{\frac{3}{4} x}{\frac{3}{4}} = \frac{35}{\frac{3}{4}}$ (Dividing by $\frac{3}{4}$)

$\frac{4}{3} \cdot \frac{3}{4} x = \frac{4}{3} \cdot 35$ (Multiplying by $\frac{4}{3}$)

$1\ x = \frac{140}{3}$

$x = \frac{140}{3}$

Check:

$\frac{3}{4} x = 35$	
$\frac{3}{4} \cdot \frac{140}{3}$	35
$\frac{140}{4}$	
35	

35. Division by zero is not possible.
Thus, the division $\frac{6}{0}$ is not possible.

37. We never divide by zero.
Thus, the division $\frac{0}{0}$ is not possible.

39. $\frac{2 - 2}{4} = \frac{0}{4} = 0$

Thus, the division $\frac{2 - 2}{4}$ is possible.

41. $\frac{8 - 8}{t - t} = \frac{0}{0}$

We never divide by zero.
Thus, the division $\frac{8 - 8}{t - t}$ is not possible.

43. $\frac{7}{8} \div \frac{14}{15} = \frac{7}{8} \cdot \frac{15}{14} = \frac{7 \cdot 15}{8 \cdot 2 \cdot 7} = \frac{15}{8 \cdot 2} = \frac{15}{16}$

45. $9u + 4t + 11u + 7u + 28$

$= (9 + 11 + 7)u + 4t + 28$

$= 27u + 4t + 28$

47. $0.1265x = 1065.636$

$\frac{0.1265x}{0.1265} = \frac{1065.636}{0.1265}$

$x = 8424$

49. Answers may vary. $4x = 3x + x$

Exercise Set 1.9

1. Two-thirds of what number is forty-eight?

$\frac{2}{3} \quad \cdot \quad x \quad = \quad 48$

We solve the equation.

$\frac{2}{3} x = 48$

$\frac{\frac{2}{3} x}{\frac{2}{3}} = \frac{48}{\frac{2}{3}}$ (Dividing by $\frac{2}{3}$)

$\frac{3}{2} \cdot \frac{2}{3} x = \frac{3}{2} \cdot 48$ (Multiplying by $\frac{3}{2}$)

$1\ x = 3 \cdot 24$

$x = 72$

To check, we find out if $\frac{2}{3}$ of 72 is 48.

$\frac{2}{3} \cdot 72 = 2 \cdot 24 = 48$

The answer is 72.

3. What number plus five is twenty-two?

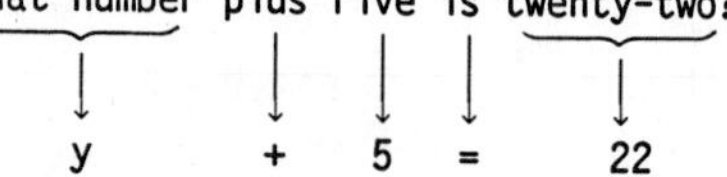

$y + 5 = 22$

We solve the equation.

$$y + 5 = 22$$
$$y + 5 - 5 = 22 - 5 \quad \text{(Subtracting 5)}$$
$$y = 17$$

To check, we add 17 to 5.

$17 + 5 = 22$

The answer is 17.

5. What number is four more than five?

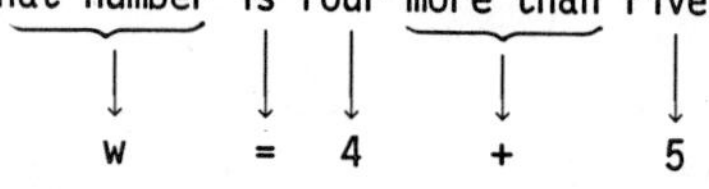

$w = 4 + 5$

We solve.

$w = 4 + 5$

$w = 9$

The answer is 9.

7. Area of Lake Superior is four times area of Lake Ontario.

$78{,}114 = 4 \cdot A$

We solve.

$$78{,}114 = 4A$$
$$\frac{1}{4} \cdot 78{,}114 = \frac{1}{4} \cdot 4A \quad \text{(Dividing by 4, or multiplying by } \tfrac{1}{4}\text{)}$$
$$\frac{78{,}114}{4} = 1A$$
$$19{,}528.5 = A$$

To check, we find out if four times 19,528.5 is 78,114.

$4 \times 19{,}528.5 = 78{,}114$

The answer is 19,528.5 km^2.

9. Two-fifths of Preitter's typing speed is Zlow's typing speed.

$\frac{2}{5} \cdot t = 35$

We solve.

$$\frac{2}{5} t = 35$$
$$\frac{\frac{2}{5} t}{\frac{2}{5}} = \frac{35}{\frac{2}{5}} \quad \text{(Dividing by } \tfrac{2}{5}\text{)}$$
$$\frac{5}{2} \cdot \frac{2}{5} t = \frac{5}{2} \cdot 35 \quad \text{(Multiplying by } \tfrac{5}{2}\text{)}$$
$$t = \frac{175}{2}, \text{ or } 87\tfrac{1}{2}$$

9. (continued)

To check, we find out if $\frac{2}{5}$ of this number is 35.

$\frac{2}{5} \cdot \frac{175}{2} = \frac{175}{5} = 35$

The answer is $87\frac{1}{2}$ words per minute.

11. Boiling point of methyl alcohol plus 13.5°C is boiling point of ethyl alcohol.

$m + 13.5 = 78.3$

We solve.

$$m + 13.5 = 78.3$$
$$m + 13.5 - 13.5 = 78.3 - 13.5 \quad \text{(Subtracting 13.5)}$$
$$m = 64.8$$

To check, we add 64.8 and 13.5.

$64.8 + 13.5 = 78.3$

The answer is 64.8°C.

13. Electricity used by tube set is 1.6 times electricity used by solid-state set.

$640 = 1.6 \cdot s$

We solve.

$$640 = 1.6s$$
$$\frac{640}{1.6} = s \quad \text{(Dividing by 1.6, or multiplying by } \tfrac{1}{1.6}\text{)}$$
$$400 = s$$

To check, we multiply 1.6 times 400.

$1.6 \times 400 = 640$

The answer is 400 kilowatt hours.

15. Ohio cost was 1.8 times California cost.

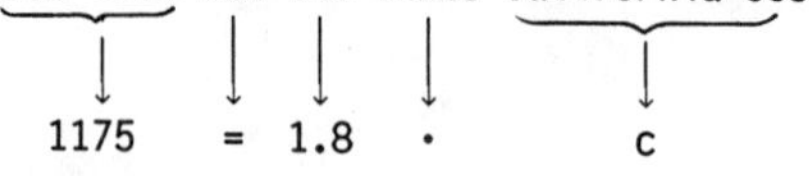

$1175 = 1.8 \cdot c$

We solve.

$$1175 = 1.8c$$
$$\frac{1175}{1.8} = c \quad \text{(Dividing by 1.8, or multiplying by } \tfrac{1}{1.8}\text{)}$$
$$652.78 \approx c \quad \text{(Rounding to the nearest cent)}$$

($\approx$ means approximately equal to)

To check, we multiply \$652.78 by 1.8.

$1.8(652.78) \approx \$1175$

The answer is \$652.78.

17. $1 \text{ inch} = 2.54 \text{ cm}$

$\frac{1}{2.54} \text{ inch} = 1 \text{ cm}$ (Dividing by 2.54)

$1 \text{ m} = 100 \text{ cm}$
$= 100 \cdot 1 \text{ cm}$
$= 100 \cdot \frac{1}{2.54} \text{ inch}$ (Substituting)
$= \frac{100}{2.54} \text{ inches}$
$\approx 39.37 \text{ inches}$

19. We need to know the salary now.

Exercise Set 1.10

1. $76\% = 76 \times 0.01 = 0.76$

3. $54.7\% = 54.7 \times 0.01 = 0.547$

5. $20\% = 20 \times \frac{1}{100} = \frac{20}{100}$, or $\frac{1}{5}$

7. $78.6\% = 78.6 \times \frac{1}{100} = \frac{78.6}{100}$
$= \frac{78.6}{100} \times \frac{10}{10}$
$= \frac{786}{1000}$, or $\frac{393}{500}$

9. $4.54 = 4.54 \times 1 = 4.54 \times (100 \times 0.01)$
$= (4.54 \times 100) \times 0.01$
$= 454 \times 0.01$
$= 454\%$

11. $0.998 = 0.998 \times 1 = 0.998 \times (100 \times 0.01)$
$= (0.998 \times 100) \times 0.01$
$= 99.8 \times 0.01$
$= 99.8\%$

13. $\frac{1}{8} = \frac{1}{8} \times 1 = \frac{1}{8} \times (100 \times \frac{1}{100})$
$= (\frac{1}{8} \times 100) \times \frac{1}{100}$
$= \frac{100}{8} \times \frac{1}{100}$
$= \frac{100}{8}\%$, or 12.5%

15. The denominator, 25, is a factor of 100.
$\frac{17}{25} = \frac{17}{25} \times \frac{4}{4} = \frac{68}{100} = 68 \times \frac{1}{100} = 68\%$

17. What is 65% of 840?
$\downarrow \quad \downarrow \quad \downarrow \quad \downarrow \quad \downarrow$
$x \quad = \quad 65\% \quad \cdot \quad 840$

Solve:
$x = (65 \times 0.01) \times 840$
$x = 0.65 \times 840$
$x = 546$

The solution is 546, so 546 is 65% of 840.

19. 24 percent of what is 20.4?
$\downarrow \quad \downarrow \quad \downarrow \quad \downarrow \quad \downarrow \quad \downarrow$
$24 \quad \% \quad \cdot \quad y \quad = 20.4$

Solve:
$24 \times 0.01 \times y = 20.4$
$0.24 \times y = 20.4$
$y = \frac{20.4}{0.24}$
$y = 85$

24% of 85 is 20.4, so the solution is 85.

21. What percent of 80 is 100?
$\downarrow \quad \downarrow \quad \downarrow \quad \downarrow \quad \downarrow \quad \downarrow$
$y \quad \% \quad \cdot \quad 80 = \quad 100$

Solve:
$y \times 0.01 \times 80 = 100$
$y \times 0.80 = 100$
$y = \frac{100}{0.80}$
$y = 125$

The solution is 125, so 125% of 80 is 100.

23. 76 is what percent of 88?
$\downarrow \quad \downarrow \quad \downarrow \quad \downarrow \quad \downarrow \quad \downarrow$
$76 \quad = \quad x \quad \% \quad \cdot 88$

Solve:
$x \times 0.01 \times 88 = 76$
$x(0.88) = 76$
$x = \frac{76}{0.88}$
$x \approx 86.4$

86.4% were correct.

25. \$208 is 26% of what amount?
$\downarrow \quad \downarrow \quad \downarrow \quad \downarrow \quad \downarrow$
$208 \quad = 26\% \quad \cdot \quad x$

Solve:
$26\% \cdot x = 208$
$0.26x = 208$
$x = \frac{208}{0.26}$
$x = 800$

The monthly income was \$800.

27. What is 8% of \$428.86?

$x = 8\% \cdot 428.86$

Solve:

$x = 0.08 \times 428.86$

$= 34.3088$

≈ 34.31

The tax is \$34.31. The total cost is \$428.86 + \$34.31, or \$463.17.

29. Original investment plus interest earned is present value.

$x + 9\%x = 8502$

Solve:

$x + 9\%x = 8502$

$1x + 0.09x = 8502$

$1.09x = 8502$

$x = \frac{8502}{1.09}$

$x = 7800$

The original investment was \$7800.

31. What is (8% - 7.4%) of \$9600?

$x = (8\% - 7.4\%) \cdot 9600$

Solve:

$x = 0.6\% \times 9600$

$x = 0.006(9600)$

$x = 57.60$

The additional earning power was \$57.60.

33. $1 - 76\% = 100\% - 76\% = (100 - 76)\% = 24\%$

35. 10,000 is what percent of 300,000?

$10{,}000 = x \; \% \cdot 300{,}000$

Solve:

$10{,}000 = x \cdot 0.01 \cdot 300{,}000$

$10{,}000 = x(3000)$

$\frac{10{,}000}{3000} = x$

$\frac{10}{3} = x$, or $x = 3\frac{1}{3}$

The average person knows $3\frac{1}{3}\%$, or $3.\overline{3}\%$, of the words in the English language.

Exercise Set 1.10A

1. $38\% = 38 \times 0.01 = 0.38$

3. $65.4\% = 65.4 \times 0.01 = 0.654$

5. $8.24\% = 8.24 \times 0.01 = 0.0824$

7. $0.012\% = 0.012 \times 0.01 = 0.00012$

9. $0.73\% = 0.73 \times 0.01 = 0.0073$

11. $125\% = 125 \times 0.01 = 1.25$

13. $30\% = 30 \times \frac{1}{100} = \frac{30}{100}$, or $\frac{3}{10}$

15. $13.5\% = 13.5 \times \frac{1}{100} = \frac{13.5}{100}$

$= \frac{13.5}{100} \cdot \frac{10}{10} = \frac{135}{1000}$, or $\frac{27}{200}$

17. $3.2\% = 3.2 \times \frac{1}{100} = \frac{3.2}{100}$

$= \frac{3.2}{100} \cdot \frac{10}{10} = \frac{32}{1000}$, or $\frac{4}{125}$

19. $120\% = 120 \times \frac{1}{100} = \frac{120}{100}$, or $\frac{6}{5}$

21. $0.35\% = 0.35 \times \frac{1}{100} = \frac{0.35}{100}$

$= \frac{0.35}{100} \cdot \frac{100}{100} = \frac{35}{10{,}000}$, or $\frac{7}{2000}$

23. $0.042\% = 0.042 \times \frac{1}{100} = \frac{0.042}{100}$

$= \frac{0.042}{100} \cdot \frac{1000}{1000}$

$= \frac{42}{100{,}000}$, or $\frac{21}{50{,}000}$

25. $0.62 = 0.62 \times (100 \times 0.01)$

$= (0.62 \times 100) \times 0.01$

$= 62 \times 0.01$

$= 62\%$

27. $0.623 = 0.623 \times (100 \times 0.01)$

$= (0.623 \times 100) \times 0.01$

$= 62.3 \times 0.01$

$= 62.3\%$

29. $7.2 = 7.2 \times (100 \times 0.01)$

$= (7.2 \times 100) \times 0.01$

$= 720 \times 0.01$

$= 720\%$

31. $2 = 2 \times (100 \times 0.01)$
$= (2 \times 100) \times 0.01$
$= 200 \times 0.01$
$= 200\%$

33. $0.072 = 0.072 \times (100 \times 0.01)$
$= (0.072 \times 100) \times 0.01$
$= 7.2 \times 0.01$
$= 7.2\%$

35. $0.0057 = 0.0057 \times (100 \times 0.01)$
$= (0.0057 \times 100) \times 0.01$
$= 0.57 \times 0.01$
$= 0.57\%$

37. $\frac{17}{100} = 17 \times \frac{1}{100} = 17\%$

39. $\frac{7}{10} = \frac{7}{10} \cdot \frac{10}{10} = \frac{70}{100} = 70 \times \frac{1}{100} = 70\%$

41. $\frac{7}{20} = \frac{7}{20} \cdot \frac{5}{5} = \frac{35}{100} = 35 \times \frac{1}{100} = 35\%$

43. $\frac{1}{2} = \frac{1}{2} \cdot \frac{50}{50} = \frac{50}{100} = 50 \times \frac{1}{100} = 50\%$

45. $\frac{3}{5} = \frac{3}{5} \cdot \frac{20}{20} = \frac{60}{100} = 60 \times \frac{1}{100} = 60\%$

47. $\frac{1}{3} = \frac{1}{3} \times (100 \times \frac{1}{100})$
$= (\frac{1}{3} \times 100) \times \frac{1}{100}$
$= \frac{100}{3} \times \frac{1}{100}$
$= \frac{100}{3}\%$, or $33\frac{1}{3}\%$, or $33.\overline{3}\%$

49. What is 38% of 250?
↓ ↓ ↓ ↓ ↓
$x = 38\% \cdot 250$
Solve:
$x = 0.38(250)$
$x = 95$
The solution is 95, so 95 is 38% of 250.

51. What percent of 80 is 20?
↓ ↓ ↓ ↓ ↓ ↓
$x \quad \% \quad \cdot \quad 80 = 20$
Solve:
$x \times 0.01 \times 80 = 20$
$x(0.80) = 20$
$x = \frac{20}{0.80}$
$x = 25$
The solution is 25, so 25% of 80 is 20.

53. 25 percent of what is 16?
↓ ↓ ↓ ↓ ↓ ↓
$25 \quad \% \quad \cdot \quad x = 16$
Solve:
$25\%x = 16$
$0.25x = 16$
$x = \frac{16}{0.25}$
$x = 64$
The solution is 64, so 25% of 64 is 16.

Exercise Set 2.1

1. -1286 ft corresponds to 1286 ft below sea level.
   29,028 ft corresponds to 29,028 ft above sea level.

3. -3 sec corresponds to 3 sec before liftoff.
   128 sec corresponds to 128 sec after liftoff.

5. Since 6 is to the right of 0, we have $6 > 0$.

7. Since -9 is to the left of 5, we have $-9 < 5$.

9. Since -6 is to the left of 6, we have $-6 < 6$.

11. Since -8 is to the left of -5, we have $-8 < -5$.

13. Since -5 is to the right of -11, we have $-5 > -11$.

15. Since -6 is to the left of -5, we have $-6 < -5$.

17. The distance of -3 from 0 is 3, so $|-3| = 3$.

19. The distance of 10 from 0 is 10, so $|10| = 10$.

21. The distance of 0 from 0 is 0, so $|0| = 0$.

23. The distance of -24 from 0 is 24, so $|-24| = 24$.

25. The distance of 53 from 0 is 53, so $|53| = 53$.

27. The distance of -8 from 0 is 8, so $|-8| = 8$.

29. When $x = -6$, $-x = -(-6)$.
    The reflection, or inverse, of -6 is 6.
    Thus, $-(-6) = 6$.

31. When $x = 6$, $-x = -(6)$.
    The reflection, or inverse, of 6 is -6.
    Thus, $-(6) = -6$.

33. When $x = -7$,
    $-(-x) = -[-(-7)]$
    $= -[7]$ [The reflection, or inverse, of -7 is 7, $-(-7) = 7$.]
    $= -7$ (The reflection, or inverse, of 7 is -7.)

35. When $x = 1$,
    $-(-x) = -[-(1)]$
    $= -[-1]$ (The reflection, or inverse, of 1 is -1.)
    $= 1$ (The reflection, or inverse, of -1 is 1.)

37. When $x = -12$, $-x = -(-12)$.
    The reflection, or inverse, of -12 is 12.
    Thus, $-(-12) = 12$.

39. When $x = 70$, $-x = -(70)$.
    The reflection, or inverse, of 70 is -70.
    Thus, $-(70) = -70$.

41. When $x = 0$,
    $-(-x) = -[-(0)]$
    $= -[0]$ (The reflection, or inverse, of 0 is 0.)
    $= 0$

43. When $x = -34$,
    $-(-x) = -[-(-34)]$
    $= -[34]$ (The reflection, or inverse, of -34 is 34.)
    $= -34$ (The reflection, or inverse, of 34 is -34.)

45. The additive inverse of -1 is 1. $-(-1) = 1$

47. The additive inverse of 7 is -7. $-(7) = -7$

49. The additive inverse of -14 is 14. $-(-14) = 14$

51. The additive inverse of 0 is 0. $-(0) = 0$

53. $\frac{21}{5} \cdot \frac{1}{7} = \frac{21 \cdot 1}{5 \cdot 7} = \frac{3 \cdot 7}{5 \cdot 7} = \frac{3}{5}$

55. The expression $6 \cdot (4 + 8) = 6 \cdot 4 + 6 \cdot 8$ illustrates the distributive law.

57. $|-7| = 7$ and $|7| = 7$
    The solutions of $|x| = 7$ are 7 and -7.

59. The inverse of -x is x. $-(-x) = x$

    $-(-(-x))$
    $= -(x)$
    $= -x$

    $-(-(-(-x)))$
    $= -(-(x))$
    $= -(-x)$
    $= x$

Exercise Set 2.2

1. $-9 + 2$ Find the difference of the absolute values: $9 - 2 = 7$. Since -9 has the greater absolute value and it is negative, make the answer negative. $-9 + 2 = -7$

3. $-9 + 5$ Find the difference of the absolute values: $9 - 5 = 4$. Since -9 has the greater absolute value and it is negative, make the answer negative. $-9 + 5 = -4$

5. $-6 + 6 = 0$
For any integer a, $-a + a = 0$.

7. $-8 + (-5)$ Two negatives
Add the absolute values: $8 + 5 = 13$
Make the answer negative: $-8 + (-5) = -13$

9. $-5 + (-11)$ Two negatives
Add the absolute values: $5 + 11 = 16$
Make the answer negative: $-5 + (-11) = -16$

11. $-6 + (-5)$ Two negatives
Add the absolute values: $6 + 5 = 11$
Make the answer negative: $-6 + (-5) = -11$

13. $9 + (-9) = 0$
For any integer a, $a + (-a) = 0$.

15. $-2 + 2 = 0$
For any integer a, $-a + a = 0$.

17. $0 + 8 = 8$
For any integer a, $0 + a = a$.

19. $0 + (-8) = -8$
For any integer a, $0 + a = a$.

21. $-25 + 0 = -25$
For any integer a, $a + 0 = a$.

23. $0 + (-17) = -17$
For any integer a, $0 + a = a$.

25. $17 + (-17) = 0$
For any integer a, $a + (-a) = 0$.

27. $-18 + 18 = 0$
For any integer a, $-a + a = 0$.

29. $8 + (-5)$ Find the difference of the absolute values: $8 - 5 = 3$. Since 8 has the greater absolute value and it is positive, leave the answer positive. $8 + (-5) = 3$

31. $-4 + (-5)$ Two negatives
Add the absolute values: $4 + 5 = 9$
Make the answer negative: $-4 + (-5) = -9$

33. $0 + (-5) = -5$
For any integer a, $0 + a = a$.

35. $13 + (-6)$ Find the difference of the absolute values: $13 - 6 = 7$. Since 13 has the greater absolute value and it is positive, leave the answer positive. $13 + (-6) = 7$

37. $-10 + 7$ Find the difference of the absolute values: $10 - 7 = 3$. Since -10 has the greater absolute value and it is negative, make the answer negative. $-10 + 7 = -3$

39. $-6 + 6 = 0$
For any integer a, $-a + a = 0$.

41. $11 + (-16)$ Find the difference of the absolute values: $16 - 11 = 5$. Since -16 has the greater absolute value and it is negative, make the answer negative. $11 + (-16) = -5$

43. $-15 + (-6)$ Two negatives
Add the absolute values: $15 + 6 = 21$
Make the answer negative: $-15 + (-6) = -21$

45. $11 + (-9)$ Find the difference of the absolute values: $11 - 9 = 2$. Since 11 has the greater absolute value and it is positive, leave the answer positive. $11 + (-9) = 2$

47. $-20 + (-6)$ Two negatives
Add the absolute values: $20 + 6 = 26$
Make the answer negative: $-20 + (-6) = -26$

49. $-15 + (-7)$ Two negatives
Add the absolute values: $15 + 7 = 22$
Make the answer negative: $-15 + (-7) = -22$

51. $40 + (-8)$ Find the difference of the absolute values: $40 - 8 = 32$. Since 40 has the greater absolute value and it is positive, leave the answer positive. $40 + (-8) = 32$

53. $-25 + 25 = 0$
For any integer a, $-a + a = 0$.

55. $63 + (-18)$ Find the difference of the absolute values: $63 - 18 = 45$. Since 63 has the greater absolute value and it is positive, leave the answer positive. $63 + (-18) = 45$

57. $$\begin{array}{r} -35 \\ -63 \\ -27 \\ -14 \\ -59 \\ \hline -198 \end{array}$$
(All negatives: Add the absolute values and make the answer negative.)

59.

	Add the positives:	Add the negatives:
27	27	-54
-54	65	-32
-32	46	-86
65	138	
46		

Add the results: 138 + (-86)

Find the difference of the absolute values: 138 - 86 = 52. Since 138 has the greater absolute values and it is positive, leave the answer positive.

138 + (-86) = 52

61. 7(3z + y + 2)

= 7·3z + 7·y + 7·2

= 21z + 7z + 14

63. What is 25% of 14?

x = 25% · 14

x = 0.25(14)

x = 3.5

65.

	Add the positives:	Add the negatives:
-64,374	53,690	-64,374
-27,159	39,087	-27,159
53,690	41,646	-11,953
39,087	134,423	-103,486
41,646		
-11,953		

Add the results: 134,423 + (-103,486)

Find the difference of the absolute values: 134,423 - 103,486 = 30,937. Since 134,423 has the greater absolute value and it is positive, leave the answer positive.

134,423 + (-103,486) = 30,937

67. The inverse of a positive number is negative. -x is negative when x is positive.

Exercise Set 2.3

1. 3 - 7 = 3 + (-7) = -4

3. 0 - 7 = 0 + (-7) = -7

5. -8 - (-2) = -8 + 2 = -6

7. -18 - (-18) = -18 + 18 = 0

9. 12 - 16 = 12 + (-16) = -4

11. 20 - 27 = 20 + (-27) = -7

13. -9 - (-3) = -9 + 3 = -6

15. -40 - (-40) = -40 + 40 = 0

17. 7 - 7 = 7 + (-7) = 0

19. 7 - (-7) = 7 + 7 = 14

21. 8 - (-3) = 8 + 3 = 11

23. -6 - 8 = -6 + (-8) = -14

25. -4 - (-9) = -4 + 9 = 5

27. 2 - 9 = 2 + (-9) = -7

29. -6 - (-5) = -6 + 5 = -1

31. 8 - (-10) = 8 + 10 = 18

33. 0 - 5 = 0 + (-5) = -5

35. -5 - (-2) = -5 + 2 = -3

37. -7 - 14 = -7 + (-14) = -21

39. 0 - (-5) = 0 + 5 = 5

41. -8 - 0 = -8 + 0 = -8

43. 7 - (-5) = 7 + 5 = 12

45. 2 - 25 = 2 + (-25) = -23

47. -42 - 26 = -42 + (-26) = -68

49. -71 - 2 = -71 + (-2) = -73

51. 24 - (-92) = 24 + 92 = 116

53. -50 - (-50) = -50 + 50 = 0

55. 18 - (-15) - 3 - (-5) + 2

= 18 + 15 + (-3) + 5 + 2

= 40 + (-3)

= 37

57. -31 + (-28) - (-14) - 17

= -31 + (-28) + 14 + (-17)

= -76 + 14

= -62

59. -34 - 28 + (-33) - 44

= -34 + (-28) + (-33) + (-44)

= -139

61. We subtract -34° from 28°.

28 - (-34) = 28 + 34 = 62

The lowest temperature ever recorded in Los Angeles is 62°F higher than the lowest temperature ever recorded in Minneapolis.

63. $A = \ell \cdot w$

$A = 36\text{ ft} \cdot 12\text{ ft}$

$A = 432\text{ ft}^2$

65. $123{,}907 - 433{,}789$

$= 123{,}907 + (-433{,}789)$

$= -309{,}882$

67. False. $a - 0 = a$ $\quad 0 - a = 0 + (-a) = -a$

$a = -a$ is only true when $a = 0$. It is not true for all other integers.

$12 - 0 = 12$, $\quad 0 - 12 = -12$, $\quad 12 \neq -12$

Exercise Set 2.4

1. $-8 \cdot 2$

Multiply their absolute values: $8 \cdot 2 = 16$

The answer is negative: $-8 \cdot 2 = -16$

The product of a negative and a positive integer is negative.

3. $-7 \cdot 6 = -42$ 5. $8 \cdot (-3) = -24$ 7. $-9 \cdot 8 = -72$

9. $-8 \cdot (-2)$

Multiply their absolute values: $8 \cdot 2 = 16$

The answer is positive: $-8 \cdot (-2) = 16$

The product of two negative integers is positive.

11. $-7 \cdot (-6) = 42$ 13. $-8 \cdot (-3) = 24$

15. $-9 \cdot (-8) = 72$ 17. $15 \cdot (-8) = -120$

19. $-25 \cdot (-40) = 1000$ 21. $-6 \cdot (-15) = 90$

23. $-25 \cdot (-8) = 200$

25. $36 \div (-6) = -6$

When we divide a positive integer by a negative integer, the answer is negative.

27. $\frac{26}{-2} = -13$ 29. $\frac{-16}{8} = -2$

31. $\frac{-48}{-12} = 4$

When we divide two negative integers, the answer is positive.

33. $\frac{-72}{9} = -8$ 35. $-100 \div (-50) = 2$

37. $-108 \div 9 = -12$ 39. $\frac{200}{-25} = -8$

41. $\frac{54}{48} = \frac{6 \cdot 9}{6 \cdot 8} = \frac{9}{8}$ 43. $3.1x = 387.5$

$x = \frac{387.5}{3.1}$

$x = 125$

45. $(-4)^2 = -4 \cdot (-4) = 16$

47. $-4 \cdot 3 + 10 \cdot (-2) = -12 + (-20) = -32$

49. $(-1)^{23}$ has 23 factors of -1. The product of an odd number of negative numbers is negative.

$(-1)^{23} = -1$

Exercise Set 2.5

1. The additive inverse of -4.7 is 4.7 because $-4.7 + 4.7 = 0$.

3. The additive inverse of $\frac{7}{2}$ is $-\frac{7}{2}$ because $\frac{7}{2} + (-\frac{7}{2}) = 0$.

5. The additive inverse of -7 is 7 because $-7 + 7 = 0$.

7. The additive inverse of -26.9 is 26.9 because $-26.9 + 26.9 = 0$.

9. $-(-\frac{1}{3}) = \frac{1}{3}$

The inverse of a negative number is positive.

11. $-(\frac{7}{6}) = -\frac{7}{6}$

The inverse of a positive number is negative.

13. $-(-9.3) = 9.3$

The inverse of a negative number is positive.

15. $-(90.3) = -90.3$

The inverse of a positive number is negative.

17. When $x = 12.4$, $-x = -(12.4) = -12.4$.

19. When $x = -\frac{9}{10}$, $-x = -(-\frac{9}{10}) = \frac{9}{10}$.

21. When $x = -34.8$, $-x = -(-34.8) = 34.8$.

23. When $x = 567$, $-x = -(567) = -567$.

25. The distance of 19.2 from 0 is 19.2, so $|19.2| = 19.2$.

27. The distance of $-\frac{2}{3}$ from 0 is $\frac{2}{3}$, so $|-\frac{2}{3}| = \frac{2}{3}$.

29. The distance of -89.3 from 0 is 89.3, so $|-89.3| = 89.3$.

31. The distance of $\frac{14}{3}$ from 0 is $\frac{14}{3}$, so $\left|\frac{14}{3}\right| = \frac{14}{3}$.

33. $-6.5 + 4.7 = -1.8$

35. $-2.8 + (-5.3) = -8.1$

37. $-\frac{3}{5} + \frac{2}{5} = \frac{-3 + 2}{5} = \frac{-1}{5} = -\frac{1}{5}$

39. $-\frac{3}{7} + (-\frac{5}{7}) = \frac{-3 + (-5)}{7} = \frac{-8}{7} = -\frac{8}{7}$

41. $-\frac{5}{8} + \frac{1}{4} = -\frac{5}{8} + \frac{2}{8} = \frac{-5 + 2}{8} = \frac{-3}{8} = -\frac{3}{8}$

43. $-\frac{3}{7} + (-\frac{2}{5}) = -\frac{15}{35} + (-\frac{14}{35}) = \frac{-15 + (-14)}{35}$
$= \frac{-29}{35} = -\frac{29}{35}$

45. $-\frac{3}{5} + (-\frac{2}{15}) = -\frac{9}{15} + (-\frac{2}{15}) = \frac{-9 + (-2)}{15}$
$= \frac{-11}{15} = -\frac{11}{15}$

47. $-5.7 + (-7.2) + 6.6$
$= -12.9 + 6.6$
$= -6.3$

49. $-8.5 + 7.9 + (-3.7)$
$= -12.2 + 7.9$
$= -4.3$

51. $-9 - (-5) = -9 + 5 = -4$

53. $\frac{3}{8} - \frac{5}{8} = \frac{3}{8} + (-\frac{5}{8}) = \frac{3 + (-5)}{8} = \frac{-2}{8} = -\frac{1}{4}$

55. $\frac{3}{4} - \frac{2}{3} = \frac{3}{4} + (-\frac{2}{3}) = \frac{9}{12} + (-\frac{8}{12})$
$= \frac{9 + (-8)}{12} = \frac{1}{12}$

57. $-\frac{3}{4} - \frac{2}{3} = -\frac{3}{4} + (-\frac{2}{3}) = -\frac{9}{12} + (-\frac{8}{12})$
$= \frac{-9 + (-8)}{12}$
$= \frac{-17}{12} = -\frac{17}{12}$

59. $-\frac{5}{8} - (-\frac{3}{4}) = -\frac{5}{8} + \frac{3}{4} = -\frac{5}{8} + \frac{6}{8} = \frac{-5 + 6}{8} = \frac{1}{8}$

61. $6.1 - (-13.8) = 6.1 + 13.8 = 19.9$

63. $-3.2 - 5.8 = -3.2 + (-5.8) = -9$

65. $0.99 - 1 = 0.99 + (-1) = -0.01$

67. Substitute 4 for x and 12 for y.
$\frac{3x}{y} = \frac{3\cdot 4}{12} = \frac{12}{12} = 1$

69. $\frac{19}{25} = \frac{19}{25}\cdot\frac{4}{4} = \frac{76}{100} = 0.76$ or

```
       0.7 6
  25 ⟌1 9.0 0
      1 7 5
        1 5 0
        1 5 0
            0
```

Thus, $\frac{19}{25} = 0.76$.

71. a) -8.6 is to the right of -9.0, so $-8.6 > -9.0$.
b) $-2\frac{1}{8} = -\frac{17}{8} = -\frac{34}{16}$ $\qquad \frac{-35}{16} = -\frac{35}{16}$
$-\frac{34}{16}$ is to the right of $-\frac{35}{16}$,
so $-2\frac{1}{8} > \frac{-35}{16}$

73. $-(-3.8 + 7.2) = -(3.4) = -3.4$

75. $-\left|-\frac{1}{3} + \frac{5}{6}\right| = -\left|-\frac{2}{6} + \frac{5}{6}\right| = -\left|\frac{3}{6}\right| = -(\frac{1}{2}) = -\frac{1}{2}$

Exercise Set 2.6

1. $9\cdot(-8)$
Multiply their absolute values: $9\cdot 8 = 72$
The answer is negative: $9\cdot(-8) = -72$
The product of a positive and a negative number is negative.

3. $4\cdot(-3.1) = -12.4$

5. $-6\cdot(-4)$
Multiply their absolute values: $6\cdot 4 = 24$
The answer is positive: $-6\cdot(-4) = 24$
The product of two negative numbers is positive.

7. $-7\cdot(-3.1) = 21.7$

9. $\frac{2}{3}\cdot(-\frac{3}{5}) = -\frac{6}{15} = -\frac{2}{5}$

11. $-\frac{3}{8}\cdot(-\frac{2}{9}) = \frac{6}{72} = \frac{1}{12}$

13. $-6.3 \times 2.7 = -17.01$

15. $-\frac{5}{9}\cdot\frac{3}{4} = -\frac{15}{36} = -\frac{5}{12}$

17. $6\cdot(-5)\cdot 3$
$= 6\cdot 3\cdot(-5)$
$= 18\cdot(-5)$
$= -90$

19. $7\cdot(-4)\cdot(-3)\cdot5$
$= 7\cdot5\cdot(-4)\cdot(-3)$
$= 35\cdot12$
$= 420$

21. $-\frac{2}{3} \cdot \frac{1}{2} \cdot(-\frac{6}{7})$
$= \frac{1}{2} \cdot (-\frac{2}{3}) \cdot (-\frac{6}{7})$
$= \frac{1}{2} \cdot \frac{12}{21}$
$= \frac{12}{42}$
$= \frac{2}{7}$

23. $-3\cdot(-4)\cdot(-5)$
$= 12\cdot(-5)$
$= -60$

25. $-2\cdot(-5)\cdot(-3)\cdot(-5)$
$= 10\cdot15$
$= 150$

27. The reciprocal of -5 is $-\frac{1}{5}$ because $-5 \cdot (-\frac{1}{5}) = \frac{5}{5} = 1$.

29. The reciprocal of $\frac{1}{4}$ is 4 because $\frac{1}{4} \cdot 4 = \frac{4}{4} = 1$.

31. The reciprocal of $-\frac{7}{5}$ is $-\frac{5}{7}$ because $-\frac{7}{5} \cdot (-\frac{5}{7}) = \frac{35}{35} = 1$.

33. The reciprocal of $-\frac{4}{11}$ is $-\frac{11}{4}$ because $-\frac{4}{11} \cdot (-\frac{11}{4}) = \frac{44}{44} = 1$.

35. $\frac{3}{4} \div (-\frac{2}{3}) = \frac{3}{4} \cdot (-\frac{3}{2}) = -\frac{9}{8}$

37. $-\frac{5}{4} \div (-\frac{3}{4}) = -\frac{5}{4} \cdot (-\frac{4}{3}) = \frac{20}{12} = \frac{5}{3}$

39. $-\frac{2}{7} \div (-\frac{4}{9}) = -\frac{2}{7} \cdot (-\frac{9}{4}) = \frac{18}{28} = \frac{9}{14}$

41. $-\frac{3}{8} \div (-\frac{8}{3}) = -\frac{3}{8} \cdot (-\frac{3}{8}) = \frac{9}{64}$

43. $\frac{5}{7} \div (-\frac{5}{7}) = \frac{5}{7} \cdot (-\frac{7}{5}) = -\frac{35}{35} = -1$

45. $-6.6 \div 3.3 = \frac{-6.6}{3.3} = \frac{-6.6}{3.3} \times \frac{10}{10}$
$= \frac{-66}{33}$
$= -2$

47. $\frac{-11}{-13} = \frac{11}{13}$

49. $\frac{23}{-14} = -\frac{23}{14}$

51. $A = \pi r^2$
Substitute 15 cm for r and 3.14 for π.
$A = 3.14(15 \text{ cm})^2$
$= 3.14 \times 15 \text{ cm} \times 15 \text{ cm}$
$= 3.14 \times 225 \times \text{cm}^2$
$= 706.5 \text{ cm}^2$

53. $24x + 32y + 64$
$= 8\cdot3x + 8\cdot4y + 8\cdot8$
$= 8(3x + 4y + 8)$

55. $(-1)^4 = (-1)(-1)(-1)(-1) = 1\cdot1 = 1$
The product of an even number of negative numbers is positive.

57. $(-2)^3 = (-2)(-2)(-2) = 4(-2) = -8$
The product of an odd number of negative numbers is negative.

Exercise Set 2.7

1. $-3(3 - 7) = -3(-4) = 12$
or
$-3(3 - 7) = -3\cdot3 - (-3)\cdot7 = -9 - (-21)$
$= -9 + 21 = 12$

3. $8(9 - 10) = 8(-1) = -8$
or
$8(9 - 10) = 8\cdot9 - 8\cdot10 = 72 - 80$
$= 72 + (-80) = -8$

5. $8x - 1.4y = 8x + (-1.4y)$
The terms are $8x$ and $-1.4y$.

7. $-5x + 3y - 14z = -5x + 3y + (-14z)$
The terms are $-5x$, $3y$, and $-14z$.

9. $8x - 24 = 8\cdot x - 8\cdot3 = 8(x - 3)$

11. $32 - 4x = 4\cdot8 - 4\cdot x = 4(8 - x)$

13. $8x + 10y - 22 = 2\cdot4x + 2\cdot5y - 2\cdot11$
$= 2(4x + 5y - 11)$

15. $ax - 7a = a\cdot x - a\cdot 7 = a(x - 7)$

17. $ax + ay - az = a\cdot x + a\cdot y - a\cdot z$
$= a(x + y - z)$

19. $7(x - 2) = 7\cdot x - 7\cdot 2 = 7x - 14$

21. $-7(y - 2) = -7\cdot y - (-7)\cdot 2 = -7y + 14$

23. $-3(7 - t) = -3\cdot 7 - (-3)\cdot t = -21 + 3t$

25. $-4(x + 3y) = -4\cdot x + (-4)\cdot 3y = -4x - 12y$

27. $7(-2x - 4y + 3) = 7\cdot(-2x) - 7\cdot 4y + 7\cdot 3$
$= -14x - 28y + 21$

29. $11x - 3x = (11 - 3)x = 8x$

31. $17y - y = 17y - 1y = (17 - 1)y = 16y$

33. $x - 12x = 1x - 12x = (1 - 12)x = -11x$

35. $x - 0.83x = 1x - 0.83x = (1 - 0.83)x = 0.17x$

37. $9x + 2y - 5x = (9 - 5)x + 2y = 4x + 2y$

39. $11x + 2y + 7 - 4x - 18 - y$
$= 11x + 2y + 7 - 4x - 18 - 1y$
$= (11 - 4)x + (2 - 1)y + (7 - 18)$
$= 7x + y - 11$

41. $2.7x + 2.3y - 1.9x - 1.8y$
$= (2.7 - 1.9)x + (2.3 - 1.8)y$
$= 0.8x + 0.5y$

43. $\frac{1}{5}x + \frac{4}{5}y + \frac{2}{5}x - \frac{1}{5}y$
$= (\frac{1}{5} + \frac{2}{5})x + (\frac{4}{5} - \frac{1}{5})y$
$= \frac{3}{5}x + \frac{3}{5}y$

45. $P = 2\ell + 2w$
$P = 2\cdot 25\text{ ft} + 2\cdot 14\text{ ft}$
$= 50\text{ ft} + 28\text{ ft}$
$= 78\text{ ft}$

47. $\frac{3}{4}x = \frac{9}{2}$
$\frac{4}{3}\cdot\frac{3}{4}x = \frac{4}{3}\cdot\frac{9}{2}$
$x = \frac{36}{6}$
$x = 6$

49. $\frac{1}{2}ah + \frac{1}{2}bh$
$= \frac{1}{2}h\cdot a + \frac{1}{2}h\cdot b$
$= \frac{1}{2}h(a + b)$

51. $\frac{5x - 15}{5} + \frac{2x + 6}{2}$
$= \frac{5(x - 3)}{5} + \frac{2(x + 3)}{2}$
$= x - 3 + x + 3$
$= 2x$

Exercise Set 2.8

1. $-(2x + 7) = -2x - 7$

3. $-(5x - 8) = -5x + 8$

5. $-(4a - 3b + 7c) = -4a + 3b - 7c$

7. $-(6x + 8y + 5) = -6x - 8y - 5$

9. $-(3x + 5y - 6) = -3x - 5y + 6$

11. $-(-8x - 6y - 43) = 8x + 6y + 43$

13. $9x - (4x + 3) = 9x - 4x - 3 = 5x - 3$

15. $2a + (5a - 9) = 2a + 5a - 9 = 7a - 9$

17. $2x + 7x - (4x + 6) = 2x + 7x - 4x - 6$
$= 5x - 6$

19. $2x - 4y - (7x - 2y) = 2x - 4y - 7x + 2y$
$= -5x - 2y$

21. $3x - y - (3x - 2y + 5z) = 3x - y - 3x + 2y - 5z$
$= y - 5z$

23. $2x - 4y - 3(7x - 2y)$
$= 2x - 4y + [-3(7x - 2y)]$
$= 2x - 4y + (-21x + 6y)$
$= 2x - 4y - 21x + 6y$
$= -19x + 2y$

25. $4a - b - 5(5a - 7b + 8c)$
$= 4a - b + [-5(5a - 7b + 8c)]$
$= 4a - b + (-25a + 35b - 40c)$
$= 4a - b - 25a + 35b - 40c$
$= -21a + 34b - 40c$

27. $[(-24) \div (-3)] \div (-\frac{1}{2})$
$= 8 \div (-\frac{1}{2})$
$= 8\cdot(-2)$
$= -16$

29. $8\cdot[9 - 2(5 - 3)]$
$= 8\cdot[9 - 2\cdot2]$
$= 8\cdot[9 - 4]$
$= 8\cdot5$
$= 40$

31. $[4(9 - 6) + 11] - [14 - (6 + 4)]$
$= [4\cdot3 + 11] - [14 - 10]$
$= [12 + 11] - [4]$
$= 23 - 4$
$= 19$

33. $[3(8 - 4) + 12] - [10 - 2(3 + 5)]$
$= [3\cdot4 + 12] - [10 - 2\cdot8]$
$= [12 + 12] - [10 - 16]$
$= 24 - [-6]$
$= 24 + 6$
$= 30$

35. $[10(x + 3) - 4] + [2(x - 17) + 6]$
$= [10x + 30 - 4] + [2x - 34 + 6]$
$= [10x + 26] + [2x - 28]$
$= 10x + 26 + 2x - 28$
$= 12x - 2$

37. $[4(2x - 5) + 7] + [3(x + 3) + 5x]$
$= [8x - 20 + 7] + [3x + 9 + 5x]$
$= [8x - 13] + [8x + 9]$
$= 8x - 13 + 8x + 9$
$= 16x - 4$

39. $[7(x + 5) - 19] - [4(x - 6) + 10]$
$= [7x + 35 - 19] - [4x - 24 + 10]$
$= [7x + 16] - [4x - 14]$
$= 7x + 16 - 4x + 14$
$= 3x + 30$

41. $3\{[7(x - 2) + 4] - [2(2x - 5) + 6]\}$
$= 3\{[7x - 14 + 4] - [4x - 10 + 6]\}$
$= 3\{[7x - 10] - [4x - 4]\}$
$= 3\{7x - 10 - 4x + 4\}$
$= 3\{3x - 6\}$
$= 9x - 18$

43. $4\{[5(x - 3) + 2] - 3[2(x + 5) - 9]\}$
$= 4\{[5x - 15 + 2] - 3[2x + 10 - 9]\}$
$= 4\{[5x - 13] - 3[2x + 1]\}$
$= 4\{5x - 13 - 6x - 3\}$
$= 4\{-x - 16\}$
$= -4x - 64$

45. $6y + 2x - 3a + c$
$= 6y + (2x - 3a + c)$
$= 6y - (-2x + 3a - c)$

47. $\{x - [a - (a - x)] + [x - a]\} - 3x$
$= \{x - [a - a + x] + [x - a]\} - 3x$
$= \{x - x + x - a\} - 3x$
$= x - a - 3x$
$= -2x - a$

Exercise Set 2.9

1. $[9 - 2(5 - 1)]$
$= [9 - 2\cdot4]$
$= 9 - 8$
$= 1$

3. $8[7 - 6(4 - 2)]$
$= 8[7 - 6\cdot2]$
$= 8[7 - 12]$
$= 8[-5]$
$= -40$

5. $[4(9 - 6) + 11] - [14 - (6 - 4)]$
$= [4\cdot3 + 11] - [14 - 2]$
$= [12 + 11] - [12]$
$= 23 - 12$
$= 11$

7. $[32 \div (-4)] \div (-2)$
$= -8 \div (-2)$
$= 4$

9. $16(-24) + 50$
$= -384 + 50$
$= -334$

11. $2^4 + 2^3 - 10\cdot20$
$= 16 + 8 - 200$
$= 24 - 200$
$= -176$

13. $5^3 + 26\cdot71 - (16 + 25\cdot3)$
$= 5^3 + 26\cdot71 - (16 + 75)$
$= 5^3 + 26\cdot71 - 91$
$= 125 + 1846 - 91$
$= 1971 - 91$
$= 1880$

15. $(12\cdot 3 - 7\cdot 4)^2$
$= (36 - 28)^2$
$= 8^2$
$= 64$

17. $3000\cdot(1 + 0.16)^3$
$= 3000(1.16)^3$
$= 3000(1.560896)$
$= 4682.688$

19. $(20\cdot 4 + 12\cdot 7)^2 - (34\cdot 56)^3$
$= (80 + 84)^2 - (1904)^3$
$= 164^2 - 1904^3$
$= 26,896 - 6,902,411,264$
$= -6,902,384,368$

21. $a^2 + 2ab + b^2$
= A^2 + 2*A*B + B^2

23. $\frac{2(3 - b)}{c}$ = 2*(3 - B)/C

25. $\frac{a}{b} - \frac{c}{d}$ = A/B - C/D

27. 2*A + 7 = $2a + 7$

29. 3*A^2 - 5 = $3a^2 - 5$

31. (A + B)^2 = $(a + b)^2$

33. Original investment + Interest earned = Present value

$x + 9\%x = \$5995$

We solve:

$x + 0.09x = 5995$

$1.09x = 5995$

$x = \frac{5995}{1.09}$

$x = 5500$

The original investment was $5500.

35. Two times y plus three times x, $2y + 3x$

37. $(-x)^2 = (-x)(-x) = (-1\cdot x)(-1\cdot x)$
$= (-1)(-1)xx$
$= x^2$

Thus, $(-x)^2 = x^2$ is true.

Exercise Set 3.1

1.
$$\begin{aligned} x + 2 &= 6 \\ x + 2 + (-2) &= 6 + (-2) \\ x + 0 &= 4 \\ x &= 4 \end{aligned}$$

Check: $x + 2 = 6$

$4 + 2 \mid 6$

6

3.
$$\begin{aligned} x + 15 &= -5 \\ x + 15 + (-15) &= -5 + (-15) \\ x + 0 &= -20 \\ x &= -20 \end{aligned}$$

Check: $x + 15 = -5$

$-20 + 15 \mid -5$

-5

5.
$$\begin{aligned} r + \frac{2}{3} &= 1 \\ r + \frac{2}{3} + (-\frac{2}{3}) &= 1 + (-\frac{2}{3}) \\ r + 0 &= \frac{3}{3} + (-\frac{2}{3}) \\ r &= \frac{1}{3} \end{aligned}$$

Check: $r + \frac{2}{3} = 1$

$\frac{1}{3} + \frac{2}{3} \mid 1$

$\frac{3}{3}$

1

7.
$$\begin{aligned} x - \frac{5}{6} &= \frac{7}{8} \\ x - \frac{5}{6} + \frac{5}{6} &= \frac{7}{8} + \frac{5}{6} \\ x + 0 &= \frac{21}{24} + \frac{20}{24} \\ x &= \frac{41}{24} \end{aligned}$$

Check: $x - \frac{5}{6} = \frac{7}{8}$

$\frac{41}{24} - \frac{5}{6} \mid \frac{7}{8}$

$\frac{41}{24} - \frac{20}{24}$

$\frac{21}{24}$

$\frac{7}{8}$

9.
$$\begin{aligned} 8 + y &= 12 \\ -8 + 8 + y &= -8 + 12 \\ 0 + y &= 4 \\ y &= 4 \end{aligned}$$

Check: $8 + y = 12$

$8 + 4 \mid 12$

12

11.
$$\begin{aligned} \frac{1}{3} + a &= \frac{5}{6} \\ -\frac{1}{3} + \frac{1}{3} + a &= -\frac{1}{3} + \frac{5}{6} \\ 0 + a &= -\frac{2}{6} + \frac{5}{6} \\ a &= \frac{3}{6} \\ a &= \frac{1}{2} \end{aligned}$$

Check: $\frac{1}{3} + a = \frac{5}{6}$

$\frac{1}{3} + \frac{1}{2} \mid \frac{5}{6}$

$\frac{2}{6} + \frac{3}{6}$

$\frac{5}{6}$

13.
$$\begin{aligned} x - 2.3 &= -7.4 \\ x - 2.3 + 2.3 &= -7.4 + 2.3 \\ x &= -5.1 \end{aligned}$$

Check: $x - 2.3 = -7.4$

$-5.1 - 2.3 \mid -7.4$

-7.4

15.
$$\begin{aligned} -2.6 + x &= 8.3 \\ -2.6 + x + 2.6 &= 8.3 + 2.6 \\ x &= 10.9 \end{aligned}$$

Check: $-2.6 + x = 8.3$

$-2.6 + 10.9 \mid 8.3$

8.3

17.
$$\begin{aligned} m + \frac{5}{6} &= -\frac{11}{12} \\ m + \frac{5}{6} + (-\frac{5}{6}) &= -\frac{11}{12} + (-\frac{5}{6}) \\ m &= -\frac{11}{12} + (-\frac{10}{12}) \\ m &= -\frac{21}{12} \\ m &= -\frac{7}{4} \end{aligned}$$

Check: $m + \frac{5}{6} = -\frac{11}{12}$

$-\frac{7}{4} + \frac{5}{6} \mid -\frac{11}{12}$

$-\frac{21}{12} + \frac{10}{12}$

$-\frac{11}{12}$

19.
$$\begin{aligned} -6 &= -2 + y \\ -6 + 2 &= -2 + y + 2 \\ -4 &= y \end{aligned}$$

Check: $-6 = -2 + y$

$-6 \mid -2 + (-4)$

-6

21.
$$\begin{aligned} 10 &= y - 6 \\ 10 + 6 &= y - 6 + 6 \\ 16 &= y \end{aligned}$$

Check: $10 = y - 6$

$10 \mid 16 - 6$

10

23.
$$\begin{aligned} -9.7 &= -4.7 + y \\ 4.7 + (-9.7) &= 4.7 + (-4.7) + y \\ -5 &= y \end{aligned}$$

Check: $-9.7 = -4.7 + y$

$-9.7 \mid -4.7 + (-5)$

-9.7

25.
$$\begin{aligned} 5\frac{1}{6} + x &= 7 \\ -5\frac{1}{6} + 5\frac{1}{6} + x &= -5\frac{1}{6} + 7 \\ x &= 1\frac{5}{6}, \text{ or } \frac{11}{6} \end{aligned}$$

Check: $5\frac{1}{6} + x = 7$

$5\frac{1}{6} + 1\frac{5}{6} \mid 7$

7

27.
$$\begin{aligned} p + \frac{2}{3} &= 7\frac{1}{3} \\ p + \frac{2}{3} + (-\frac{2}{3}) &= 7\frac{1}{3} + (-\frac{2}{3}) \\ p &= 6\frac{2}{3}, \text{ or } \frac{20}{3} \end{aligned}$$

Check: $p + \frac{2}{3} = 7\frac{1}{3}$

$6\frac{2}{3} + \frac{2}{3} \mid 7\frac{1}{3}$

$7\frac{1}{3}$

29.
$$\begin{aligned} 22\frac{1}{7} &= 30 + t \\ -30 + 22\frac{1}{7} &= -30 + 30 + t \\ -7\frac{6}{7} &= t \\ \text{or } t &= -\frac{55}{7} \end{aligned}$$

Check: $22\frac{1}{7} = 30 + t$

$22\frac{1}{7} \mid 30 + (-7\frac{6}{7})$

$22\frac{1}{7}$

31. $-3 + (-8) = -11$

33. $-\frac{2}{3} \cdot \frac{5}{8} = -\frac{10}{24} = -\frac{5 \cdot 2}{12 \cdot 2} = -\frac{5}{12}$

35.
$$\begin{aligned} -356.788 &= -699.034 + t \\ 699.034 + (-356.788) &= 699.034 + (-699.034) + t \\ 342.246 &= t \end{aligned}$$

37.
$$\begin{aligned} x + 4 &= 5 + x \\ -x + x + 4 &= -x + 5 + x \\ 4 &= 5 \end{aligned}$$

Since $4 = 5$ is false, there is no solution.

Exercise Set 3.2

1.
$$\begin{aligned} 6x &= 36 \\ \frac{1}{6} \cdot 6x &= \frac{1}{6} \cdot 36 \\ 1 \cdot x &= 6 \\ x &= 6 \end{aligned}$$

Check: $\begin{array}{c|c} \multicolumn{2}{c}{6x = 36} \\ \hline 6 \cdot 6 & 36 \\ 36 & \end{array}$

3.
$$\begin{aligned} 9x &= -36 \\ \frac{1}{9} \cdot 9x &= \frac{1}{9} \cdot (-36) \\ 1x &= -4 \\ x &= -4 \end{aligned}$$

Check: $\begin{array}{c|c} \multicolumn{2}{c}{9x = -36} \\ \hline 9(-4) & -36 \\ -36 & \end{array}$

5.
$$\begin{aligned} -12x &= 72 \\ -\frac{1}{12} \cdot (-12x) &= -\frac{1}{12} \cdot 72 \\ 1 \cdot x &= -6 \\ x &= -6 \end{aligned}$$

Check: $\begin{array}{c|c} \multicolumn{2}{c}{-12x = 72} \\ \hline -12(-6) & 72 \\ 72 & \end{array}$

7.
$$\begin{aligned} \frac{1}{7} t &= 9 \\ 7 \cdot \frac{1}{7} t &= 7 \cdot 9 \\ 1 \cdot t &= 63 \\ t &= 63 \end{aligned}$$

Check: $\begin{array}{c|c} \multicolumn{2}{c}{\frac{1}{7} t = 9} \\ \hline \frac{1}{7} \cdot 63 & 9 \\ 9 & \end{array}$

9.
$$\begin{aligned} \frac{1}{5} &= \frac{1}{3} t \\ 3 \cdot \frac{1}{5} &= 3 \cdot \frac{1}{3} t \\ \frac{3}{5} &= 1 \cdot t \\ \frac{3}{5} &= t \end{aligned}$$

Check: $\begin{array}{c|c} \multicolumn{2}{c}{\frac{1}{5} = \frac{1}{3} t} \\ \hline \frac{1}{5} & \frac{1}{3} \cdot \frac{3}{5} \\ & \frac{1}{5} \end{array}$

11.
$$\begin{aligned} -2.7y &= 54 \\ -\frac{1}{2.7} \cdot (-2.7y) &= -\frac{1}{2.7} \cdot 54 \\ 1 \cdot y &= -\frac{54}{2.7} \\ y &= -\frac{540}{27} \qquad \left(-\frac{54}{2.7} \cdot \frac{10}{10} = -\frac{540}{27}\right) \\ y &= -20 \end{aligned}$$

Check: $\begin{array}{c|c} \multicolumn{2}{c}{-2.7y = 54} \\ \hline -2.7(-20) & 54 \\ 54 & \end{array}$

13.
$$\begin{aligned} \frac{3}{4} x &= 27 \\ \frac{4}{3} \cdot \frac{3}{4} x &= \frac{4}{3} \cdot 27 \\ 1 \cdot x &= \frac{108}{3} \\ x &= 36 \end{aligned}$$

Check: $\begin{array}{c|c} \multicolumn{2}{c}{\frac{3}{4} x = 27} \\ \hline \frac{3}{4} \cdot 36 & 27 \\ \frac{108}{4} & \\ 27 & \end{array}$

15.
$$\begin{aligned} -\frac{3}{5} r &= -\frac{9}{10} \\ -\frac{5}{3} \cdot \left(-\frac{3}{5} r\right) &= -\frac{5}{3} \cdot \left(-\frac{9}{10}\right) \\ r &= \frac{45}{30} \\ r &= \frac{3}{2} \end{aligned}$$

Check: $\begin{array}{c|c} \multicolumn{2}{c}{-\frac{3}{5} r = -\frac{9}{10}} \\ \hline -\frac{3}{5} \cdot \frac{3}{2} & -\frac{9}{10} \\ -\frac{9}{10} & \end{array}$

17.
$$\begin{aligned} 12 &= \frac{6}{5} y \\ \frac{5}{6} \cdot 12 &= \frac{5}{6} \cdot \frac{6}{5} y \\ \frac{60}{6} &= y \\ 10 &= y \end{aligned}$$

Check: $\begin{array}{c|c} \multicolumn{2}{c}{12 = \frac{6}{5} y} \\ \hline 12 & \frac{6}{5} \cdot 10 \\ & \frac{60}{5} \\ & 12 \end{array}$

19.
$$\begin{aligned} -3.3y &= 6.6 \\ -\frac{1}{3.3} \cdot (-3.3y) &= -\frac{1}{3.3} \cdot (6.6) \\ y &= -\frac{6.6}{3.3} \\ y &= -\frac{66}{33} \\ y &= -2 \end{aligned}$$

Check: $\begin{array}{c|c} \multicolumn{2}{c}{-3.3y = 6.6} \\ \hline -3.3(-2) & 6.6 \\ 6.6 & \end{array}$

21.
$$\begin{aligned} 38.7m &= 309.6 \\ \frac{1}{38.7} (38.7m) &= \frac{1}{38.7} (309.6) \\ m &= \frac{309.6}{38.7} \\ m &= \frac{3096}{387} \\ m &= 8 \end{aligned}$$

Check: $\begin{array}{c|c} \multicolumn{2}{c}{38.7m = 309.6} \\ \hline 38.7(8) & 309.6 \\ 309.6 & \end{array}$

23. $20.07 = \frac{3}{2} y$

$\frac{2}{3}\cdot(20.07) = \frac{2}{3} \cdot \frac{3}{2} y$

$\frac{40.14}{3} = y$

$13.38 = y$

Check: $20.07 = \frac{3}{2} y$

20.07	$\frac{3}{2}(13.38)$
	$\frac{40.14}{2}$
	20.07

25. $-\frac{3}{2} r = -\frac{27}{4}$

$-\frac{2}{3}\cdot(-\frac{3}{2} r) = -\frac{2}{3}\cdot(-\frac{27}{4})$

$r = \frac{54}{12}$

$r = \frac{9}{2}$

Check: $-\frac{3}{2} r = -\frac{27}{4}$

$-\frac{3}{2} \cdot \frac{9}{2}$	$-\frac{27}{4}$
$-\frac{27}{4}$	

27. $-\frac{2}{3} y = -10.6$

$-\frac{3}{2}\cdot(-\frac{2}{3} y) = -\frac{3}{2}\cdot(-10.6)$

$y = \frac{31.8}{2}$

$y = 15.9$

Check: $-\frac{2}{3} y = -10.6$

$-\frac{2}{3}(15.9)$	-10.6
$-\frac{31.8}{3}$	
-10.6	

29. $-x = 100$

$-1\cdot x = 100$

$-1\cdot(-1\cdot x) = -1\cdot 100$

$1\cdot x = -100$

$x = -100$

Check: $-x = 100$

$-(-100)$	100
100	

31. $-9 = -\frac{x}{6}$

$-9 = -1 \cdot \frac{x}{6}$

$-1\cdot(-9) = -1\cdot(-1 \cdot \frac{x}{6})$

$9 = \frac{x}{6}$

$6\cdot 9 = 6 \cdot \frac{x}{6}$

$54 = x$

Check: $-9 = -\frac{x}{6}$

-9	$-\frac{54}{6}$
	-9

33. $x - 8x = 1x - 8x = (1 - 8)x = -7x$

35. The distance of -3.2 from 0 is 3.2. Thus, $|-3.2| = 3.2$.

37. $0\cdot x = 0$

$0 = 0$

Since $0 = 0$ is true, $0\cdot x = 0$ is true for all rational numbers.

39. $4\cdot|x| = 48$

$\frac{1}{4} \cdot 4 \cdot |x| = \frac{1}{4} \cdot 48$

$|x| = 12$

$x = 12$ or $x = -12$ $(|12| = 12, |-12| = 12)$

The solutions are 12 and -12.

Exercise Set 3.3

1. $5x + 6 = 31$

$5x + 6 + (-6) = 31 + (-6)$

$5x = 25$

$\frac{1}{5}\cdot 5x = \frac{1}{5}\cdot 25$

$x = 5$

Check: $5x + 6 = 31$

$5\cdot 5 + 6$	31
$25 + 6$	
31	

3. $4x - 6 = 34$

$4x - 6 + 6 = 34 + 6$

$4x = 40$

$\frac{1}{4}\cdot 4x = \frac{1}{4}\cdot 40$

$x = 10$

Check: $4x - 6 = 34$

$4\cdot 10 - 6$	34
$40 - 6$	
34	

5. $7x + 2 = -54$

$7x + 2 + (-2) = -54 + (-2)$

$7x = -56$

$\frac{1}{7}\cdot 7x = \frac{1}{7}\cdot(-56)$

$x = -8$

Check: $7x + 2 = -54$

$7(-8) + 2$	-54
$-56 + 2$	
-54	

7. $5x + 7x = 72$

$12x = 72$

$\frac{1}{12}\cdot 12x = \frac{1}{12}\cdot 72$

$x = 6$

Check: $5x + 7x = 72$

$5\cdot 6 + 7\cdot 6$	72
$30 + 42$	
72	

9. $4y - 2y = 10$

$2y = 10$

$\frac{1}{2}\cdot 2y = \frac{1}{2}\cdot 10$

$y = 5$

Check: $4y - 2y = 10$

$4\cdot 5 - 2\cdot 5$	10
$20 - 10$	
10	

11. $10.2y - 7.3y = -58$

$2.9y = -58$

$\frac{1}{2.9}\cdot 2.9y = \frac{1}{2.9}\cdot(-58)$

$y = -\frac{58}{2.9}$

$y = -20$

11. (continued)

Check:
$$\begin{array}{r|l} \multicolumn{2}{c}{10.2y - 7.3y = -58} \\ \hline 10.2(-20) - 7.3(-20) & -58 \\ -204 + 146 & \\ -58 & \end{array}$$

13.
$$\begin{aligned} x + \tfrac{1}{3}x &= 8 \\ (1 + \tfrac{1}{3})x &= 8 \\ \tfrac{4}{3}x &= 8 \\ \tfrac{3}{4}\cdot\tfrac{4}{3}x &= \tfrac{3}{4}\cdot 8 \\ x &= 6 \end{aligned}$$

Check:
$$\begin{array}{r|l} \multicolumn{2}{c}{x + \tfrac{1}{3}x = 8} \\ \hline 6 + \tfrac{1}{3}\cdot 6 & 8 \\ 6 + 2 & \\ 8 & \end{array}$$

15.
$$\begin{aligned} x + 0.08x &= 9072 \\ 1x + 0.08x &= 9072 \\ 1.08x &= 9072 \\ \frac{1}{1.08}\cdot 1.08x &= \frac{1}{1.08}\cdot 9072 \\ x &= \frac{9072}{1.08} \\ x &= 8400 \end{aligned}$$

Check:
$$\begin{array}{r|l} \multicolumn{2}{c}{x + 0.08x = 9072} \\ \hline 8400 + 0.08(8400) & 9072 \\ 8400 + 672 & \\ 9072 & \end{array}$$

17.
$$\begin{aligned} 8y - 35 &= 3y \\ 8y &= 3y + 35 \\ 8y - 3y &= 35 \\ 5y &= 35 \\ y &= \frac{35}{5} \\ y &= 7 \end{aligned}$$

Check:
$$\begin{array}{r|l} \multicolumn{2}{c}{8y - 35 = 3y} \\ \hline 8\cdot 7 - 35 & 3\cdot 7 \\ 56 - 35 & 21 \\ 21 & \end{array}$$

19.
$$\begin{aligned} 4x - 7 &= 3x \\ 4x &= 3x + 7 \\ 4x - 3x &= 7 \\ x &= 7 \end{aligned}$$

Check:
$$\begin{array}{r|l} \multicolumn{2}{c}{4x - 7 = 3x} \\ \hline 4\cdot 7 - 7 & 3\cdot 7 \\ 28 - 7 & 21 \\ 21 & \end{array}$$

21.
$$\begin{aligned} x + 1 &= 16 - 4x \\ x + 4x &= 16 - 1 \\ 5x &= 15 \\ x &= \frac{15}{5} \\ x &= 3 \end{aligned}$$

Check:
$$\begin{array}{r|l} \multicolumn{2}{c}{x + 1 = 16 - 4x} \\ \hline 3 + 1 & 16 - 4\cdot 3 \\ 4 & 16 - 12 \\ & 4 \end{array}$$

23.
$$\begin{aligned} 2x - 1 &= 4 + x \\ 2x - x &= 4 + 1 \\ x &= 5 \end{aligned}$$

Check:
$$\begin{array}{r|l} \multicolumn{2}{c}{2x - 1 = 4 + x} \\ \hline 2\cdot 5 - 1 & 4 + 5 \\ 10 - 1 & 9 \\ 9 & \end{array}$$

25.
$$\begin{aligned} 5x + 2 &= 3x + 6 \\ 5x - 3x &= 6 - 2 \\ 2x &= 4 \\ x &= 2 \end{aligned}$$

Check:
$$\begin{array}{r|l} \multicolumn{2}{c}{5x + 2 = 3x + 6} \\ \hline 5\cdot 2 + 2 & 3\cdot 2 + 6 \\ 10 + 2 & 6 + 6 \\ 12 & 12 \end{array}$$

27.
$$\begin{aligned} 5 - 2x &= 3x - 7x + 25 \\ 5 - 2x &= -4x + 25 \\ 4x - 2x &= 25 - 5 \\ 2x &= 20 \\ x &= \frac{20}{2} \\ x &= 10 \end{aligned}$$

Check:
$$\begin{array}{r|l} \multicolumn{2}{c}{5 - 2x = 3x - 7x + 25} \\ \hline 5 - 2\cdot 10 & 3\cdot 10 - 7\cdot 10 + 25 \\ 5 - 20 & 30 - 70 + 25 \\ -15 & -40 + 25 \\ & -15 \end{array}$$

29.
$$\begin{aligned} 4 + 3x - 6 &= 3x + 2 - x \\ 3x - 2 &= 2x + 2 \\ 3x - 2x &= 2 + 2 \\ x &= 4 \end{aligned}$$

Check:
$$\begin{array}{r|l} \multicolumn{2}{c}{4 + 3x - 6 = 3x + 2 - x} \\ \hline 4 + 3\cdot 4 - 6 & 3\cdot 4 + 2 - 4 \\ 4 + 12 - 6 & 12 + 2 - 4 \\ 16 - 6 & 14 - 4 \\ 10 & 10 \end{array}$$

31.
$$\begin{aligned} 4y - 4 + y &= 6y + 20 - 4y \\ 5y - 4 &= 2y + 20 \\ 5y - 2y &= 20 + 4 \\ 3y &= 24 \\ y &= \frac{24}{3} \\ y &= 8 \end{aligned}$$

Check:
$$\begin{array}{r|l} \multicolumn{2}{c}{4y - 4 + y = 6y + 20 - 4y} \\ \hline 4\cdot 8 - 4 + 8 & 6\cdot 8 + 20 - 4\cdot 8 \\ 32 - 4 + 8 & 48 + 20 - 32 \\ 28 + 8 & 68 - 32 \\ 36 & 36 \end{array}$$

33. $\frac{5}{2}x + \frac{1}{2}x = 3x + \frac{3}{2} + \frac{5}{2}x$

$$\begin{aligned} \frac{1}{2}x &= 3x + \frac{3}{2} && \text{(Adding } -\tfrac{5}{2}x\text{)} \\ 2(\tfrac{1}{2}x) &= 2(3x + \tfrac{3}{2}) && \text{(Clearing fractions)} \\ x &= 6x + 3 \\ -3 &= 6x - x \\ -3 &= 5x \\ -\frac{3}{5} &= x \end{aligned}$$

Check:

$\frac{5}{2}x + \frac{1}{2}x$	$= 3x + \frac{3}{2} + \frac{5}{2}x$
$\frac{5}{2}(-\frac{3}{5}) + \frac{1}{2}(-\frac{3}{5})$	$3(-\frac{3}{5}) + \frac{3}{2} + \frac{5}{2}(-\frac{3}{5})$
$-\frac{15}{10} - \frac{3}{10}$	$-\frac{9}{5} + \frac{3}{2} - \frac{3}{2}$
$-\frac{18}{10}$	$-\frac{9}{5}$
$-\frac{9}{5}$	

35.

$$\begin{aligned} 2.1x + 45.2 &= 3.2 - 8.4x \\ 10(2.1x + 45.2) &= 10(3.2 - 8.4x) \\ & \text{(Clearing decimals)} \\ 21x + 452 &= 32 - 84x \\ 21x + 84x &= 32 - 452 \\ 105x &= -420 \\ x &= \frac{-420}{105} \\ x &= 4 \end{aligned}$$

Check:

$2.1x + 45.2$	$= 3.2 - 8.4x$
$2.1(-4) + 45.2$	$3.2 - 8.4(-4)$
$-8.4 + 45.2$	$3.2 + 33.6$
36.8	36.8

37.

$$\begin{aligned} \frac{1}{5}t - 0.4 + \frac{2}{5}t &= 0.6 - \frac{1}{10}t \\ 10(\tfrac{1}{5}t - 0.4 + \tfrac{2}{5}t) &= 10(0.6 - \tfrac{1}{10}t) \\ & \text{(Clearing decimals and fractions)} \\ 2t - 4 + 4t &= 6 - t \\ 6t - 4 &= 6 - t \\ 6t + t &= 6 + 4 \\ 7t &= 10 \\ t &= \frac{10}{7} \end{aligned}$$

39. $\frac{80}{96} = \frac{5\cdot16}{6\cdot16} = \frac{5}{6}\cdot\frac{16}{16} = \frac{5}{6}$

41. When $x = -14$,
$-(-x) = -[-(-14)] = -[14] = -14$

43.

$$\begin{aligned} 0.008 + 9.62x - 42.8 &= 0.944x + 0.0083 - x \\ 9.62x - 42.792 &= -0.056x + 0.0083 \\ 9.62x + 0.056x &= 0.0083 + 42.792 \\ 9.676x &= 42.8003 \\ x &= \frac{42.8003}{9.676} \\ x &\approx 4.42 \end{aligned}$$

45.

$$\begin{aligned} 0 &= y - (-14) - (-3y) \\ 0 &= y + 14 + 3y \\ 0 &= 4y + 14 \\ -14 &= 4y \\ \frac{-14}{4} &= y \\ -\frac{7}{2} &= y \end{aligned}$$

Exercise Set 3.4

1.

$$\begin{aligned} 3(2y - 3) &= 27 \\ 6y - 9 &= 27 \\ 6y &= 27 + 9 \\ 6y &= 36 \\ y &= \frac{36}{6} \\ y &= 6 \end{aligned}$$

Check:

$3(2y - 3)$	$= 27$
$3(2\cdot6 - 3)$	27
$3(12 - 3)$	
$3\cdot9$	
27	

3.

$$\begin{aligned} 40 &= 5(3x + 2) \\ 40 &= 15x + 10 \\ 40 - 10 &= 15x \\ 30 &= 15x \\ \frac{30}{15} &= x \\ 2 &= x \end{aligned}$$

Check:

$40 =$	$5(3x + 2)$
40	$5(3\cdot2 + 2)$
	$5(6 + 2)$
	$5\cdot8$
	40

5.

$$\begin{aligned} 2(3 + 4m) - 9 &= 45 \\ 6 + 8m - 9 &= 45 \\ 8m - 3 &= 45 \\ 8m &= 45 + 3 \\ 8m &= 48 \\ m &= \frac{48}{8} \\ m &= 6 \end{aligned}$$

Check:

$2(3 + 4m) - 9$	$= 45$
$2(3 + 4\cdot6) - 9$	45
$2(3 + 24) - 9$	
$2\cdot27 - 9$	
$54 - 9$	
45	

7.

$$\begin{aligned} 5r - (2r + 8) &= 16 \\ 5r - 2r - 8 &= 16 \\ 3r - 8 &= 16 \\ 3r &= 16 + 8 \\ 3r &= 24 \\ r &= \frac{24}{3} \\ r &= 8 \end{aligned}$$

Check:

$5r - (2r + 8)$	$= 16$
$5\cdot8 - (2\cdot8 + 8)$	16
$40 - (16 + 8)$	
$40 - 24$	
16	

9.
$$\begin{aligned} 3g - 3 &= 3(7 - g) \\ 3g - 3 &= 21 - 3g \\ 3g + 3g &= 21 + 3 \\ 6g &= 24 \\ g &= \frac{24}{6} \\ g &= 4 \end{aligned}$$

Check:

$3g - 3$	$= 3(7 - g)$
$3 \cdot 4 - 3$	$3(7 - 4)$
$12 - 3$	$3 \cdot 3$
9	9

11.
$$\begin{aligned} 6 - 2(3x - 1) &= 2 \\ 6 - 6x + 2 &= 2 \\ 8 - 6x &= 2 \\ 8 - 2 &= 6x \\ 6 &= 6x \\ \frac{6}{6} &= x \\ 1 &= x \end{aligned}$$

Check:

$6 - 2(3x - 1)$	$= 2$
$6 - 2(3 \cdot 1 - 1)$	2
$6 - 2(3 - 1)$	
$6 - 2 \cdot 2$	
$6 - 4$	
2	

13.
$$\begin{aligned} 5(d + 4) &= 7(d - 2) \\ 5d + 20 &= 7d - 14 \\ 20 + 14 &= 7d - 5d \\ 34 &= 2d \\ \frac{34}{2} &= d \\ 17 &= d \end{aligned}$$

15.
$$\begin{aligned} 3(x - 2) &= 5(x + 2) \\ 3x - 6 &= 5x + 10 \\ 3x - 5x &= 10 + 6 \\ -2x &= 16 \\ x &= -\frac{16}{2} \\ x &= -8 \end{aligned}$$

Check:

$3(x - 2)$	$= 5(x + 2)$
$3(-8 - 2)$	$5(-8 + 2)$
$3(-10)$	$5(-6)$
-30	-30

17.
$$\begin{aligned} 8(2t + 1) &= 4(7t + 7) \\ 16t + 8 &= 28t + 28 \\ 16t - 28t &= 28 - 8 \\ -12t &= 20 \\ t &= -\frac{20}{12} \\ t &= -\frac{5}{3} \end{aligned}$$

19.
$$\begin{aligned} 3(r - 6) + 2 &= 4(r + 2) - 21 \\ 3r - 18 + 2 &= 4r + 8 - 21 \\ 3r - 16 &= 4r - 13 \\ 13 - 16 &= 4r - 3r \\ -3 &= r \end{aligned}$$

Check:

$3(r - 6) + 2$	$= 4(r + 2) - 21$
$3(-3 - 6) + 2$	$4(-3 + 2) - 21$
$3(-9) + 2$	$4(-1) - 21$
$-27 + 2$	$-4 - 21$
-25	-25

21.
$$\begin{aligned} 19 - (2x + 3) &= 2(x + 3) + x \\ 19 - 2x - 3 &= 2x + 6 + x \\ 16 - 2x &= 3x + 6 \\ 16 - 6 &= 3x + 2x \\ 10 &= 5x \\ \frac{10}{5} &= x \\ 2 &= x \end{aligned}$$

23.
$$\begin{aligned} \frac{1}{4}(8y + 4) - 17 &= -\frac{1}{2}(4y - 8) \\ 2y + 1 - 17 &= -2y + 4 \\ 2y - 16 &= -2y + 4 \\ 2y + 2y &= 4 + 16 \\ 4y &= 20 \\ y &= \frac{20}{4} \\ y &= 5 \end{aligned}$$

25.
$$C = 2\pi r$$
$$C = 2 \cdot \frac{22}{7} \cdot \frac{7}{2} \quad \text{(Substituting)}$$
$$C = 22 \text{ ft}$$

27.
$$\begin{aligned} &7x - 21 - 14y \\ &= 7 \cdot x - 7 \cdot 3 - 7 \cdot 2y \\ &= 7(x - 3 - 2y) \end{aligned}$$

29.
$$\begin{aligned} 475(54x + 7856) + 9762 &= 402(83x + 975) \\ 25{,}650x + 3{,}731{,}600 + 9762 &= 33{,}366x + 391{,}950 \\ 25{,}650x + 3{,}741{,}362 &= 33{,}366x + 391{,}950 \\ 3{,}741{,}362 - 391{,}950 &= 33{,}366x - 25{,}650x \\ 3{,}349{,}412 &= 7716x \\ \frac{3{,}349{,}412}{7716} &= x \\ \frac{837{,}353}{1929} &= x \\ 434.087 &\approx x \end{aligned}$$

31.
$$\begin{aligned} 30{,}000 + 20{,}000x &= 55{,}000(1 + 12{,}500x) \\ 30{,}000 + 20{,}000x &= 55{,}000 + 687{,}500{,}000x \\ 30{,}000 - 55{,}000 &= 687{,}500{,}000x - 20{,}000x \\ -25{,}000 &= 687{,}480{,}000x \\ -\frac{25{,}000}{687{,}480{,}000} &= x \\ -\frac{25}{687{,}480} &= x \\ -\frac{5}{137{,}496} &= x \\ -0.0000364 &\approx x \end{aligned}$$

Exercise Set 3.5

1. Let x represent the lowest temperature ever recorded in Los Angeles.

   Lowest temperature in Minneapolis is 62° less than lowest temperature in Los Angeles.

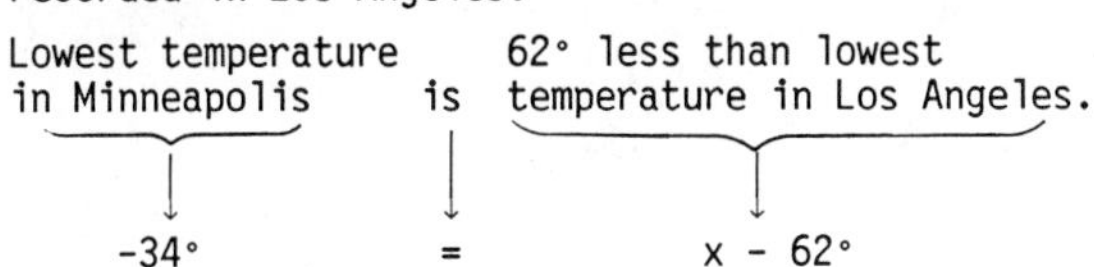

   $-34° = x - 62°$

   We solve the equation.

   $-34 = x - 62$

   $28 = x$

   We check: $28° - 62° = -34°$

   The lowest temperature ever recorded in Los Angeles is 28°.

3. Let x be a certain number. Then 6x is 6 times that number.

   18 subtracted from six times a certain number is 96.

   $6x - 18 = 96$

   We solve the equation.

   $6x - 18 = 96$

   $6x = 96 + 18$

   $6x = 114$

   $x = \frac{114}{6}$

   $x = 19$

   Six times 19 is 114. Subtracting 18 from 114 is 96. This checks.

   The number is 19.

5. Let x be a certain number. Then 2x is the double of the number, and $\frac{2}{5}x$ is two-fifths of the number.

   16 added to the double of a number is two-fifths of the original number.

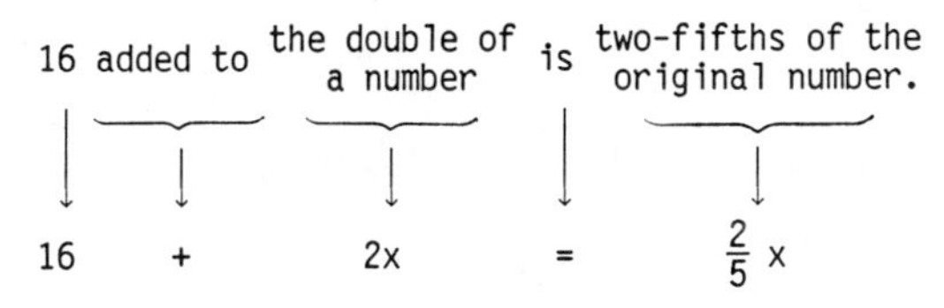

   $16 + 2x = \frac{2}{5}x$

   We solve the equation.

   $16 + 2x = \frac{2}{5}x$

   $5(16 + 2x) = 5(\frac{2}{5}x)$, LCM = 5.

   $80 + 10x = 2x$

   $10x - 2x = -80$

   $8x = -80$

   $x = -\frac{80}{8}$

   $x = -10$

   The double of -10 is -20. Sixteen added to -20 is -4. Two-fifths of -10 is -4. This checks.

   The original number is -10.

7. This problem is stated explicitly enough that we can go right to the translation.

   A number plus two-fifths of the number is 56.

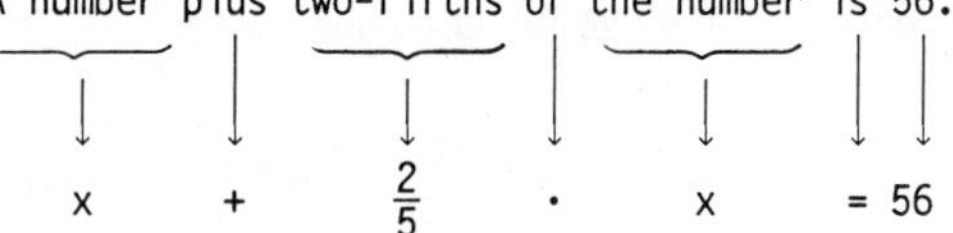

   $x + \frac{2}{5} \cdot x = 56$

   We solve the equation.

   $x + \frac{2}{5}x = 56$

   $(1 + \frac{2}{5})x = 56$

   $\frac{7}{5}x = 56$

   $\frac{5}{7} \cdot \frac{7}{5}x = \frac{5}{7} \cdot 56$

   $x = 40$

   Two-fifths of 40 is 16. The number 40 plus 16 is 56. This checks.

   The number is 40.

9. First draw a picture.

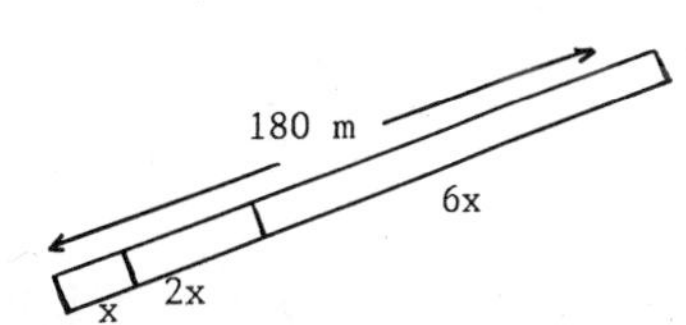

   We use x for the first length, 2x for the second length, and 3·2x, or 6x, for the third length.

   Length of 1st piece plus Length of 2nd piece plus Length of 3rd piece is 180

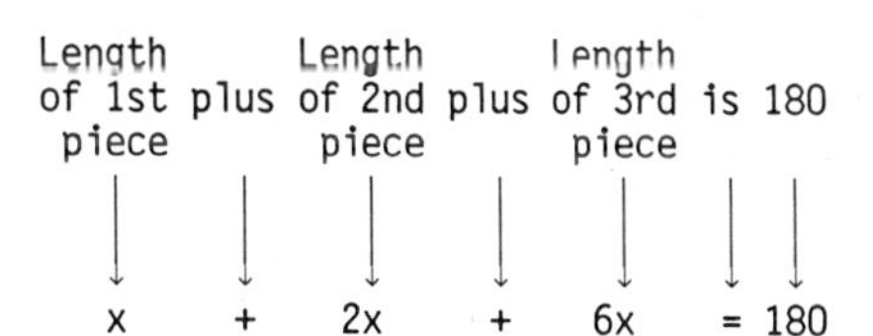

   $x + 2x + 6x = 180$

   We solve the equation.

   $x + 2x + 6x = 180$

   $9x = 180$

   $x = \frac{180}{9}$

   $x = 20$

   If the first piece is 20 m long, then the second is 2·20 m, or 40 m and the third is 6·20 m, or 120 m. The lengths of these pieces add up to 180 m (20 + 40 + 120 = 180). This checks.

   The first piece measures 20 m. The second measures 40 m, and the third measures 120 m.

11. We let x be the smaller odd integer. Then x + 2 is the next larger consecutive odd integer.

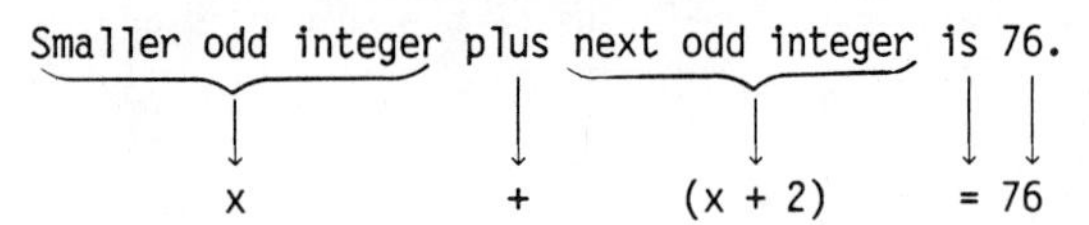

We solve the equation.

$$x + (x + 2) = 76$$
$$2x + 2 = 76$$
$$2x = 76 - 2$$
$$2x = 74$$
$$x = \frac{74}{2}$$
$$x = 37$$

If the smaller odd integer is 37, then the larger is 37 + 2, or 39. They are consecutive odd integers. Their sum, 37 + 39, is 76. This checks.

The integers are 37 and 39.

13. We let x be the smaller even integer. Then x + 2 is the next larger consecutive even integer.

Smaller even integer plus next even integer is 114.

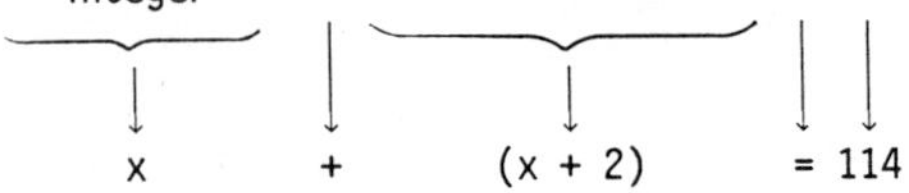

We solve the equation.

$$x + (x + 2) = 114$$
$$2x + 2 = 114$$
$$2x = 114 - 2$$
$$2x = 112$$
$$x = \frac{112}{2}$$
$$x = 56$$

If the smaller even integer is 56, then the larger is 56 + 2, or 58. They are consecutive even integers. Their sum, 56 + 58, is 114. This checks.

The integers are 56 and 58.

15. Let x be the smallest integer. Then x + 1 and x + 1 + 1, or x + 2, are the next two consecutive integers.

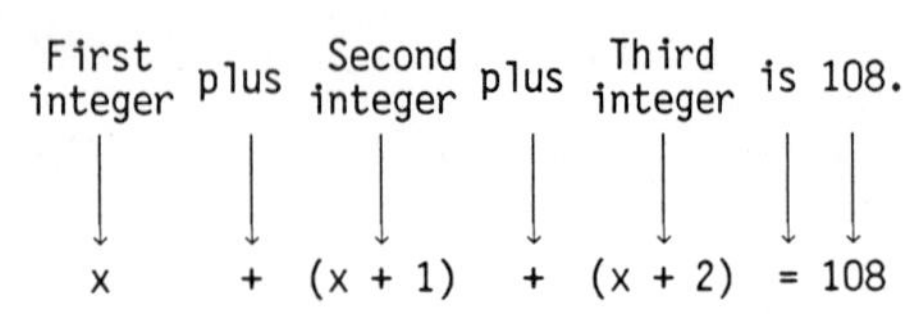

15. (continued)

We solve the equation.

$$x + (x + 1) + (x + 2) = 108$$
$$3x + 3 = 108$$
$$3x = 108 - 3$$
$$3x = 105$$
$$x = \frac{105}{3}$$
$$x = 35$$

If the smallest integer is 35, then the second is 35 + 1, or 36, and the third is 35 + 1 + 1, or 37. They are consecutive integers. Their sum, 35 + 36 + 37, is 108. This checks.

The integers are 35, 36, and 37.

17. Let x be the smallest odd integer. Then x + 2 and x + 2 + 2, or x + 4, are the next two consecutive odd integers.

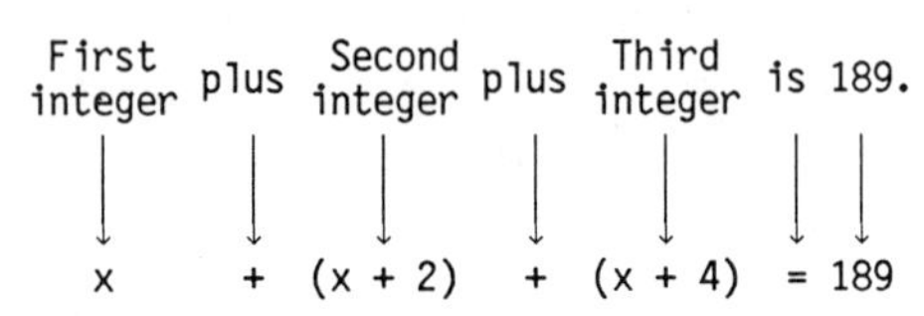

We solve the equation.

$$x + (x + 2) + (x + 4) = 189$$
$$3x + 6 = 189$$
$$3x = 189 - 6$$
$$3x = 183$$
$$x = \frac{183}{3}$$
$$x = 6$$

If the smallest odd integer is 61, then the next consecutive odd integer is 61 + 2, or 63. The third is 61 + 4, or 65. They are consecutive odd integers. Their sum, 61 + 63 + 65, is 189. This checks.

The integers are 61, 63, and 65.

19. First draw a picture.

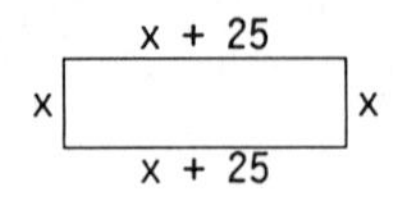

Let x represent the width. Then x + 25 represents the length.

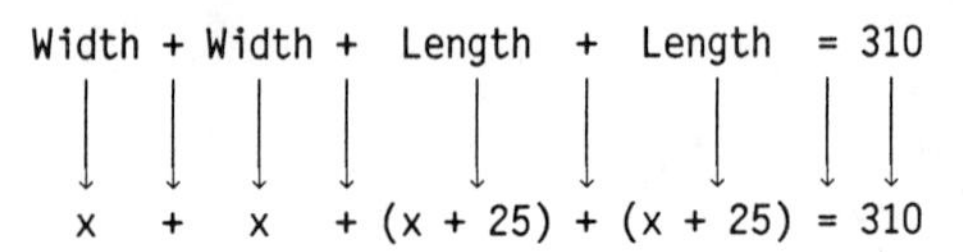

19. (continued)

We solve the equation.

$$x + x + (x + 25) + (x + 25) = 310$$
$$4x + 50 = 310$$
$$4x = 310 - 50$$
$$4x = 260$$
$$x = \frac{260}{4}$$
$$x = 65$$

The length which is 25 more than the width is 65 + 25, or 90.

The perimeter is 65 + 65 + 90 + 90, which is 310. This checks.

The width is 65 m, and the length is 90 m.

21. First draw a picture.

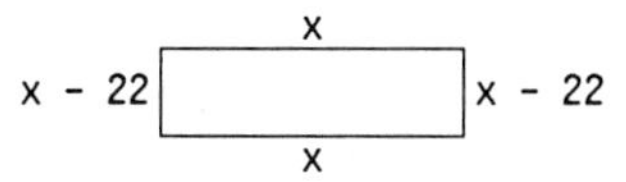

x

x - 22 [rectangle] x - 22

x

Let x represent the length. Then x - 22 represents the width.

Width + Width + Length + Length is 152.

$$(x - 22) + (x - 22) + x + x = 152$$

We solve the equation.

$$(x - 22) + (x - 22) + x + x = 152$$
$$4x - 44 = 152$$
$$4x = 152 + 44$$
$$4x = 196$$
$$x = \frac{196}{4}$$
$$x = 49$$

The width which is 22 less than the length is 49 - 22, or 27.

The perimeter is 27 + 27 + 49 + 49, which is 152. This checks.

The width is 27 m, and the length is 49 m.

23.

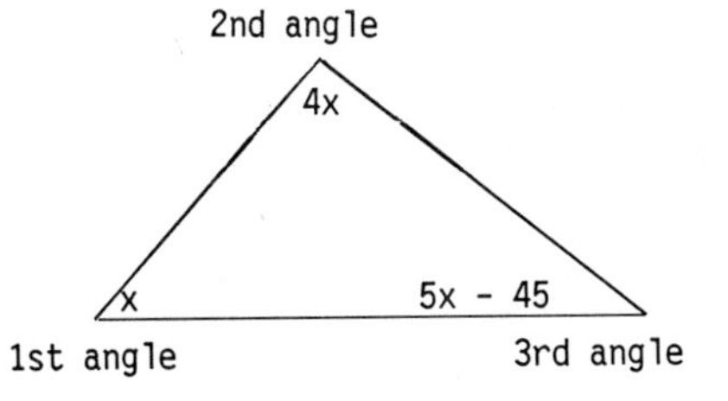

23. (continued)

We let x represent the measure of the first angle, 4x the second angle, and (x + 4x) - 45, or 5x - 45, the third angle. The measures of the angles of any triangle add up to 180°.

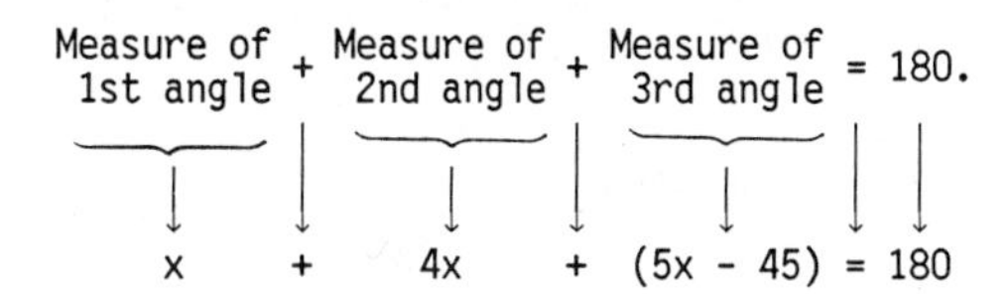

We solve the equation.

$$x + 4x + (5x - 45) = 180$$
$$10x - 45 = 180$$
$$10x = 180 + 45$$
$$10x = 225$$
$$x = \frac{225}{10}$$
$$x = 22.5$$

Possible answers for the angle measures are as follows:

1st angle: $x = 22.5°$
2nd angle: $4x = 4(22.5) = 90°$
3rd angle: $5x - 45 = 5(22.5) - 45 = 112.5 - 45 = 67.5°$

Consider 22.5°, 90°, and 67.5°. The second is four times the first, and the third is 45° less than five times the first. The sum is 180°. These numbers check.

The measure of the first angle is 22.5°.

25. Let x represent the original investment. Interest earned in 1 year is found by taking 7% of the original investment. We let 7%·x represent the interest. The amount in the account at the end of the year is the sum of the original investment and the interest earned.

Original investment plus interest earned is $4708.

$$x + 7\% \cdot x = 4708$$

We solve the equation.

$$x + 7\%x = 4708$$
$$1 \cdot x + 0.07x = 4708$$
$$1.07x = 4708$$
$$x = \frac{4708}{1.07}$$
$$x = 4400$$

7% of $4400 is $308. Adding this to $4400 we get $4708. This checks.

The original investment was $4400.

27. Let x represent the marked price. Then, $40\%\cdot x$ represents the reduction. The sale price is found by subtracting the amount of reduction from the marked price.

Marked price minus reduction is \$9.60.

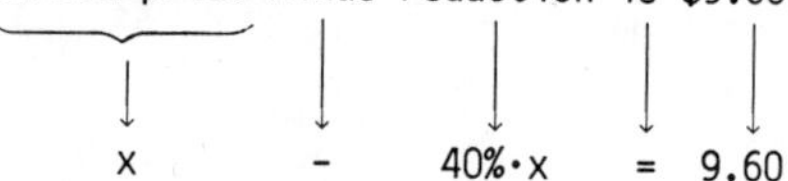

$x \quad - \quad 40\%\cdot x \quad = \quad 9.60$

We solve the equation.

$x - 40\%x = 9.60$

$1\cdot x - 0.40x = 9.60$

$(1 - 0.40)x = 9.60$

$0.6x = 9.60$

$x = \frac{9.60}{0.6}$

$x = 16$

40% of \$16 is \$6.40. Subtracting this from \$16 we get \$9.60. This checks.

The marked price was \$16.

29. The total cost is the daily charge plus the mileage charge. The total cost cannot exceed \$200

Daily rate plus Cost per mile times Number of miles driven is Budget

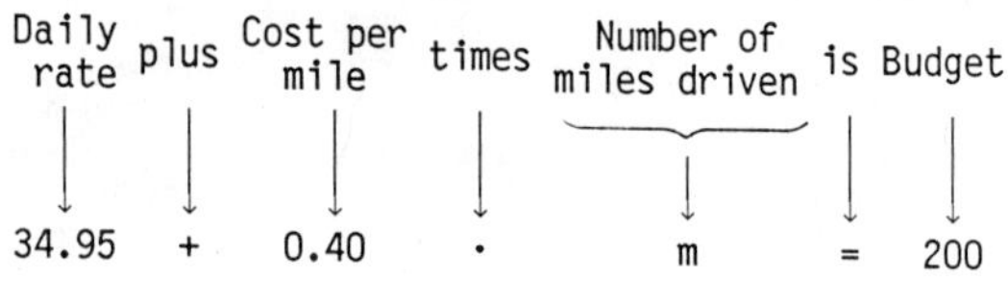

$34.95 \quad + \quad 0.40 \quad \cdot \quad m \quad = \quad 200$

We let m represent the number of miles driven.

$34.95 + 0.40m = 200$

$0.40m = 200 - 34.95$

$0.40m = 165.05$

$m = \frac{165.05}{0.40}$

$m = 412.625$

The mileage cost is found by multiplying 412.625 by \$0.40 obtaining \$165.05. Then we add \$165.05 to \$34.95, the daily rate, and get \$200.

The mileage must be approximately 412.6 miles or less to stay within budget.

31.

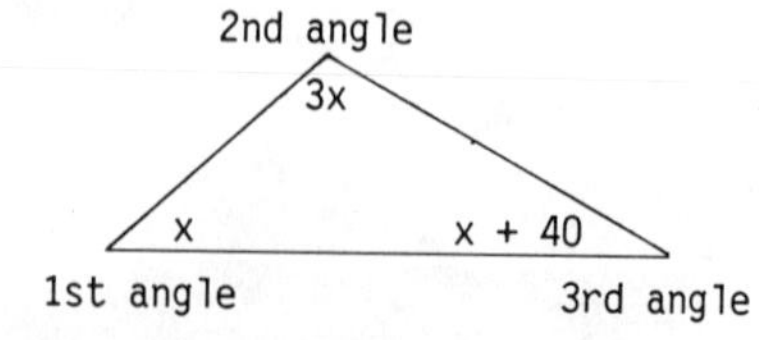

We let x represent the measure of the first angle, 3x the second angle, and x + 40 the third angle. The measures of the angles of any triangle add up to 180°.

Measure of 1st angle + Measure of 2nd angle + Measure of 3rd angle = 180

$x \quad + \quad 3x \quad + \quad (x + 40) \quad = 180$

31. (continued)

$x + 3x + (x + 40) = 180$

$5x + 40 = 180$

$5x = 180 - 40$

$5x = 140$

$x = \frac{140}{5}$

$x = 28$

Possible answers for the angle measures are as follows:

1st angle: $x = 28°$
2nd angle: $3x = 3(28) = 84°$
3rd angle: $x + 40 = 28 + 40 = 68°$

Consider 28°, 84°, and 68°. The second angle is three times the first, and the third is 40° more than the first. The sum, 28° + 84° + 68°, is 180°. These numbers check.

The measures of the angles are 28°, 84°, and 68°, respectively.

33. We let x represent the former population. Then 2%x represents the increase.

Former population plus 2% increase is 4.8 billion.

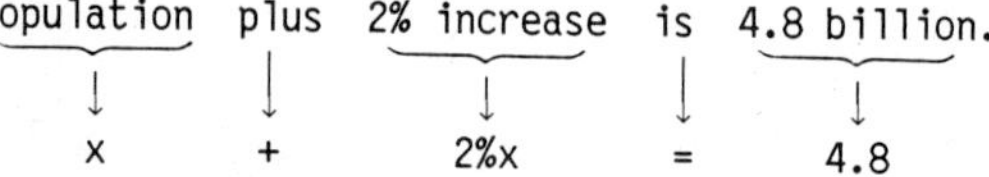

$x \quad + \quad 2\%x \quad = \quad 4.8$

We solve the equation.

$x + 2\%x = 4.8$

$1.02x = 4.8$

$x = \frac{4.8}{1.02}$

$x \approx 4.71$

To check we find 2% of 4.71 and add the result to 4.71.

$2\% \times 4.71 \approx 0.09, \quad 4.71 + 0.09 = 4.8$

The value checks.

The former population was approximately 4.71 billion.

35. $R = -0.028t + 20.8$

R stands for the record in seconds, and t stands for the number of years since 1920.

Substitute 19.0 for R and solve for t.

$R = -0.028t + 20.8$

$19.0 = -0.028t + 20.8$

$0.028t = 20.8 - 19.0$

$0.028t = 1.8$

$t = \frac{1.8}{0.028}$

$t \approx 64.3$

In approximately 64.3 years since 1920, the record should be 19.0 seconds.

37. 1776 plus "four score and seven" is 1863.

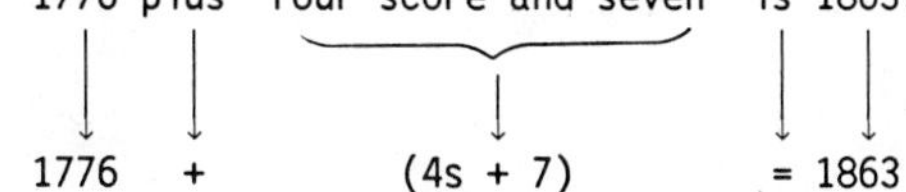

1776 + (4s + 7) = 1863

We let s represent a score.

$1776 + (4s + 7) = 1863$

$4s + 1783 = 1863$

$4s = 1863 - 1783$

$4s = 80$

$s = \frac{80}{4}$

$s = 20$

If a score is 20 years, then four score and seven represents 87 years. Adding 87 to 1776 we get 1863. This checks.

A score is 20 years.

39. Let x represent the amount the pump reads. Then 9%x represents the tax to be added.

Amount on tank plus tax is $10.

x + 9%x = 10

We solve:

$x + 9\%x = 10$

$1.09x = 10$

$x = \frac{10}{1.09}$

$x \approx 9.17$

The pump should read $9.17, not $9.10.

Exercise Set 3.6

1. $A = bh$

$A\cdot\frac{1}{h} = bh\cdot\frac{1}{h}$

$\frac{A}{h} = b$

3. $d = rt$

$d\cdot\frac{1}{t} = rt\cdot\frac{1}{t}$

$\frac{d}{t} = r$

5. $I = Prt$

$I\cdot\frac{1}{rt} = Prt\cdot\frac{1}{rt}$

$\frac{I}{rt} = P$

7. $F = ma$

$\frac{1}{m}\cdot F = \frac{1}{m}\cdot ma$

$\frac{F}{m} = a$

9. $P = 2\ell + 2w$

$P - 2\ell = 2w$

$\frac{1}{2}(P - 2\ell) = \frac{1}{2}\cdot 2w$

$\frac{P - 2\ell}{2} = w$

or $\frac{1}{2}P - \ell = w$

11. $A = \pi r^2$

$\frac{1}{\pi}\cdot A = \frac{1}{\pi}\cdot\pi r^2$

$\frac{A}{\pi} = r^2$

13. $A = \frac{1}{2}bh$

$2\cdot A = 2\cdot\frac{1}{2}bh$

$2A = bh$

$2A\cdot\frac{1}{h} = bh\cdot\frac{1}{h}$

$\frac{2A}{h} = b$

15. $E = mc^2$

$E\cdot\frac{1}{c^2} = mc^2\cdot\frac{1}{c^2}$

$\frac{E}{c^2} = m$

17. $A = \frac{a + b + c}{3}$

$3\cdot A = 3\cdot\frac{a + b + c}{3}$

$3A = a + b + c$

$3A - a - c = b$

19. $v = \frac{3k}{t}$

$t\cdot v = t\cdot\frac{3k}{t}$

$tv = 3k$

$tv\cdot\frac{1}{v} = 3k\cdot\frac{1}{v}$

$t = \frac{3k}{v}$

21. $A = \frac{1}{2}ah + \frac{1}{2}bh$

$2\cdot A = 2(\frac{1}{2}ah + \frac{1}{2}bh)$

$2A = ah + bh$

$2A - ah = bh$

$(2A - ah)\cdot\frac{1}{h} = bh\cdot\frac{1}{h}$

$\frac{2A - ah}{h} = b$

or $\frac{2A}{h} - a = b$

23. $H = \frac{D^2N}{2.5}$

$2.5\cdot H = 2.5\cdot\frac{D^2N}{2.5}$

$2.5H = D^2N$

$2.5H\frac{1}{N} = D^2N\cdot\frac{1}{N}$

$\frac{2.5H}{N} = D^2$

25. $$A = \frac{\pi r^2 S}{360}$$
$$360 \cdot A = 360 \cdot \frac{\pi r^2 S}{360}$$
$$360A = \pi r^2 S$$
$$\frac{1}{\pi r^2} \cdot 360A = \frac{1}{\pi r^2} \cdot \pi r^2 S$$
$$\frac{360A}{\pi r^2} = S$$

27. $$R = -0.0075t + 3.85$$
$$R - 3.85 = -0.0075t$$
$$\frac{R - 3.85}{-0.0075} = t$$
Note: $\frac{R - 3.85}{-0.0075} \cdot \frac{-1}{-1} = \frac{-R + 3.85}{0.0075} = \frac{3.85 - R}{0.0075}$
$$t = \frac{R - 3.85}{-0.0075} \text{ or } \frac{3.85 - R}{0.0075}$$

29. What percent of 7500 is 2500?
$$x \quad \% \quad \cdot 7500 = 2500$$
$$x \times 0.01 \times 7500 = 2500$$
$$x(75) = 2500$$
$$x = \frac{2500}{75}$$
$$x = 33.\overline{3}$$
The answer is $33.\overline{3}\%$.

31. $-45.8 - (-32.6) = -45.8 + 32.6 = -13.2$

33. $A = \ell w$
Substitute 2ℓ for ℓ and $2w$ for w and compare the A values.
$A = 2\ell \cdot 2w = 4 \cdot \ell w$
When ℓ and w are each doubled, A quadruples.

35. $A = \frac{1}{2}bh$
Substitute $b + 4$ for b and compare the A values.
$$A = \frac{1}{2}(b + 4)h$$
$$A = \frac{1}{2}(bh + 4h)$$
$$A = \frac{1}{2} \cdot bh + \frac{1}{2} \cdot 4h$$
$$A = \frac{1}{2}bh + 2h$$
When b increases by 4 units, A increases by 2h units.

Exercise Set 3.7

1. $3^{-2} = \frac{1}{3^2}$, or $\frac{1}{3 \cdot 3}$, or $\frac{1}{9}$

3. $10^{-4} = \frac{1}{10^4}$, or $\frac{1}{10 \cdot 10 \cdot 10 \cdot 10}$, or $\frac{1}{10,000}$

5. $\frac{1}{4^3} = 4^{-3}$

7. $\frac{1}{x^3} = x^{-3}$

9. $\frac{1}{a^4} = a^{-4}$

11. $\frac{1}{p^n} = p^{-n}$

13. $7^{-3} = \frac{1}{7^3}$

15. $a^{-3} = \frac{1}{a^3}$

17. $y^{-4} = \frac{1}{y^4}$

19. $z^{-n} = \frac{1}{z^n}$

21. $2^4 \cdot 2^3 = 2^{4+3} = 2^7$

23. $3^{-5} \cdot 3^8 = 3^{-5+8} = 3^3$

25. $x^{-2} \cdot x = x^{-2} \cdot x^1 = x^{-2+1} = x^{-1}$

27. $x^4 \cdot x^3 = x^{4+3} = x^7$

29. $x^{-7} \cdot x^{-6} = x^{-7+(-6)} = x^{-13}$

31. $t^8 \cdot t^{-8} = t^{8+(-8)} = t^0 = 1$

33. $\frac{7^5}{7^2} = 7^{5-2} = 7^3$

35. $\frac{x}{x^{-1}} = \frac{x^1}{x^{-1}} = x^{1-(-1)} = x^{1+1} = x^2$

37. $\frac{x^7}{x^{-2}} = x^{7-(-2)} = x^{7+2} = x^9$

39. $\frac{z^{-6}}{z^{-2}} = z^{-6-(-2)} = z^{-6+2} = z^{-4}$

41. $\frac{x^{-5}}{x^{-8}} = x^{-5-(-8)} = x^{-5+8} = x^3$

43. $\frac{m^{-9}}{m^{-9}} = m^{-9-(-9)} = m^{-9+9} = m^0 = 1$

45. $(2^3)^2 = 2^{3 \cdot 2} = 2^6$

47. $(5^2)^{-3} = 5^{2(-3)} = 5^{-6}$

49. $(x^{-3})^{-4} = x^{-3(-4)} = x^{12}$

51. $(x^4y^5)^{-3} = x^{4(-3)}y^{5(-3)} = x^{-12}y^{-15}$

53. $(x^{-6}y^{-2})^{-4} = x^{-6(-4)}y^{-2(-4)} = x^{24}y^8$

55. $(3x^3y^{-8}z^{-3})^2 = 3^2 \, x^{3 \cdot 2} \, y^{-8 \cdot 2} \, z^{-3 \cdot 2}$
$= 9 \, x^6 \, y^{-16} \, z^{-6}$

57. $A = P(1 + r)^t$
$A = 2000(1 + 12\%)^2$ (Substituting)
$= 2000(1.12)^2$
$= 2000(1.2544)$
$= \$2508.80$

59. $A = P(1 + r)^t$
$A = 10{,}400(1 + 16.5\%)^5$
$= 10{,}400(1.165)^5$
$= 10{,}400(2.14599955)$
$\approx \$22{,}318.40$

61. $-23.8(-5.5) = 130.9$

63. Let $x = 4$, $y = 5$, $m = 2$, and $n = 3$.
$x^m \cdot y^n = 4^2 \cdot 5^3 = 16 \cdot 125 = 2000$
$(xy)^{mn} = (4 \cdot 5)^{2 \cdot 3} = 20^6 = 64{,}000{,}000$
$x^m \cdot y^n \neq (xy)^{mn}$
Thus, $x^m \cdot y^n = (xy)^{mn}$ is not always true.

65. $[\frac{x}{y}]^n = \frac{x^n}{y^n}$, True

Exercise Set 3.8

1. $4{,}200{,}000 = 4{,}200{,}000 \times (10^{-6} \times 10^6)$
$= (4{,}200{,}000 \times 10^{-6}) \times 10^6$
$= 4.2 \times 10^6$

3. $4.8 \text{ billion} = 4.8 \times 1{,}000{,}000{,}000$
$= 4.8 \times 10^9$

5. $78{,}000{,}000{,}000$
$= 78{,}000{,}000{,}000 \times (10^{-10} \times 10^{10})$
$= (78{,}000{,}000{,}000 \times 10^{-10}) \times 10^{10}$
$= 7.8 \times 10^{10}$

7. $907{,}000{,}000{,}000{,}000{,}000$
$= 907{,}000{,}000{,}000{,}000{,}000 \times (10^{-17} \times 10^{17})$
$= (907{,}000{,}000{,}000{,}000{,}000 \times 10^{-17}) \times 10^{17}$
$= 9.07 \times 10^{17}$

9. 0.00000374
$= 0.00000374 \times (10^6 \times 10^{-6})$
$= (0.00000374 \times 10^6) \times 10^{-6}$
$= 3.74 \times 10^{-6}$

11. 0.000000018
$= 0.000000018 \times (10^8 \times 10^{-8})$
$= (0.000000018 \times 10^8) \times 10^{-8}$
$= 1.8 \times 10^{-8}$

13. $10{,}000{,}000$
$= 10{,}000{,}000 \times (10^{-7} \times 10^7)$
$= (10{,}000{,}000 \times 10^{-7}) \times 10^7$
$= 1 \times 10^7$
$= 10^7$

15. 0.000000001
$= 0.000000001 \times (10^9 \times 10^{-9})$
$= (0.000000001 \times 10^9) \times 10^{-9}$
$= 1 \times 10^{-9}$
$= 10^{-9}$

17. We move the decimal point 8 places to the right.
$7.84 \times 10^8 = 784{,}000{,}000$

19. We move the decimal point 10 places to the left.
$8.764 \times 10^{-10} = 0.0000000008764$

21. We move the decimal point 8 places to the right.
$10^8 = 1.0 \times 10^8 = 100{,}000{,}000$

23. We move the decimal point 4 places to the left.
$10^{-4} = 1.0 \times 10^{-4} = 0.0001$

25. $(3 \times 10^4)(2 \times 10^5)$
$= (3 \times 2)(10^4 \times 10^5)$
$= 6 \times 10^9$

27. $(5.2 \times 10^5)(6.5 \times 10^{-2})$
$= (5.2 \times 6.5)(10^5 \times 10^{-2})$
$= 33.8 \times 10^3$
$= (3.38 \times 10^1) \times 10^3$
$= 3.38 \times (10^1 \times 10^3)$
$= 3.38 \times 10^4$

29. $(9.9 \times 10^{-6})(8.23 \times 10^{-8})$
$= (9.9 \times 8.23)(10^{-6} \times 10^{-8})$
$= 81.477 \times 10^{-14}$
$= (8.1477 \times 10^1) \times 10^{-14}$
$= 8.1477 \times (10^1 \times 10^{-14})$
$= 8.1477 \times 10^{-13}$

31. $\frac{8.5 \times 10^8}{3.4 \times 10^{-5}}$
$= \frac{8.5}{3.4} \times \frac{10^8}{10^{-5}}$
$= 2.5 \times 10^{13}$

33. $(3.0 \times 10^6) \div (6.0 \times 10^9)$

$$= \frac{3.0 \times 10^6}{6.0 \times 10^9}$$

$$= \frac{3.0}{6.0} \times \frac{10^6}{10^9}$$

$$= 0.5 \times 10^{-3}$$

$$= (5 \times 10^{-1}) \times 10^{-3}$$

$$= 5 \times (10^{-1} \times 10^{-3})$$

$$= 5 \times 10^{-4}$$

35. $\frac{7.5 \times 10^{-9}}{2.5 \times 10^{12}}$

$$= \frac{7.5}{2.5} \times \frac{10^{-9}}{10^{12}}$$

$$= 3 \times 10^{-21}$$

37. $\frac{20,000}{300,000} = \frac{2 \times 10^4}{3 \times 10^5} = 0.6\overline{6} \times 10^{-1}$

$$= (6.\overline{6} \times 10^{-1}) \times 10^{-1}$$

$$= 6.\overline{6} \times (10^{-1} \times 10^{-1})$$

$$= 6.\overline{6} \times 10^{-2}$$

39. $\frac{2}{3} - \frac{3}{4} = \frac{8}{12} - \frac{9}{12} = -\frac{1}{12}$

41. When $x = -24$, $-x = -(-24) = 24$.

43. $\{2.1 \times 10^6[(2.5\times10^{-3})\div(5.0\times10^{-5})]\} \div (3.0\times10^{17})$

$$= \frac{(2.1 \times 10^6)(\frac{2.5 \times 10^{-3}}{5.0 \times 10^{-5}})}{3.0 \times 10^{17}}$$

$$= \frac{(2.1 \times 10^6)(0.5 \times 10^2)}{3.0 \times 10^{17}}$$

$$= \frac{2.1 \times 0.5}{3.0} \times \frac{10^6 \times 10^2}{10^{17}}$$

$$= 0.35 \times 10^{-9}$$

$$= (3.5 \times 10^{-1}) \times 10^{-9}$$

$$= 3.5 \times (10^{-1} \times 10^{-9})$$

$$= 3.5 \times 10^{-10}$$

45. The reciprocal of 4.0×10^{10} is $\frac{1}{4.0 \times 10^{10}}$.

$$\frac{1}{4.0 \times 10^{10}} = \frac{1}{4} \times \frac{1}{10^{10}} = 0.25 \times 10^{-10}$$

$$= (2.5 \times 10^{-1}) \times 10^{-10}$$

$$= 2.5 \times (10^{-1} \times 10^{-10})$$

$$= 2.5 \times 10^{-11}$$

Exercise Set 4.1

1. $x = 4;\ -5x + 2 = -5\cdot4 + 2$
$= -20 + 2$
$= -18$

3. $x = 4;\ 2x^2 - 5x + 7 = 2\cdot4^2 - 5\cdot4 + 7$
$= 2\cdot16 - 20 + 7$
$= 32 - 20 + 7$
$= 12 + 7$
$= 19$

5. $x = 4;\ x^3 - 5x^2 + x = 4^3 - 5\cdot4^2 + 4$
$= 64 - 5\cdot16 + 4$
$= 64 - 80 + 4$
$= 68 - 80$
$= -12$

7. $a = 18;$
$0.4a^2 - 40a + 1039 = 0.4(18^2) - 40\cdot18 + 1039$
$= 0.4(324) - 40\cdot18 + 1039$
$= 129.6 - 720 + 1039$
$= 1168.6 - 720$
$= 448.6$

The average number of daily accidents in the United States involving 18 year-old drivers is approximately 449.

9. $x = -1;\ 3x + 5 = 3(-1) + 5$
$= -3 + 5$
$= 2$

11. $x = -1;\ x^2 - 2x + 1 = (-1)^2 - 2(-1) + 1$
$= 1 + 2 + 1$
$= 4$

13. $x = -1;\ -3x^3 + 7x^2 - 3x - 2$
$= -3(-1)^3 + 7(-1)^2 - 3(-1) - 2$
$= -3(-1) + 7(1) - 3(-1) - 2$
$= 3 + 7 + 3 - 2$
$= 13 - 2$
$= 11$

15. $2 - 3x + x^2 = 2 + (-3x) + x^2$
The terms are 2, $-3x$, and x^2.

17. $5x^3 + 6x^2 - 3x^2$
Like terms: $6x^2$ and $-3x^2$

19. $2x^4 + 5x - 7x - 3x^4$
Like terms: $2x^4$ and $-3x^4$
Like terms: $5x$ and $-7x$

21. $2x - 5x$
$= (2 - 5)x$
$= -3x$

23. $x - 9x$
$= 1x - 9x$
$= (1 - 9)x$
$= -8x$

25. $5x^3 + 6x^3 + 4$
$= (5 + 6)x^3 + 4$
$= 11x^3 + 4$

27. $5x^3 + 6x - 4x^3 - 7x$
$= (5 - 4)x^3 + (6 - 7)x$
$= 1x^3 + (-1)x$
$= x^3 - x$

29. $6b^5 + 3b^2 - 2b^5 - 3b^2$
$= (6 - 2)b^5 + (3 - 3)b^2$
$= 4b^5 + 0b^2$
$= 4b^5$

31. $\frac{1}{4}x^5 - 5 + \frac{1}{2}x^5 - 2x - 37$
$= (\frac{1}{4} + \frac{1}{2})x^5 - 2x + (-5 - 37)$
$= \frac{3}{4}x^5 - 2x - 42$

33. $6x^2 + 2x^4 - 2x^2 - x^4 - 4x^2$
$= 6x^2 + 2x^4 - 2x^2 - 1x^4 - 4x^2$
$= (6 - 2 - 4)x^2 + (2 - 1)x^4$
$= 0x^2 + 1x^4$
$= 0 + x^4$
$= x^4$

35. $\frac{1}{4}x^3 - x^2 - \frac{1}{6}x^2 + \frac{3}{8}x^3 + \frac{5}{16}x^3$
$= \frac{1}{4}x^3 - 1x^2 - \frac{1}{6}x^2 + \frac{3}{8}x^3 + \frac{5}{16}x^3$
$= (\frac{1}{4} + \frac{3}{8} + \frac{5}{16})x^3 + (-1 - \frac{1}{6})x^2$
$= (\frac{4}{16} + \frac{6}{16} + \frac{5}{16})x^3 + (-\frac{6}{6} - \frac{1}{6})x^2$
$= \frac{15}{16}x^3 - \frac{7}{6}x^2$

37. $3(s + t + 8) = 3\cdot s + 3\cdot t + 3\cdot8 = 3s + 3t + 24$

39. $9x + 2y - 4x - 2y = (9 - 4)x + (2 - 2)y = 5x$

41. $3x^2 + 2x - 2 + 3x^0 = 3x^2 + 2x - 2 + 3\cdot1$
$= 3x^2 + 2x - 2 + 3$
$= 3x^2 + 2x + 1$

<u>43</u>. $\frac{9}{2}x^8 + \frac{1}{9}x^2 + \frac{1}{2}x^9 + \frac{9}{2}x^1 + \frac{9}{2}x^9 + \frac{8}{9}x^2 + \frac{1}{2}x - \frac{1}{2}x^8$

$= (\frac{1}{2} + \frac{9}{2})x^9 + (\frac{9}{2} - \frac{1}{2})x^8 + (\frac{1}{9} + \frac{8}{9})x^2 + (\frac{9}{2} + \frac{1}{2})x$

$= \frac{10}{2}x^9 + \frac{8}{2}x^8 + \frac{9}{9}x^2 + \frac{10}{2}x$

$= 5x^9 + 4x^8 + x^2 + 5x$

Exercise Set 4.2

<u>1</u>. $x^5 + x + 6x^3 + 1 + 2x^2$

$= x^5 + x^1 + 6x^3 + 1x^0 + 2x^2$

$= x^5 + 6x^3 + 2x^2 + x^1 + 1x^0$

$= x^5 + 6x^3 + 2x^2 + x + 1$

<u>3</u>. $5x^3 + 15x^9 + x - x^2 + 7x^8$

$= 5x^3 + 15x^9 + x^1 - x^2 + 7x^8$

$= 15x^9 + 7x^8 + 5x^3 - x^2 + x^1$

$= 15x^9 + 7x^8 + 5x^3 - x^2 + x$

<u>5</u>. $8y^3 - 7y^2 + 9y^6 - 5y^8 + y^7$

$= -5y^8 + y^7 + 9y^6 + 8y^3 - 7y^2$

<u>7</u>. $3x^4 - 5x^6 - 2x^4 + 6x^6$

$= x^4 + x^6$

$= x^6 + x^4$

<u>9</u>. $-2x + 4x^3 - 7x + 9x^3 + 8$

$= -9x + 13x^3 + 8$

$= 13x^3 - 9x + 8$

<u>11</u>. $3x + 3x + 3x - x^2 - 4x^2$

$= 9x - 5x^2$

$= -5x^2 + 9x$

<u>13</u>. $-x + \frac{3}{4} + 15x^4 - x - \frac{1}{2} - 3x^4$

$= -2x + \frac{1}{4} + 12x^4$

$= 12x^4 - 2x + \frac{1}{4}$

<u>15</u>. $-7x^3 + 6x^2 + 3x + 7 = -7x^3 + 6x^2 + 3x^1 + 7x^0$

The degree of $-7x^3$ is 3.

The degree of $6x^2$ is 2.

The degree of $3x$ is 1.

The degree of 7 is 0.

The degree of the polynomial is 3, the largest exponent.

<u>17</u>. $x^2 - 3x + x^6 - 9x^4 = x^2 - 3x^1 + x^6 - 9x^4$

The degree of x^2 is 2.

The degree of $-3x$ is 1.

The degree of x^6 is 6.

The degree of $-9x^4$ is 4.

The degree of the polynomial is 6, the largest exponent.

<u>19</u>. $-3x + 6$

The coefficient of $-3x$, the first term, is -3.

The coefficient of 6, the second term, is 6.

<u>21</u>. $6x^3 + 7x^2 - 8x - 2 = 6x^3 + 7x^2 + (-8x) + (-2)$

The coefficient of $6x^3$, the first term, is 6.

The coefficient of $7x^2$, the second term, is 7.

The coefficient of $-8x$, the third term, is -8.

The coefficient of -2, the fourth term, is -2.

<u>23</u>. In the polynomial $x^3 - 27$, there are no x^2 or x terms. The x^2 term (or second degree term) and the x term (or first degree term) are missing.

<u>25</u>. In the polynomial $x^4 - x$, there are no x^3, x^2, or x^0 terms. The x^3 term (or third degree term), the x^2 term (or second degree term), and the x^0 term (or zero degree term) are missing.

<u>27</u>. No terms are missing in the polynomial $2x^3 - 5x^2 + x - 3$.

<u>29</u>. The polynomial $x^2 - 10x + 25$ is a <u>trinomial</u> because it has only three terms.

<u>31</u>. The polynomial $x^3 - 7x^2 + 2x - 4$ is <u>none of these</u> because it has more than three terms.

<u>33</u>. The polynomial $4x^2 - 25$ is a <u>binomial</u> because it has only two terms.

<u>35</u>. The polynomial $40x$ is a <u>monomial</u> because it has only one term.

<u>37</u>. When x is 6,

$-(-x) = -(-6) = 6$.

<u>39</u>. $(-6)(-5) = 30$

<u>41</u>. $(5m^5)^2 = 5^2m^{10} = 25m^{10}$

The degree of $25m^{10}$ is 10.

Exercise Set 4.3

1. $(3x + 2) + (-4x + 3)$
$= (3 - 4)x + (2 + 3)$
$= -x + 5$

3. $(-6x + 2) + (x^2 + x - 3)$
$= x^2 + (-6 + 1)x + (2 - 3)$
$= x^2 - 5x - 1$

5. $(3y^5 + 6y^2 - 1) + (7y^2 + 6y - 2)$
$= 3y^5 + (6 + 7)y^2 + 6y + (-1 - 2)$
$= 3y^5 + 13y^2 + 6y - 3$

7. $(-4x^4 + 6x^2 - 3x - 5) + (6x^3 + 5x + 9)$
$= -4x^4 + 6x^3 + 6x^2 + (-3 + 5)x + (-5 + 9)$
$= -4x^4 + 6x^3 + 6x^2 + 2x + 4$

9. $(7x^3 + 6x^2 + 4x + 1) + (-7x^3 + 6x^2 - 4x + 5)$
$= (7 - 7)x^3 + (6 + 6)x^2 + (4 - 4)x + (1 + 5)$
$= 0x^3 + 12x^2 + 0x + 6$
$= 12x^2 + 6$

11. $(5x^4 - 6x^3 - 7x^2 + x - 1) + (4x^3 - 6x + 1)$
$= 5x^4 + (-6 + 4)x^3 - 7x^2 + (1 - 6)x + (-1 + 1)$
$= 5x^4 - 2x^3 - 7x^2 - 5x + 0$
$= 5x^4 - 2x^3 - 7x^2 - 5x$

13. $(9x^8 - 7x^4 + 2x^2 + 5) + (8x^7 + 4x^4 - 2x)$
$= 9x^8 + 8x^7 + (-7 + 4)x^4 + 2x^2 - 2x + 5$
$= 9x^8 + 8x^7 - 3x^4 + 2x^2 - 2x + 5$

15. $(\frac{1}{4}x^4 + \frac{2}{3}x^3 + \frac{5}{8}x^2 + 7) + (-\frac{3}{4}x^4 + \frac{3}{8}x^2 - 7)$
$= (\frac{1}{4} - \frac{3}{4})x^4 + \frac{2}{3}x^3 + (\frac{5}{8} + \frac{3}{8})x^2 + (7 - 7)$
$= -\frac{2}{4}x^4 + \frac{2}{3}x^3 + \frac{8}{8}x^2 + 0$
$= -\frac{1}{2}x^4 + \frac{2}{3}x^3 + x^2$

17. $(0.02x^5 - 0.2x^3 + x + 0.08) + (-0.01x^5 + x^4 - 0.8x - 0.02)$
$= (0.02 - 0.01)x^5 + x^4 - 0.2x^3 + (1 - 0.8)x + (0.08 - 0.02)$
$= 0.01x^5 + x^4 - 0.2x^3 + 0.2x + 0.06$

19.
$$\begin{array}{r} -3t^4 + 6t^2 + 2t - 1 \\ - 3t^2 + 2t + 1 \\ \hline -3t^4 + 3t^2 + 4t + 0 \end{array}$$
$-3t^4 + 3t^2 + 4t$

21.
$$\begin{array}{rrrrr} 3x^5 & & - 6x^3 & & + 3x \\ & - 3x^4 & + 3x^3 & + x^2 & \\ \hline 3x^5 & - 3x^4 & - 3x^3 & + x^2 & + 3x \end{array}$$

23. $(x = 1x)$
$$\begin{array}{rrrr} & - 3x^2 & + x & \\ 5x^3 & - 6x^2 & & + 1 \\ & & 3x & - 8 \\ \hline 5x^3 & - 9x^2 & + 4x & - 7 \end{array}$$

25. $(x^2 = 1x^2)$ $(1\frac{1}{2} = \frac{3}{2})$ $(6\frac{3}{4} = \frac{27}{4})$
$$\begin{array}{rrrrr} -\frac{1}{2}x^4 & - \frac{3}{4}x^3 & & + 6x & \\ & \frac{1}{2}x^3 & + x^2 & + \frac{1}{4}x & \\ \frac{3}{4}x^4 & & + \frac{1}{2}x^2 & + \frac{1}{2}x & + \frac{1}{4} \\ \hline \frac{1}{4}x^4 & - \frac{1}{4}x^3 & + \frac{3}{2}x^2 & + \frac{27}{4}x & + \frac{1}{4} \end{array}$$

27.
$$\begin{array}{rrrrr} & & - 4x^2 & & \\ 4x^4 & - 3x^3 & + 6x^2 & + 5x & \\ & 6x^3 & - 8x^2 & & + 1 \\ -5x^4 & & & & \\ & & 6x^2 & - 3x & \\ \hline -1x^4 & + 3x^3 & + 0x^2 & + 2x & + 1 \end{array}$$
$-x^4 + 3x^3 + 2x + 1$

29.
$$\begin{array}{rrrr} 3x^4 & - 6x^2 & + 7x & \\ & 3x^2 & - 3x & + 1 \\ -2x^4 & + 7x^2 & + 3x & \\ & & 5x & - 2 \\ \hline 1x^4 & + 4x^2 & + 12x & - 1 \end{array}$$
$x^4 + 4x^2 + 12x - 1$

31.
$$\begin{array}{rrrrr} 3x^5 & - 6x^4 & + 3x^3 & & - 1 \\ & 6x^4 & - 4x^3 & + 6x^2 & \\ 3x^5 & & + 2x^3 & & \\ & - 6x^4 & & - 7x^2 & \\ -5x^5 & & + 3x^3 & & + 2 \\ \hline 1x^5 & - 6x^4 & + 4x^3 & - 1x^2 & + 1 \end{array}$$
$x^5 - 6x^4 + 4x^3 - x^2 + 1$

33. $(p^4 = 1p^4)$ $(-p^3 = -1p^3)$
$$\begin{array}{rrrrr} & - p^3 & + 6p^2 & + 3p & + 5 \\ p^4 & & - 3p^2 & & + 2 \\ & & & - 5p & + 3 \\ 6p^4 & & + 4p^2 & & - 1 \\ & - p^3 & & + 6p & \\ \hline 7p^4 & - 2p^3 & + 7p^2 & + 4p & + 9 \end{array}$$

35.

$$\begin{array}{rrrrrr} & -3x^4 & +6x^3 & -6x^2 & +5x & +1 \\ 5x^5 & & -3x^3 & & -5x & \\ & 4x^4 & +7x^3 & & +3x & +1 \\ -2x^5 & & & +7x^2 & & -8 \\ \hline 3x^5 & +1x^4 & +10x^3 & +1x^2 & +3x & -6 \\ 3x^5 & +x^4 & +10x^3 & +x^2 & +3x & -6 \end{array}$$

37.

$$\begin{array}{rrrrr} 0.15x^4 & +0.10x^3 & -0.90x^2 & & \\ & -0.01x^3 & +0.01x^2 & +x & \\ 1.25x^4 & & +0.11x^2 & & +0.01 \\ & 0.27x^3 & & & +0.99 \\ -0.35x^4 & & +15.00x^2 & & -0.03 \\ \hline 1.05x^4 & +0.36x^3 & +14.22x^2 & +x & +0.97 \end{array}$$

$(-0.9 = -0.90)$

$(15 = 15.00)$

39. a)

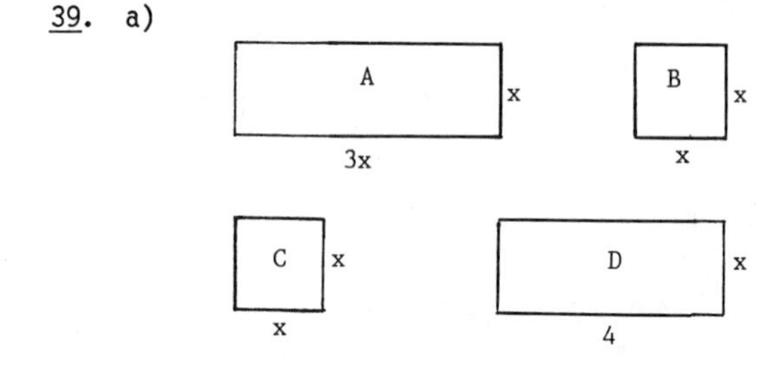

The area of a rectangle is the product of the length and the width. The sum of the areas is found as follows:

Area of A + Area of B + Area of C + Area of D

$= 3x \cdot x + x \cdot x + x \cdot x + 4 \cdot x$

$= 3x^2 + x^2 + x^2 + 4x$

$= 5x^2 + 4x$

A polynomial for the sum of the areas is $5x^2 + 4x$.

b) $x = 3$; $5x^2 + 4x = 5 \cdot 3^2 + 4 \cdot 3$

$= 5 \cdot 9 + 4 \cdot 3$

$= 45 + 12$

$= 57$

When $x = 3$, the sum of the areas is 57 square units.

$x = 8$; $5x^2 + 4x = 5 \cdot 8^2 + 4 \cdot 8$

$= 5 \cdot 64 + 4 \cdot 8$

$= 320 + 32$

$= 352$

When $x = 8$, the sum of the areas is 352 square units.

41. $x^7 \cdot x^3 = x^{7+3} = x^{10}$

43. 3478 is 37% of what number?

$3478 = 37\% \cdot x$

We solve.

$3478 = 0.37x$

$\frac{3478}{0.37} = x$

$9400 = x$

3478 is 37% of 9400.

45. a) Compare $(ax + b) + (cx + d)$ and $(cx + d) + (ax + b)$.

$(ax + b) + (cx + d)$

$= ax + b + cx + d$

$= ax + cx + b + d$

$= (a + c)x + (b + d)$

$(cx + d) + (ax + b)$

$= cx + d + ax + b$

$= ax + cx + b + d$

$= (a + c)x + (b + d)$

Since $(ax + b) + (cx + d) = (cx + d) + (ax + b)$ addition of binomials is commutative.

b) Compare $(ax^2 + bx + c) + (dx^2 + ex + f)$ and $(dx^2 + ex + f) + (ax^2 + bx + c)$.

$(ax^2 + bx + c) + (dx^2 + ex + f)$

$= ax^2 + bx + c + dx^2 + ex + f$

$= ax^2 + dx^2 + bx + ex + c + f$

$(a + d)x^2 + (b + e)x + (c + f)$

$(dx^2 + ex + f) + (ax^2 + bx + c)$

$= dx^2 + ex + f + ax^2 + bx + c$

$= ax^2 + dx^2 + bx + ex + c + f$

$= (a + d)x^2 + (b + e)x + (c + f)$

Since $(ax^2 + bx + c) + (dx^2 + ex + f) = (dx^2 + ex + f) + (ax^2 + bx + c)$, addition of trinomials is commutative.

Exercise Set 4.4

1. Two equivalent expressions for the additive inverse of $-5x$ and $-(-5x)$ and $5x$.

$-(-5x) = 5x$

3. Two equivalent expressions for the additive inverse of $-x^2 + 10x - 2$ are $-(-x^2 + 10x - 2)$ and $x^2 - 10x + 2$.

$-(-x^2 + 10x - 2) = x^2 - 10x + 2$

5. Two equivalent expressions for the additive inverse of $12x^4 - 3x^3 + 3$ are $-(12x^4 - 3x^3 + 3)$ and $-12x^4 + 3x^3 - 3$.

$-(12x^4 - 3x^3 + 3) = -12x^4 + 3x^3 - 3$

7. We change the sign of every term inside parentheses.

$-(3x - 7) = -3x + 7$

9. We change the sign of every term inside parentheses.

$-(4x^2 - 3x + 2) = -4x^2 + 3x - 2$

11. We change the sign of every term inside parentheses.

$-(-4x^4 - 6x^2 + \frac{3}{4}x - 8) = 4x^4 + 6x^2 - \frac{3}{4}x + 8$

13. $(5x^2 + 6) - (3x^2 - 8)$

$= (5x^2 + 6) + (-3x^2 + 8)$

$= 2x^2 + 14$

15. $(6x^5 - 3x^4 + x + 1) - (8x^5 + 3x^4 - 1)$

$= (6x^5 - 3x^4 + x + 1) + (-8x^5 - 3x^4 + 1)$

$= -2x^5 - 6x^4 + x + 2$

17. $(6x^2 + 2x) - (-3x^2 - 7x + 8)$

$= (6x^2 + 2x) + (3x^2 + 7x - 8)$

$= 9x^2 + 9x - 8$

19. $(\frac{5}{8}x^3 - \frac{1}{4}x - \frac{1}{3}) - (-\frac{1}{8}x^3 + \frac{1}{4}x - \frac{1}{3})$

$= (\frac{5}{8}x^3 - \frac{1}{4}x - \frac{1}{3}) + (\frac{1}{8}x^3 - \frac{1}{4}x + \frac{1}{3})$

$= \frac{6}{8}x^3 - \frac{2}{4}x + 0$

$= \frac{3}{4}x^3 - \frac{1}{2}x$

21. $(0.08x^3 - 0.02x^2 + 0.01x)-(0.02x^3 + 0.03x^2 - 1)$

$= (0.08x^3 - 0.02x^2 + 0.01x)+(-0.02x^3 - 0.03x^2 + 1)$

$= 0.06x^3 - 0.05x^2 + 0.01x + 1$

23. Subtract:

$$\begin{array}{l} x^2 + 5x + 6 \\ \underline{x^2 + 2x} \end{array}$$

Think: Add

$$\begin{array}{r} x^2 + 5x + 6 \\ \underline{-x^2 - 2x \quad\;\;} \\ 3x + 6 \end{array}$$

Subtracting the polynomial $x^2 + 2x$ is the same as adding the polynomial $-x^2 - 2x$.

25. Subtract:

$$\begin{array}{l} x^4 \qquad\quad - 3x^2 + x + 1 \\ \underline{x^4 - 4x^3 \qquad\qquad\qquad} \end{array}$$

Think: Add

$$\begin{array}{r} x^4 \qquad\quad - 3x^2 + x + 1 \\ \underline{-x^4 + 4x^3 \qquad\qquad\qquad} \\ 4x^3 - 3x^2 + x + 1 \end{array}$$

Subtracting the polynomial $x^4 - 4x^3$ is the same as adding the polynomial $-x^4 + 4x^3$.

27. Subtract:

$$\begin{array}{l} 5x^4 + 6x^3 - 9x^2 \\ \underline{-6x^4 - 6x^3 \qquad\quad + 8x - 9} \end{array}$$

Think: Add

$$\begin{array}{r} 5x^4 + 6x^3 - 9x^2 \qquad\qquad\; \\ \underline{6x^4 + 6x^3 \qquad\quad - 8x - 9} \\ 11x^4 + 12x^3 - 9x^2 - 8x - 9 \end{array}$$

Subtracting the polynomial $-6x^4 - 6x^3 + 8x + 9$ is the same as adding the polynomial $6x^4 + 6x^3 - 8x - 9$.

29. Subtract:

$$\begin{array}{r} 3x^4 + 6x^2 + 8x - 1 \\ \underline{4x^5 - 6x^4 \qquad\quad - 8x - 7} \end{array}$$

Think: Add

$$\begin{array}{r} 3x^4 + 6x^2 + 8x - 1 \\ \underline{-4x^5 + 6x^4 \qquad\quad + 8x + 7} \\ -4x^5 + 9x^4 + 6x^2 + 16x + 6 \end{array}$$

Subtracting the polynomial $4x^5 - 6x^4 - 8x - 7$ is the same as adding the polynomial $-4x^5 + 6x^4 + 8x + 7$.

31. Subtract:

$$\begin{array}{r} x^5 \qquad\qquad\qquad\quad - 1 \\ \underline{x^5 - x^4 + x^3 - x^2 + x - 1} \end{array}$$

Think: Add

$$\begin{array}{r} x^5 \qquad\qquad\qquad\quad - 1 \\ \underline{-x^5 + x^4 - x^3 + x^2 - x + 1} \\ x^4 - x^3 + x^2 - x \quad \end{array}$$

Subtracting the polynomial $x^5 - x^4 + x^3 - x^2 + x - 1$ is the same as adding the polynomial $-x^5 + x^4 - x^3 + x^2 - x + 1$.

33. $8(x - 2) = 8 \cdot x - 8 \cdot 2 = 8x - 16$

35. $3x - 3 = -4x + 4$
$3x + 4x = 4 + 3$
$7x = 7$
$x = 1$

37. $(345.099x^3 - 6.178x) - (-224.508x^3 + 8.99x)$
$= 345.099x^3 - 6.178x + 224.508x^3 - 8.99x$
$= 569.607x^3 - 15.168x$

39. $(-y^4 - 7y^3 + y^2) + (-2y^4 + 5y - 2) - (-6y^3 + y^2)$
$= -y^4 - 7y^3 + y^2 - 2y^4 + 5y - 2 + 6y^3 - y^2$
$= -3y^4 - y^3 + 5y - 2$

Exercise Set 4.5

1. $(3x)(-4) = 3(-4)x = -12x$

3. $(6x^2)(7) = (6 \cdot 7)x^2 = 42x^2$

5. $(-5x)(-6) = (-5)(-6)x = 30x$

7. $(y^2)(-2y) = -2(y^2 \cdot y) = -2y^{2+1} = -2y^3$

9. $(x^4)(x^2) = x^{4+2} = x^6$

11. $(3x^4)(2x^2) = (3 \cdot 2)(x^4 \cdot x^2) = 6x^{4+2} = 6x^6$

13. $(-4x^4)(0) = 0$ Any number multiplied by 0 is 0.

15. $(-0.1x^6)(0.2x^4) = (-0.1)(0.2)(x^6 \cdot x^4)$
$= -0.02x^{6+4}$
$= -0.02x^{10}$

17. $(2x)(4x - 6)$
$= (2x)(4x) - (2x)(6)$
$= 8x^2 - 12x$

19. $(-6x^2)(x^2 + x)$
$= (-6x^2)(x^2) + (-6x^2)(x)$
$= -6x^4 - 6x^3$

21. $(4x^4)(x^2 - 6x)$
$= (4x^4)(x^2) - (4x^4)(6x)$
$= 4x^6 - 24x^5$

23. $(x + 6)(-x + 2)$
$= (x + 6)((-x) + (x + 6)(2)$
$= [x(-x) + 6(-x)] + [x(2) + 6(2)]$
$= -x^2 - 6x + 2x + 12$
$= -x^2 - 4x + 12$

25. $(y - 5)(2y - 5)$
$= (y - 5)2y - (y - 5)5$
$= (y \cdot 2y - 5 \cdot 2y) - (y \cdot 5 - 5 \cdot 5)$
$= (2y^2 - 10y) - (5y - 25)$
$= 2y^2 - 10y - 5y + 25$
$= 2y^2 - 15y + 25$

27. $(3x - 5)(3x + 5)$
$= (3x - 5)3x + (3x - 5)5$
$= (3x \cdot 3x - 5 \cdot 3x) + (3x \cdot 5 - 5 \cdot 5)$
$= 9x^2 - 15x + 15x - 25$
$= 9x^2 - 25$

29. $(2x - \frac{1}{2})(x + \frac{3}{2})$
$= (2x - \frac{1}{2})x + (2x - \frac{1}{2})\frac{3}{2}$
$= (2x \cdot x - \frac{1}{2} \cdot x) + (2x \cdot \frac{3}{2} - \frac{1}{2} \cdot \frac{3}{2})$
$= 2x^2 - \frac{1}{2}x + 3x - \frac{3}{4}$
$= 2x^2 + \frac{5}{2}x - \frac{3}{4}$

31. $(x^2 + 6x + 1)(x + 1)$
$= (x^2 + 6x + 1)x + (x^2 + 6x + 1)1$
$= (x^2 \cdot x + 6x \cdot x + 1 \cdot x) + (x^2 \cdot 1 + 6x \cdot 1 + 1 \cdot 1)$
$= x^3 + 6x^2 + x + x^2 + 6x + 1$
$= x^3 + 7x^2 + 7x + 1$

33. $(-5q^2 - 7q + 3)(2q + 1)$
$= (-5q^2 - 7q + 3)2q + (-5q^2 - 7q + 3)1$
$= (-5q^2 \cdot 2q - 7q \cdot 2q + 3 \cdot 2q) + (-5q^2 \cdot 1 - 7q \cdot 1 + 3 \cdot 1)$
$= -10q^3 - 14q^2 + 6q - 5q^2 - 7q + 3$
$= -10q^3 - 19q^2 - q + 3$

35. $3x^2 - 6x + 2$
$x^2 - 3$
$3x^4 - 6x^3 + 2x^2$ (Multiplying by x^2)
$- 9x^2 + 18x - 6$ (Multiplying by -3)
$3x^4 - 6x^3 - 7x^2 + 18x - 6$ (Adding)

37. $2t^2 - t - 4$
$3t^2 + 2t - 1$
$6t^4 - 3t^3 - 12t^2$ (Multiplying by $3t^2$)
$4t^3 - 2t^2 - 8t$ (Multiplying by 2t)
$- 2t^2 + t + 4$ (Multiplying by -1)
$6t^4 + t^3 - 16t^2 - 7t + 4$ (Adding)

39. $x^3 + x^2 + x + 1$
$x - 1$
$x^4 + x^3 + x^2 + x$ (Multiplying by x)
$- x^3 - x^2 - x - 1$ (Multiplying by -1)
$x^4 \qquad - 1$ (Adding)

41. $y = mx + b$
$y - b = mx$ (Adding -b)
$\frac{y - b}{m} = x$ (Multiplying by $\frac{1}{m}$)

43. $(t^{-4})^5 = t^{-4(5)} = t^{-20}$

45. $(x + 3)(x + 6) + (x + 3)(x + 6)$
$= 2[(x + 3)(x + 6)]$
$= 2[(x + 3)x + (x + 3)6]$
$= 2(x^2 + 3x + 6x + 18)$
$= 2(x^2 + 9x + 18)$
$= 2x^2 + 18x + 36$

47.

$V = (12 - 2x)(12 - 2x)(x)$ ($V = \ell \cdot w \cdot h$)
$= [(12 - 2x)12 - (12 - 2x)2x]x$
$= (144 - 24x - 24x + 4x^2)x$
$= (144 - 48x + 4x^2)x$
$= 144x - 48x^2 + 4x^3$

$SA = (12 - 2x)(12 - 2x) + 4(12 - 2x)x$
$= (144 - 48x + 4x^2) + (48x - 8x^2)$
$= 144 - 4x^2$

Exercise Set 4.6

1. $4x(x + 1) = 4x^2 + 4x$

3. $-3x(x - 1) = -3x^2 + 3x$

5. $x^2(x^3 + 1) = x^5 + x^2$

7. $3x(2x^2 - 6x + 1) = 6x^3 - 18x^2 + 3x$

9. $(x + 1)(x^2 + 3)$
F O I L
$= x \cdot x^2 + x \cdot 3 + 1 \cdot x^2 + 1 \cdot 3$
$= x^3 + 3x + x^2 + 3$

11. $(x^3 + 2)(x + 1)$
F O I L
$= x^3 \cdot x + x^3 \cdot 1 + 2 \cdot x + 2 \cdot 1$
$= x^4 + x^3 + 2x + 2$

13. $(x + 2)(x - 3)$
F O I L
$= x \cdot x + x \cdot (-3) + 2 \cdot x + 2 \cdot (-3)$
$= x^2 - 3x + 2x - 6$
$= x^2 - x - 6$

15. $(3x + 2)(3x + 3)$
F O I L
$= 3x \cdot 3x + 3x \cdot 3 + 2 \cdot 3x + 2 \cdot 3$
$= 9x^2 + 9x + 6x + 6$
$= 9x^2 + 15x + 6$

17. $(5x - 6)(x + 2)$
F O I L
$= 5x \cdot x + 5x \cdot 2 + (-6) \cdot x + (-6) \cdot 2$
$= 5x^2 + 10x - 6x - 12$
$= 5x^2 + 4x - 12$

19. $(3x - 1)(3x + 1)$
F O I L
$= 3x \cdot 3x + 3x \cdot 1 + (-1) \cdot 3x + (-1) \cdot 1$
$= 9x^2 + 3x - 3x - 1$
$= 9x^2 - 1$

21. $(4x - 2)(x - 1)$
F O I L
$= 4x \cdot x + 4x \cdot (-1) + (-2) \cdot x + (-2) \cdot (-1)$
$= 4x^2 - 4x - 2x + 2$
$= 4x^2 - 6x + 2$

23. $(x - \frac{1}{4})(x + \frac{1}{4})$
F O I L
$= x \cdot x + x \cdot \frac{1}{4} + (-\frac{1}{4}) \cdot x + (-\frac{1}{4}) \cdot \frac{1}{4}$
$= x^2 + \frac{1}{4}x - \frac{1}{4}x - \frac{1}{16}$
$= x^2 - \frac{1}{16}$

25. $(t - 0.1)(t + 0.1)$

F O I L

$= t\cdot t + t\cdot(0.1) + (-0.1)\cdot t + (-0.1)(0.1)$

$= t^2 + 0.1t - 0.1t - 0.01$

$= t^2 - 0.01$

27. $(2x^2 + 6)(x + 1)$

F O I L

$= 2x^3 + 2x^2 + 6x + 6$

29. $(-2x + 1)(x + 6)$

F O I L

$= -2x^2 - 12x + x + 6$

$= -2x^2 - 11x + 6$

31. $(a + 7)(a + 7)$

F O I L

$= a^2 + 7a + 7a + 49$

$= a^2 + 14a + 49$

33. $(1 + 2x)(1 - 3x)$

F O I L

$= 1 - 3x + 2x - 6x^2$

$= 1 - x - 6x^2$

35. $(x^2 + 3)(x^3 - 1)$

F O I L

$= x^5 - x^2 + 3x^3 - 3$

37. $(x^2 - 2)(x - 1)$

F O I L

$= x^3 - x^2 - 2x + 2$

39. $(3q^2 - 2)(q^4 - 2)$

F O I L

$= 3q^6 - 6q^2 - 2q^4 + 4$

41. $(3x^5 + 2)(2x^2 + 6)$

F O I L

$= 6x^7 + 18x^5 + 4x^2 + 12$

43. $(8x^3 + 1)(x^3 + 8)$

$= 8x^6 + 64x^3 + x^3 + 8$

$= 8x^6 + 65x^3 + 8$

45. $(4x^2 + 3)(x - 3)$

F O I L

$= 4x^3 - 12x^2 + 3x - 9$

47. $(4x^4 + x^2)(x^2 + x)$

F O I L

$= 4x^6 + 4x^5 + x^4 + x^3$

49. $4.7 \times 10^{-5} = 0.000047$

51. $4y(y + 5)(2y + 8)$

$= 4y(2y^2 + 8y + 10y + 40)$

$= 4y(2y^2 + 18y + 40)$

$= 8y^3 + 72y^2 + 160y$

53. $(x + 2)(x - 5) = (x + 1)(x - 3)$

$x^2 - 5x + 2x - 10 = x^2 - 3x + x - 3$

$x^2 - 3x - 10 = x^2 - 2x - 3$

$-3x - 10 = -2x - 3$

$-3x + 2x = 10 - 3$

$-x = 7$

$x = -7$

Exercise Set 4.7

1. $(x + 4)(x - 4)$

$= x^2 - 4^2$

$= x^2 - 16$

3. $(2x + 1)(2x - 1)$

$= (2x)^2 - 1^2$

$= 4x^2 - 1$

5. $(5m - 2)(5m + 2)$

$= (5m)^2 - 2^2$

$= 25m^2 - 4$

7. $(2x^2 + 3)(2x^2 - 3)$

$= (2x^2)^2 - 3^2$

$= 4x^4 - 9$

9. $(3x^4 - 4)(3x^4 + 4)$

$= (3x^4)^2 - 4^2$

$= 9x^8 - 16$

11. $(x^6 - x^2)(x^6 + x^2)$

$= (x^6)^2 - (x^2)^2$

$= x^{12} - x^4$

13. $(a^4 + 3a)(a^4 - 3a)$

$= (a^4)^2 - (3a)^2$

$= a^8 - 9a^2$

15. $(x^{12} - 3)(x^{12} + 3)$
$= (x^{12})^2 - 3^2$
$= x^{24} - 9$

17. $(2x^8 + 3)(2x^8 - 3)$
$= (2x^8)^2 - 3^2$
$= 4x^{16} - 9$

19. $(x + 2)^2$
$= x^2 + 2\cdot x\cdot 2 + 2^2$
$= x^2 + 4x + 4$

21. $(3x^2 + 1)^2$
$= (3x^2)^2 + 2\cdot 3x^2\cdot 1 + 1^2$
$= 9x^4 + 6x^2 + 1$

23. $(x - \frac{1}{2})^2$
$= x^2 - 2\cdot x\cdot \frac{1}{2} + (\frac{1}{2})^2$
$= x^2 - x + \frac{1}{4}$

25. $(3 + x)^2$
$= 3^2 + 2\cdot 3\cdot x + x^2$
$= 9 + 6x + x^2$

27. $(y^2 + 1)^2$
$= (y^2)^2 + 2\cdot y^2\cdot 1 + 1^2$
$= y^4 + 2y^2 + 1$

29. $(2 - 3x^4)^2$
$= 2^2 - 2\cdot 2\cdot 3x^4 + (3x^4)^2$
$= 4 - 12x^4 + 9x^8$

31. $(5 + 6t^2)^2$
$= 5^2 + 2\cdot 5\cdot 6t^2 + (6t^2)^2$
$= 25 + 60t^2 + 36t^4$

33. $(3 - 2x^3)^2$
$= 3^2 - 2\cdot 3\cdot 2x^3 + (2x^3)^2$ (Using Method 1)
$= 9 - 12x^3 + 4x^6$

35. $4x(x^2 + 6x - 3)$
$= 4x^3 + 24x^2 - 12x$ (Using Method 4)

37. $(2x^2 - \frac{1}{2})(2x^2 - \frac{1}{2})$
$= (2x^2)^2 - 2\cdot 2x^2\cdot \frac{1}{2} + (\frac{1}{2})^2$ (Using Method 1)
$= 4x^4 - 2x^2 + \frac{1}{4}$

39. $(-1 + 3p)(1 + 3p)$
$= (3p - 1)(3p + 1)$
$= (3p)^2 - 1^2$ (Using Method 2)
$= 9p^2 - 1$

41. $3t^2(5t^3 - t^2 + t)$
$= 15t^5 - 3t^4 + 3t^3$ (Using Method 5)

43. $(6x^4 + 4)^2$
$= (6x^4)^2 + 2\cdot 6x^4\cdot 4 + 4^2$ (Using Method 1)
$= 36x^8 + 48x^4 + 16$

45. $(3x + 2)(4x^2 + 5)$
$= 12x^3 + 15x + 8x^2 + 10$ (Using Method 3)
$= 12x^3 + 8x^2 + 15x + 10$

47. $(8 - 6x^4)^2$
$= 8^2 - 2\cdot 8\cdot 6x^4 + (6x^4)^2$ (Using Method 1)
$= 64 - 96x^4 + 36x^8$

49. Let x represent the number of watts used by the television set. Then 10x represents the number of watts used by the lamps, and 40x represents the number of watts used by the air conditioner. The total wattage is 2550 watts. We translate and solve.

$$x + 10x + 40x = 2550$$
$$51x = 2550$$
$$x = 50$$

The television uses 50 watts. The lamps use $10\cdot 50$, or 500 watts. The air conditioner uses $40\cdot 50$, or 2000 watts. The values check since the total wattage is 50 + 500 + 2000, or 2550 watts.

51. $(67.58x + 3.225)^2$
$= (67.58x)^2 + 2(67.58x)(3.225) + (3.225)^2$
$= 4567.0564x^2 + 435.891x + 10.400625$

53. $(5t^2 - 3)^2(5t^2 + 3)^2$
$= (5t^2 - 3)(5t^2 + 3)(5t^2 - 3)(5t^2 + 3)$
$= (25t^4 - 9)(25t^4 - 9)$
$= (625t^8 - 450t^4 + 81)$

Exercise Set 5.1

1. - 5. Answers may vary.

1. $6x^3 = (6x)(x^2) = (3x^2)(2x) = (2x^2)(3x)$

3. $-9x^5 = (-3x^2)(3x^3) = (-x)(9x^4) = (3x^2)(-3x^3)$

5. $24x^4 = (6x)(4x^3) = (-3x^2)(-8x^2) = (2x^3)(12x)$

7. $x^2 - 4x$
$= x \cdot x - x \cdot 4$
$= x(x - 4)$

9. $x^3 + 6x^2$
$= x^2 \cdot x + x^2 \cdot 6$
$= x^2(x + 6)$

11. $8x^4 - 24x^2$
$= 8x^2 \cdot x^2 - 8x^2 \cdot 3$
$= 8x^2(x^2 - 3)$

13. $17x^5 + 34x^3 + 51x$
$= 17x \cdot x^4 + 17x \cdot 2x^2 + 17x \cdot 3$
$= 17x(x^4 + 2x^2 + 3)$

15. $10x^3 + 25x^2 + 15x - 20$
$= 5 \cdot 2x^3 + 5 \cdot 5x^2 + 5 \cdot 3x - 5 \cdot 4$
$= 5(2x^3 + 5x^2 + 3x - 4)$

17. $\frac{5}{3}x^6 + \frac{4}{3}x^5 + \frac{1}{3}x^4 + \frac{1}{3}x^3$
$= \frac{1}{3}x^3 \cdot 5x^3 + \frac{1}{3}x^3 \cdot 4x^2 + \frac{1}{3}x^3 \cdot x + \frac{1}{3}x^3 \cdot 1$
$= \frac{1}{3}x^3(5x^3 + 4x^2 + x + 1)$

19. $y^2 + 4y + y + 4$
$= (y^2 + 4y + (y + 4)$
$= y(y + 4) + 1(y + 4)$
$= (y + 1)(y + 4)$

21. $x^2 + 5x - 2x - 10$
$= (x^2 + 5x) + (-2x - 10)$
$= x(x + 5) - 2(x + 5)$
$= (x - 2)(x + 5)$

23. $16 - 12x - 4x + 3x^2$
$= (16 - 12x) + (-4x + 3x^2)$
$= 4(4 - 3x) - x(4 - 3x)$
$= (4 - x)(4 - 3x)$

25. $2x^3 + 6x^2 + x + 3$
$= (2x^3 + 6x^2) + (x + 3)$
$= 2x^2(x + 3) + 1(x + 3)$
$= (2x^2 + 1)(x + 3)$

27. $x^3 + 8x^2 - 3x - 24$
$= (x^3 + 8x^2) + (-3x - 24)$
$= x^2(x + 8) - 3(x + 8)$
$= (x^2 - 3)(x + 8)$

29. $12t^3 - 16t^2 + 3t - 4$
$= (12t^3 - 16t^2) + (3t - 4)$
$= 4t^2(3t - 4) + 1(3t - 4)$
$= (4t^2 + 1)(3t - 4)$

31. $4x^5 + 6x^3 + 6x^2 + 9$
$= (4x^5 + 6x^2) + (6x^3 + 9)$
$= 2x^2(2x^3 + 3) + 3(2x^3 + 3)$
$= (2x^2 + 3)(2x^3 + 3)$

33. $(x - 4)(x + 4)$
$= x^2 - 4^2$
$= x^2 - 16$

35. $2 - 5(x + 5) = 3(x - 2) - 1$
$2 - 5x - 25 = 3x - 6 - 1$
$-5x - 23 = 3x - 7$
$-16 = 8x$
$-2 = x$

37. $x^6 + x^4 + x^2 + 1$
$= x^4(x^2 + 1) + 1(x^2 + 1)$
$= (x^4 + 1)(x^2 + 1)$

39. The polynomials $5x$ and $6x^2$ are not relatively prime because they have x as a common factor.

41. The polynomials $3x$ and $9x - 1$ are relatively prime because they have no common factors other than constants.

Exercise Set 5.2

1. $x^2 - 36$
The first expression is a square: x^2
The second expression is a square: $36 = 6^2$
There is a minus sign between x^2 and 36.
$x^2 - 36$ is a difference of squares.

3. $x^2 - 35$
The second expression, 35, is not a square.
$x^2 - 35$ is not a difference of squares.

5. $16x^2 - 25$
The first expression is a square: $16x^2 = (4x)^2$
The second expression is a square: $25 = 5^2$
There is a minus sign between $16x^2$ and 25.
$16x^2 - 25$ is a difference of squares.

7. $49 + 16t^4$
There is not a minus sign between 49 and $16t^4$.
$49 + 16t^4$ is not a difference of squares.

9. $x^2 - 4$
$= x^2 - 2^2$
$= (x - 2)(x + 2)$

11. $t^2 - 9$
$= t^2 - 3^2$
$= (t - 3)(t + 3)$

13. $16a^2 - 9$
$= (4a)^2 - 3^2$
$= (4a - 3)(4a + 3)$

15. $4x^2 - 25$
$= (2x)^2 - 5^2$
$= (2x - 5)(2x + 5)$

17. $8x^2 - 98$
$= 2(4x^2 - 49)$
$= 2[(2x)^2 - 7^2]$
$= 2(2x - 7)(2x + 7)$

19. $36x - 49x^3$
$= x(36 - 49x^2)$
$= x[6^2 - (7x)^2]$
$= x(6 - 7x)(6 + 7x)$

21. $16y^2 - 25y^4$
$= y^2(16 - 25y^2)$
$= y^2[4^2 - (5y)^2]$
$= y^2(4 - 5y)(4 + 5y)$

23. $49a^4 - 81$
$= (7a^2)^2 - 9^2$
$= (7a^2 - 9)(7a^2 + 9)$

25. $a^{12} - 4a^2$
$= a^2(a^{10} - 4)$
$= a^2[(a^5)^2 - 2^2]$
$= a^2(a^5 - 2)(a^5 + 2)$

27. $81y^6 - 25$
$= (9y^3)^2 - 5^2$
$= (9y^3 - 5)(9y^3 + 5)$

29. $x^4 - 1$
$= (x^2 + 1)(x^2 - 1)$
$= (x^2 + 1)(x + 1)(x - 1)$

31. $4x^4 - 64$
$= 4(x^4 - 16)$
$= 4(x^2 + 4)(x^2 - 4)$
$= 4(x^2 + 4)(x + 2)(x - 2)$

33. $1 - y^8$
$= (1 + y^4)(1 - y^4)$
$= (1 + y^4)(1 + y^2)(1 - y^2)$
$= (1 + y^4)(1 + y^2)(1 + y)(1 - y)$

35. $x^{12} - 16$
$= (x^6 + 4)(x^6 - 4)$
$= (x^6 + 4)(x^3 + 2)(x^3 - 2)$

37. $\frac{1}{16} - y^2$
$= (\frac{1}{4})^2 - y^2$
$= (\frac{1}{4} - y)(\frac{1}{4} + y)$

39. $25 - \frac{1}{49} y^2$
$= 5^2 - (\frac{1}{7} y)^2$
$= (5 - \frac{1}{7} y)(5 + \frac{1}{7} y)$

41. $16 - t^4$
$= (4 + t^2)(4 - t^2)$
$= (4 + t^2)(2 + t)(2 - t)$

43. $(t - 9)^2$
$= t^2 - 2 \cdot t \cdot 9 + 9^2$
$= t^2 - 18t + 81$

45. $(-16)(-5) = 80$

47. $3.24x^2 - 0.81$
$= (1.8x)^2 - 0.9^2$
$= (1.8x + 0.9)(1.8x - 0.9)$

49. $1.28t^2 - 2$
$= 2(0.64t^2 - 1)$
$= 2(0.8t + 1)(0.8t - 1)$

51. $x^6 - x^4 - x^2 + 1$
$= (x^6 - x^4) + (-x^2 + 1)$
$= x^4(x^2 - 1) - 1(x^2 - 1)$
$= (x^4 - 1)(x^2 - 1)$
$= (x^2 + 1)(x^2 - 1)(x + 1)(x - 1)$
$= (x^2 + 1)(x + 1)(x - 1)(x + 1)(x - 1)$
$= (x^2 + 1)(x + 1)^2(x - 1)^2$

Exercise Set 5.3

1. $x^2 - 14x + 49$

x^2 and 49 are squares. There is no minus sign before x^2 or 49. If we multiply the squre roots, x and 7, and double the product, we get 14x, the additive inverse of the remaining term, -14x. Thus $x^2 - 14x + 49$ is a trinomial square.

3. $x^2 - 64 + 16x$, or $x^2 + 16x - 64$

x^2 and 64 are squares. There is a minus sign before 64. This polynomial is not a trinomial square.

5. $y^2 - 3y + 9$

y^2 and 9 are squares. There is no minus sign before y^2 or 9. If we multiply the square roots, y and 3, and double the product, we get 6y which is not the remaining term, -3y, or its additive inverse, 3y. Thus $y^2 - 3y + 9$ is not a trinomial square.

7. $25 + 40x + 8x^2$, or $8x^2 + 40x + 25$

Two of the terms must be squares. This polynomial is not a trinomial square because only one term, 25, is a square.

9. $x^2 - 14x + 49$
$= x^2 - 2 \cdot x \cdot 7 + 7^2$
$= (x - 7)^2$ $\quad [A^2 - 2AB + B^2 = (A - B)^2]$

11. $y^2 + 16y + 64$
$= y^2 + 2 \cdot y \cdot 8 + 8^2$
$= (y + 8)^2$ $\quad [A^2 + 2AB + B^2 = (A + B)^2]$

13. $m^2 + 1 - 2m$
$= m^2 - 2m + 1$
$= m^2 - 2 \cdot m \cdot 1 + 1^2$
$= (m - 1)^2$ $\quad [A^2 + 2AB + B^2 = (A + B)^2]$

15. $4 + 4x + x^2$
$= x^2 + 4x + 4$
$= x^2 + 2 \cdot x \cdot 2 + 2^2$
$= (x + 2)^2$ $\quad [A^2 + 2AB + B^2 = (A + B)^2]$

17. $2x^2 - 4x + 2$
$= 2(x^2 - 2x + 1)$
$= 2(x^2 - 2 \cdot x \cdot 1 + 1^2)$
$= 2(x - 1)^2$

19. $x^3 - 18x^2 + 81x$
$= x(x^2 - 18x + 81)$
$= x(x^2 - 2 \cdot x \cdot 9 + 9^2)$
$= x(x - 9)^2$

21. $20t^2 + 100t + 125$
$= 5(4t^2 + 20t + 25)$
$= 5[(2t)^2 + 2 \cdot 2t \cdot 5 + 5^2]$
$= 5(2t + 5)^2$

23. $49 - 42x + 9x^2$
$= 7^2 - 2 \cdot 7 \cdot 3x + (3x)^2$
$= (7 - 3x)^2$

25. $5y^4 + 10y^2 + 5$
$= 5(y^4 + 2y^2 + 1)$
$= 5[(y^2)^2 + 2 \cdot y^2 \cdot 1 + 1^2]$
$= 5(y^2 + 1)^2$

27. $y^6 + 26y^3 + 169$
$= (y^3)^2 + 2 \cdot y^3 \cdot 13 + 13^2$
$= (y^3 + 13)^2$

29. $16x^{10} - 8x^5 + 1$
$= (4x^5)^2 - 2 \cdot 4x^5 \cdot 1 + 1^2$
$= (4x^5 - 1)^2$

31. $(x + 6)(x - 4)$
$= x^2 - 4x + 6x - 24$
$= x^2 + 2x - 24$

33. Let x represent the number of liters of oxygen which can be dissolved in 100 liters of water at 20°C.
$5 = 1.6x$ (Translating)
$\frac{5}{1.6} = x$
$3.125L = x$

35. $8.1x^2 - 6.4$
$= 0.1(81x^2 - 64)$
$= 0.1(9x + 8)(9x - 8)$

37. $4(a + 5)^2 + 20(a + 5) + 25$
$= [2(a + 5)]^2 + 2 \cdot 2(a + 5) \cdot 5 + 5^2$
$= [2(a + 5) + 5]^2$
$= (2a + 10 + 5)^2$
$= (2a + 15)^2$

Exercise Set 5.4

1. $x^2 + 8x + 15$

The leading term is x^2. The first term of each binomial factor is x: (x + _)(x + _). Since the constant term is positive and the coefficient of the middle term is positive, we look for a factorization of 15 in which both factors are positive. Their sum must be 8, the coefficient of the middle term.

Pairs of factors	Sums of factors
1, 15	16
* 3, 5	8

The numbers we want are 3 and 5.

$x^2 + 8x + 15 = (x + 3)(x + 5)$

3. $x^2 - x - 2$

The leading term is x^2. The first term of each binomial factor is x: (x + _)(x + _). Since the constant term is negative, we look for a factorization of -2 in which one factor is positive and one factor is negative. Their sum must be -1, the coefficient of the middle term.

Pairs of factors	Sums of factors
* 1, -2	-1
-1, 2	1

The numbers we want are 1 and -2.

$x^2 - x - 2 = (x + 1)(x - 2)$

5. $x^2 + 7x + 12$

The leading term is x^2. The first term of each binomial factor is x: (x + _)(x + _). Since the constant term is positive and the coefficient of the middle term is positive, we look for a factorization of 12 in which both factors are positive. Their sum must be 7, the coefficient of the middle term.

Pairs of factors	Sums of factors
1, 12	13
2, 6	8
* 3, 4	7

The numbers we want are 3 and 4.

$x^2 + 7x + 12 = (x + 3)(x + 4)$

7. $x^2 + 2x - 15$

Since the constant term is negative, we look for a factorization of -15 in which one factor is positive and one factor is negative. Their sum must be 2, the coefficient of the middle term.

Pairs of factors	Sums of factors
1, -15	-14
-1, 15	14
3, -5	-2
* -3, 5	2

The numbers we want are -3 and 5.

$x^2 + 2x - 15 = (x - 3)(x + 5)$

9. $y^2 + 9y + 8$

Since the constant term is positive and the coefficient of the middle term is positive, we look for a factorization of 8 in which both factors are positive. Their sum must be 9.

Pairs of factors	Sums of factors
* 1, 8	9
2, 4	6

The numbers we want are 1 and 8.

$y^2 + 9y + 8 = (y + 1)(y + 8)$

11. $x^4 + 5x^2 + 6$

The leading term is x^4. The first term of each binomial factor is x^2: (x^2 + _)(x^2 + _). Since the constant term is positive and the coefficient of the middle term is positive, we look for a factorization of 6 in which both factors are positive. Their sum must be 5.

Pairs of factors	Sums of factors
1, 6	7
* 2, 3	5

The numbers we want are 2 and 3.

$x^4 + 5x^2 + 6 = (x^2 + 2)(x^2 + 3)$

13. $x^2 + 3x - 28$

Since the constant term is negative, we look for a factorization of -28 in which one factor is positive and one factor is negative. Their sum must be 3.

Pairs of factors	Sums of factors
1, -28	-27
-1, 28	27
2, -14	-12
-2, 14	12
4, -7	-3
* -4, 7	3

The numbers we want are -4 and 7.

$x^2 + 3x - 28 = (x - 4)(x + 7)$

15. $16 + 8x + x^2$, or $x^2 + 8x + 16$

Since the constant term is positive and the coefficient of the middle term is positive, we look for a factorization of 16 in which both factors are positive. Their sum must be 8.

Pairs of factors	Sums of factors
1, 16	17
2, 8	10
* 4, 4	8

The numbers we want are 4 and 4.

$x^2 + 8x + 16 = (x + 4)(x + 4)$, or $(x + 4)^2$

The answer can also be expressed as $(4 + x)(4 + x)$, or $(4 + x)^2$.

Note that the trinomial is a trinomial square.

17. $a^2 - 12a + 11$

Since the constant term is positive and the coefficient of the middle term is negative, we look for a factorization of 11 in which both factors are negative. Their sum must be -12. The numbers we want are -1 and -11:

$-1(-11) = 11$ and $-1 + (-11) = -12$

$a^2 - 12a + 11 = (a - 1)(a - 11)$

19. $x^2 - \frac{2}{5}x + \frac{1}{25}$

Since the constant term is positive and the coefficient of the middle term is negative, we look for a factorization of $\frac{1}{25}$ in which both factors are negative. Their sum must be $-\frac{2}{5}$.

Pairs of factors	Sums of factors
$-1, -\frac{1}{25}$	$-\frac{26}{25}$
* $-\frac{1}{5}, -\frac{1}{5}$	$-\frac{2}{5}$

The numbers we want are $-\frac{1}{5}$ and $-\frac{1}{5}$.

$x^2 - \frac{2}{5}x + \frac{1}{25} = (x - \frac{1}{5})(x - \frac{1}{5})$, or $(x - \frac{1}{5})^2$

Note that the trinomial is a trinomial square.

21. $y^2 - 0.2y - 0.08$

Since the constant term is negative, we look for a factorization of -0.08 in which one factor is positive and one factor is negative. Their sum must be -0.2.

Pairs of factors	Sums of factors
0.1, -0.8	-0.7
-0.1, 0.8	0.7
* 0.2, -0.4	-0.2
-0.2, 0.4	0.2

The numbers we want are 0.2 and -0.4.

$y^2 - 0.2y - 0.08 = (y + 0.2)(y - 0.4)$

23. $y^2 + 11y + 28$

Since the constant term is positive and the coefficient of the middle term is positive, we look for a factorization of 28 in which both factors are positive. Their sum must be 11.

Pairs of factors	Sums of factors
1, 28	29
2, 14	16
* 4, 7	11

The numbers we want are 4 and 7.

$y^2 + 11y + 28 = (y + 4)(y + 7)$

25. $30 + 11a + a^2$, or $a^2 + 11a + 30$

Since the constant term is positive and the coefficient of the middle term is positive, we look for a factorization of 30 in which both factors are positive. Their sum must be 11.

Pairs of factors	Sums of factors
1, 30	31
2, 15	17
3, 10	13
* 5, 6	11

The numbers we want are 5 and 6.

$a^2 + 11a + 30 = (a + 5)(a + 6)$

The answer can also be expressed as $(5 + a)(6 + a)$.

27. $x^2 - x - 42$

Since the constant term is negative, we look for a factorization of -42 in which one factor is positive and one factor is negative. Their sum must be -1.

Pairs of factors	Sums of factors
1, -42	-41
-1, 42	41
2, -21	-19
-2, 21	19
3, -14	-11
-3, 14	11
* 6, -7	-1
-6, 7	1

The numbers we want are 6 and -7.

$x^2 - x - 42 = (x + 6)(x - 7)$

29. $x^2 - 2x - 99$

Since the constant term is negative, we look for a factorization of -99 in which one factor is positive and one factor is negative. Their sum must be -2.

Pairs of factors	Sums of factors
1, -99	-98
-1, 99	98
3, -33	-30
-3, 33	30
* 9, -11	-2
-9, 11	2

The numbers we want are 9 and -11.

$x^2 - 2x - 99 = (x + 9)(x - 11)$

31. $6x - 72 + x^2$, or $x^2 + 6x - 72$

Since the constant term is negative, we look for a factorization of -72 in which one factor is positive and one factor is negative. Their sum must be 6.

Pairs of factors	Sums of factors
1, -72	-71
-1, 72	71
2, -36	-34
-2, 36	34
3, -24	-21
-3, 24	21
4, -18	-14
-4, 18	14
6, -12	-6
* -6, 12	6
8, -9	-1
-8, 9	1

The numbers we want are -6 and 12.

$x^2 + 6x - 72 = (x - 6)(x + 12)$

33. $x^2 + 20x + 100$

Since the constant term is positive and the coefficient of the middle term is positive, we look for a factorization of 100 in which both factors are positive. Their sum must be 20.

Pairs of factors	Sums of factors
1, 100	101
2, 50	52
4, 25	29
5, 20	25
* 10, 10	20

The numbers we want are 10 and 10.

$x^2 + 20x + 100 = (x + 10)(x + 10)$, or $(x + 10)^2$

Note that the trinomial is a trinomial square.

35. $8x(2x^2 - 6x + 1)$

$= 8x \cdot 2x^2 - 8x \cdot 6x + 8x \cdot 1$

$= 16x^3 - 48x^2 + 8x$

37. $4x + 9 = 17$

$4x = 8$

$x = 2$

39. $x^2 - \frac{1}{2}x - \frac{3}{16}$

We look for two factors, one positive and one negative, whose product is $-\frac{3}{16}$ and whose sum is $-\frac{1}{2}$.

They are $-\frac{3}{4}$ and $\frac{1}{4}$.

$-\frac{3}{4} \cdot \frac{1}{4} = -\frac{3}{16}$ and $-\frac{3}{4} + \frac{1}{4} = -\frac{2}{4} = -\frac{1}{2}$.

$x^2 - \frac{1}{2}x - \frac{3}{16} = (x - \frac{3}{4})(x + \frac{1}{4})$

41. $y^2 + my + 50$

We look for pairs of factors whose product is 50. The sum of each pair is represented by m.

Pairs of factors whose product is 50	Sums of factors
1, 50	51
-1, -50	-51
2, 25	27
-2, -25	-27
5, 10	15
-5, -10	-15

The polynomial $y^2 + my + 50$ can be factored if m is 51, -51, 27, -27, 15, or -15.

Exercise Set 5.5

1. $3x^2 + 4x + 1$

a) First look for a common factor. There is none (other than 1).

b) Multiply the leading coefficient and the constant, 3 and 1: $3 \cdot 1 = 3$.

c) Try to factor 3 so that the sum of the factors is 4. The numbers we want are 1 and 3: $1 \cdot 3 = 3$ and $1 + 3 = 4$.

d) Split the middle term: $4x = 1x + 3x$

e) Factor by grouping:

$$\begin{aligned} 3x^2 + 4x + 1 &= 3x^2 + x + 3x + 1 \\ &= x(3x + 1) + 1(3x + 1) \\ &= (x + 1)(3x + 1) \end{aligned}$$

3. $12x^2 + 28x - 24$

Method 1:

a) We first factor out the common factor, 4.

$12x^2 + 28x - 24 = 4(3x^2 + 7x - 6)$

b) Now we factor the trinomial $3x^2 + 7x - 6$. Multiply the leading coefficient and the constant, 3 and -6: $3(-6) = -18$.

c) Try to factor -18 so that the sum of the factors is 7.

	Pairs of factors	Sums of factors
	-1, 18	17
	1, -18	-17
*	-2, 9	7
	2, -9	-7
	-3, 6	3
	3, -6	-3

d) Split the middle term: $7x = -2x + 9x$

e) Factor by grouping:

$$\begin{aligned} 3x^2 + 7x - 6 &= 3x^2 - 2x + 9x - 6 \\ &= x(3x - 2) + 3(3x - 2) \\ &= (x + 3)(3x - 2) \end{aligned}$$

We must include the common factor to get a factorization of the original trinomial.

$12x^2 + 28x - 24 = 4(x + 3)(3x - 2)$

Method 2:

$12x^2 + 28x - 24$

First we look for a common factor. The number 4 is a common factor, so we factor it out.

$4(3x^2 + 7x - 6)$

Next we factor the trinomial $3x^2 + 7x - 6$. We look for pairs of numbers whose product is 3. It is common practice that both of these factors should be positive. Thus, the only pair to consider is 1 and 3.

3. (continued)

We have this possibility:

$(x \quad)(3x \quad)$

Next we look for pairs of numbers whose product is -6. These are:

-1, 6	2, -3
1, -6	-2, 3

Then we list possibilities for factorization using these pairs of numbers.

$(x - 1)(3x + 6)$	$(x + 2)(3x - 3)$
$(x + 6)(3x - 1)$	$(x - 3)(3x + 2)$
$(x + 1)(3x - 6)$	$(x - 2)(3x + 3)$
$(x - 6)(3x + 1)$	$(x + 3)(3x - 2)$

We multiply and find that the desired factorization is $(x + 3)(3x - 2)$. The complete factorization is $4(x + 3)(3x - 2)$.

5. $2x^2 - x - 1$

Method 1:

a) First look for a common factor. There is none (other than 1).

b) Multiply the leading coefficient and the constant, 2 and -1: $2(-1) = -2$.

c) Try to factor -2 so that the sum of the factors is -1. The numbers we want are -2 and 1: $-2 \cdot 1 = -2$ and $-2 + 1 = -1$.

d) Split the middle term: $-x = -2x + 1x$

e) Factor by grouping:

$$\begin{aligned} 2x^2 - x - 1 &= 2x^2 - 2x + x - 1 \\ &= 2x(x - 1) + 1(x - 1) \\ &= (2x + 1)(x - 1) \end{aligned}$$

Method 2:

$2x^2 - x - 1$

First we look for a common factor. There is none (other than 1.)

Next we look for pairs of numbers whose product is 2. There is only one pair, 2 and 1, since we agree these factors should be positive.

We have this possibility:

$(2x \quad)(x \quad)$

Next we look for pairs of numbers whose product is -1. There is only one pair, -1 and 1.

Then we list possibilities for factorization using this pair of numbers.

$(2x - 1)(x + 1) \qquad (2x + 1)(x - 1)$

We multiply and find that the desired factorization is $(2x + 1)(x - 1)$.

7. $9x^2 + 18x - 16$

a) First look for a common factor. There is none (other than 1).

b) Multiply the leading coefficient and the constant, 9 and -16: $9(-16) = -144$.

c) Try to factor -144 so that the sum of the factors is 18.

Pairs of factors	Sums of factors
-1, 144	143
1, -144	-143
-2, 72	70
2, -72	-70
-3, 48	45
3, -48	-45
-4, 36	32
4, -36	-32
* -6, 24	18
6, -24	-18
-8, 18	10
8, -18	-10
-9, 16	7
9, -16	-7
-12, 12	0

d) Split the middle term: $18x = -6x + 24x$

e) Factor by grouping:

$$\begin{aligned} 9x^2 + 18x - 16 &= 9x^2 - 6x + 24x - 16 \\ &= 3x(3x - 2) + 8(3x - 2) \\ &= (3x + 8)(3x - 2) \end{aligned}$$

9. $15x^2 - 25x - 10$

Method 1:

a) We first factor out the common factor, 5.

$15x^2 - 25x - 10 = 5(3x^2 - 5x - 2)$

b) Now we factor the trinomial $3x^2 - 5x - 2$. Multiply the leading coefficient and the constant, 3 and -2: $3(-2) = -6$.

c) Try to factor -6 so that the sum of the factors is -5. The numbers we want are -6 and 1: $-6 \cdot 1 = -6$ and $-6 + 1 = -5$.

d) Split the middle term: $-5x = -6x + x$

e) Factor by grouping:

$$\begin{aligned} 3x^2 - 5x - 2 &= 3x^2 - 6x + x - 2 \\ &= 3x(x - 2) + (x - 2) \\ &= (3x + 1)(x - 2) \end{aligned}$$

We must include the common factor to get a factorization of the original trinomial.

$15x^2 - 25x - 10 = 5(3x + 1)(x - 2)$

9. (continued)

Method 2:

$15x^2 - 25x - 10$

First we factor out the common factor, 5.

$5(3x^2 - 5x - 2)$

Next we factor $3x^2 - 5x - 2$. We look for pairs of numbers whose product is 3. There is only one pair, 3 and 1, since we agree these factors should be positive.

We have the possibility:

$(3x \quad)(x \quad)$

Next we look for pairs of numbers whose product is -2. These are 2, -1 and -2, 1.

Then we list possibilities for factorization using these pairs of numbers.

$(3x + 2)(x - 1)$ $(3x - 2)(x + 1)$
$(3x - 1)(x + 2)$ $(3x + 1)(x - 2)$

We multiply and find that the desired factorization is $(3x + 1)(x - 2)$. The complete factorization is $5(3x + 1)(x - 2)$.

11. $12x^2 + 31x + 20$

Method 1:

a) First look for a common factor. There is none (other than 1).

b) Multiply the leading coefficient and the constant, 12 and 20: $12 \cdot 20 = 240$.

c) Try to factor 240 so that the sum of the factors is 31. We only need to consider positive factors.

Pairs of factors	Sums of factors
1, 240	241
2, 120	122
3, 80	83
4, 60	64
5, 48	53
6, 40	46
8, 30	38
10, 24	34
12, 20	32
* 15, 16	31

d) Split the middle term: $31x = 15x + 16x$

e) Factor by grouping:

$$\begin{aligned} 12x^2 + 31x + 20 &= 12x^2 + 15x + 16x + 20 \\ &= 3x(4x + 5) + 4(4x + 5) \\ &= (3x + 4)(4x + 5) \end{aligned}$$

11. (continued)

Method 2:

$12x^2 + 31x + 20$

First we look for a common factor. There is none (other than 1).

Next we look for pairs of numbers whose product is 12. Since it is common practice that both factors be positive, we only consider

1, 12   2, 6   and   3, 4.

We have these possibilities:

$(x \quad)(12x \quad)$
$(2x \quad)(6x \quad)$
$(3x \quad)(4x \quad)$

Next we look for pairs of numbers whose product is 20. These are

1, 20	2, 10	4, 5
-1, -20	-2, -10	-4, -5

Using these pairs of numbers we have 36 possible factorizations. It is very time consuming to make the complete list and multiply each. With practice (trial and error) we will be able to mentally shorten this list and consider just a few. Since the coefficients of all three terms are positive, we do not need to consider possibilities that contain -1, -20 -2, -10 and -4, -5.

This shortens the list to only 18 possibilities. Since the absolute value of the coefficient of the middle term (in this problem it is 31) is not large compared to the coefficients of the first and third terms, it saves time to check the following possibilities:

$(3x + 4)(4x + 5)$ $(2x + 4)(6x + 5)$
$(3x + 5)(4x + 4)$ $(2x + 5)(6x + 4)$
$(3x + 2)(4x + 10)$ $(2x + 2)(6x + 10)$
$(3x + 10)(4x + 2)$ $(2x + 10)(6x + 2)$

We multiply and find that the desired factorization is $(3x + 4)(4x + 5)$. If none of the above had given the desired factorization, then we would have to begin checking the other 10 possibilities.

13. $14x^2 + 19x - 3$

a) First look for a common factor. There is none (other than 1).

b) Multiply the leading coefficient and the constant, 14 and -3: $14(-3) = -42$.

c) Try to factor -42 so that the sum of the factors is 19.

Pairs of factors	Sums of factors
-1, 42	41
1, -42	-41
* -2, 21	19
2, -21	-19
-3, 14	11
3, -14	-11
-6, 7	1
6, -7	-1

13. (continued)

d) Split the middle term: $19x = -2x + 21x$

e) Factor by grouping:

$$14x^2 + 19x - 3 = 14x^2 - 2x + 21x - 3$$
$$= 2x(7x - 1) + 3(7x - 1)$$
$$= (2x + 3)(7x - 1)$$

15. $9x^4 + 18x^2 + 8$

a) First look for a common factor. There is none (other than 1).

b) Multiply the leading coefficient and the constant, 9 and 8: $9 \cdot 8 = 72$.

c) Try to factor 72 so that the sum of the factors is 18. We only need to consider positive factors.

Pairs of factors	Sums of factors
1, 72	73
2, 36	38
3, 24	27
4, 18	22
* 6, 12	18
8, 9	17

d) Split the middle term: $18x^2 = 6x^2 + 12x^2$.

e) Factor by grouping:

$$9x^4 + 18x^2 + 8 = 9x^4 + 6x^2 + 12x^2 + 8$$
$$= 3x^2(3x^2 + 2) + 4(3x^2 + 2)$$
$$= (3x^2 + 4)(3x^2 + 2)$$

17. This polynomial is a trinomial square.

Recall: $A^2 - 2AB + B^2 = (A - B)^2$

$$9x^2 - 42x + 49$$
$$= (3x)^2 - 2 \cdot 3x \cdot 7 + 7^2$$
$$= (3x - 7)^2$$

Method 1 and Method 2 could also be used.

19. $6x^3 + 4x^2 - 10x$

a) We find factor out the common factor, 2x.
$6x^3 + 4x^2 - 10x = 2x(3x^2 + 2x - 5)$

b) Now we factor the trinomial $3x^2 + 2x - 5$. Multiply the leading coefficient and the constant, 3 and -5: $3(-5) = -15$.

c) Try to factor -15 so that the sum of the factors is 2.

Pairs of factors	Sums of factors
-1, 15	14
1, -15	-14
* -3, 5	2
3, -5	-2

d) Split the middle term: $2x = -3x + 5x$

e) Factor by grouping:

$3x^2 + 2x - 5 = 3x^2 - 3x + 5x - 5$
$= 3x(x - 1) + 5(x - 1)$
$= (3x + 5)(x - 1)$

We must include the common factor to get a factorization of the original trinomial.

$6x^3 + 4x^2 - 10x = 2x(3x + 5)(x - 1)$

21. $12a - 5 + 9a^2 = 9a^2 + 12a - 5$

a) First look for a common factor. There is none (other than 1).

b) Multiply the leading coefficient and the constant, 9 and -5: $9(-5) = -45$.

c) Try to factor -45 so that the sum of the factors is 12.

Pairs of factors	Sums of factors
-1, 45	44
1, -45	-44
* -3, 15	12
3, -15	-12
-5, 9	4
5, -9	-4

d) Split the middle term: $12a = -3a + 15a$

e) Factor by grouping:

$9a^2 + 12a - 5 = 9a^2 - 3a + 15a - 5$
$= 3a(3a - 1) + 5(3a - 1)$
$= (3a + 5)(3a - 1)$

23. This polynomial is a trinomial square.

Recall: $A^2 + 2AB + B^2 = (A + B)^2$

$25x^2 + 40x + 16$
$= (5x)^2 + 2\cdot 5x\cdot 4 + 4^2$
$= (5x + 4)^2$

Method 1 and Method 2 could also be used.

25. $7 + 23m + 6m^2 = 6m^2 + 23m + 7$

a) First look for a common factor. There is none (other than 1).

b) Multiply the leading coefficient and the constant, 6 and 7: $6\cdot 7 = 42$.

c) Try to factor 42 so that the sum of the factors is 23. We only need to consider positive factors.

Pairs of factors	Sums of factors
1, 42	43
* 2, 21	23
3, 14	17
6, 7	13

d) Split the middle term: $23m = 2m + 21m$

e) Factor by grouping:

$6m^2 + 23m + 7 = 6m^2 + 2m + 21m + 7$
$= 2m(3m + 1) + 7(3m + 1)$
$= (2m + 7)(3m + 1)$
or $(7 + 2m)(1 + 3m)$

27. $4x + 6x^2 - 10 = 6x^2 + 4x - 10$

a) We first factor out the common factor, 2.

$6x^2 + 4x - 10 = 2(3x^2 + 2x - 5)$

b) Now we factor the trinomial $3x^2 + 2x - 5$. Multiply the leading coefficient and the constant, 3 and -5: $3(-5) = -15$

c) Try to factor -15 so that the sum of the factors is 2. The numbers we want are 5 and -3: $5(-3) = -15$ and $5 + (-3) = 2$.

d) Split the middle term: $2x = 5x - 3x$

e) Factor by grouping:

$3x^2 + 2x - 5 = 3x^2 + 5x - 3x - 5$
$= x(3x + 5) - 1(3x + 5)$
$= (x - 1)(3x + 5)$

We must include the common factor to get a factorization of the original trinomial.

$6x^2 + 4x - 10 = 2(x - 1)(3x + 5)$

29. $12x^3 + 31x^2 + 20x$

We first factor out the common factor, x.

$12x^3 + 31x^2 + 20x = x(12x^2 + 31x + 20)$

In Exercise 11 we factored $12x^2 + 31x + 20$.

$12x^2 + 31x + 20 = (3x + 4)(4x + 5)$

We must include the common factor to get a factorization of the original trinomial.

$12x^3 + 31x^2 + 20 = x(3x + 4)(4x + 5)$

31. $14x^4 + 19x^3 - 3x^2$

We first factor out the common factor, x^2.

$14x^4 + 19x^3 - 3x^2 = x^2(14x^2 + 19x - 3)$

In Exercise 13 we factored $14x^2 + 19x - 3$.

$14x^2 + 19x - 3 = (2x + 3)(7x - 1)$

We must include the common factor to get a factorization of the original trinomial.

$14x^4 + 19x^3 - 3x^2 = x^2(2x + 3)(7x - 1)$

33. $x^2 + 3x - 7$

We look for a factorization of -7 such that the sum of the factors is 3. There is no such factorization. Thus the polynomial cannot be factored.

35. $\frac{y^{-12}}{y^{-8}} = y^{-12-(-8)} = y^{-12+8} = y^{-4}$

37. $3(2y + 3) = 21$

$6y + 9 = 21$

$6y = 12$

$y = 2$

39. $27x^3 - 63x^2 - 147x + 343$

First look for a common factor. There is none (other than 1).

Factor by grouping:

$(27x^3 - 63x^2) - (147x - 343)$

$= 9x^2(3x - 7) - 49(3x - 7)$

$= (9x^2 - 49)(3x - 7)$

The binomial factor $9x^2 - 49$ is a difference of squares. It can be factored further.

$9x^2 - 49 = (3x + 7)(3x - 7)$

$27x^3 - 63x^2 - 147x + 343 = (3x + 7)(3x - 7)(3x - 7)$ or $(3x + 7)(3x - 7)^2$

41. $3x^{6a} - 2x^{3a} - 1$

$= 3(x^{3a})^2 - 2(x^{3a}) - 1$

$= (3x^{3a} + 1)(x^{3a} - 1)$

Exercise Set 5.6

1. $2x^2 - 128$

$= 2(x^2 - 64)$

$= 2(x - 8)(x + 8)$

3. $a^2 + 25 - 10a$

$= a^2 - 10a + 25$

$= (a - 5)^2$

5. $2x^2 - 11x + 12$

There is no common factor (other than 1). This polynomial has three terms, but it is not a trinomial square. Multiply the leading coefficient and the constant, 2 and 12: $2 \cdot 12 = 24$. Try to factor 24 so that the sum of the factors is -11. The numbers we want are -3 and -8: $-3(-8) = 24$ and $-3 + (-8) = -11$. Split the middle term and factor by grouping.

$2x^2 - 11x + 12$

$= 2x^2 - 3x - 8x + 12$

$= x(2x - 3) - 4(2x - 3)$

$= (x - 4)(2x - 3)$

7. $x^3 + 24x^2 + 144x$

$= x(x^2 + 24x + 144)$

$= x(x + 12)^2$

9. $x^2 + 3x + 2x + 6$

$= x(x + 3) + 2(x + 3)$

$= (x + 2)(x + 3)$

11. $24x^2 - 54$

$= 6(4x^2 - 9)$

$= 6(2x + 3)(2x - 3)$

13. $20x^3 - 4x^2 - 72x$

$= 4x(5x^2 - x - 18)$

$= 4x(5x + 9)(x - 2)$

15. $x^2 + 4$ is a sum of squares. It cannot be factored.

17. $x^4 + 7x^2 - 3x^2 - 21$

$= x^2(x^2 + 7) - 3(x^2 + 7)$

$= (x^2 - 3)(x^2 + 7)$

19. $x^5 - 14x^4 + 49x^3$

$= x^3(x^2 - 14x + 49)$

$= x^3(x - 7)^2$

21. $20 - 6x - 2x^2$ or $20 - 6x - 2x^2$

$= 2(10 - 3x - x^2)$ $= -2x^2 - 6x + 20$

$= 2(5 + x)(2 - x)$ $= -2(x^2 + 3x - 10)$

$= -2(x + 5)(x - 2)$

$20 - 6x - 2x^2 = 2(5 + x)(2 - x)$, or $-2(x + 5)(x - 2)$

23. $x^2 + 3x + 1$

We look for two factors whose product is 1 and whose sum is 3. There are none. The polynomial cannot be factored.

25. $4x^4 - 64$

$= 4(x^4 - 16)$

$= 4(x^2 + 4)(x^2 - 4)$

$= 4(x^2 + 4)(x + 2)(x - 2)$

27. $1 - y^8$
$= (1 + y^4)(1 - y^4)$
$= (1 + y^4)(1 + y^2)(1 - y^2)$
$= (1 + y^4)(1 + y^2)(1 + y)(1 - y)$

29. $x^5 - 4x^4 - 3x^3$
$= x^3(x^2 - 4x - 3)$

31. $36a^2 - 15a + \frac{25}{16}$
$= (6a)^2 - 2\cdot 6a\cdot \frac{5}{4} + (\frac{5}{4})^2$
$= (6a - \frac{5}{4})^2$

33. $a^4 - 2a^2 + 1$
$= (a^2)^2 - 2\cdot a^2\cdot 1 + 1^2$
$= (a^2 - 1)^2$
$= (a^2 - 1)(a^2 - 1)$
$= (a + 1)(a - 1)(a + 1)(a - 1)$, or $(a + 1)^2(a - 1)^2$

35. Population of London is 95% of population of New York.

$7{,}028{,}000 = 95\% \cdot x$

$7{,}028{,}000 = 0.95x$
$7{,}397{,}895 \approx x$

The population of New York is approximately 7,397,895.

37. $\frac{11}{6} - (-\frac{11}{18}) = \frac{33}{18} + \frac{11}{18} = \frac{44}{18} = \frac{22}{9}$

39. $12.25x^2 - 7x + 1$
$= \frac{49}{4}x^2 - 7x + 1$
$= (\frac{7}{2}x)^2 - 2\cdot\frac{7}{2}x\cdot 1 + 1^2$
$= (\frac{7}{2}x - 1)^2$; or $(3.5x - 1)^2$

41. $5x^2 + 13x + 7.2$
There is no common factor (other than 1). Multiply the leading coefficient and the constant, 5 and 7.2: $5(7.2) = 36$. Try to factor 36 so that the sum of the factors is 13. The numbers we want are 4 and 9: $4\cdot 9 = 36$ and $4 + 9 = 13$. Split the middle term and factor by grouping.

$5x^2 + 13x + 7.2 = 5x^2 + 4x + 9x + \frac{36}{5}$
$= x(5x + 4) + \frac{9}{5}(5x + 4)$
$= (x + \frac{9}{5})(5x + 4)$
$= (x + 1.8)(5x + 4)$

43. $y^2(y - 1) - 2y(y - 1) + (y - 1)$
$= (y^2 - 2y + 1)(y - 1)$
$= (y - 1)^2(y - 1)$
$= (y - 1)^3$

Exercise Set 5.7

1. $(x + 8)(x + 6) = 0$
$x + 8 = 0$ or $x + 6 = 0$
$x = -8$ or $x = -6$

Check:

For -8

$(x + 8)(x + 6) = 0$	
$(-8 + 8)(-8 + 6)$	0
$0\cdot(-2)$	
0	

For -6

$(x + 8)(x + 6) = 0$	
$(-6 + 8)(-6 + 6)$	0
$2\cdot 0$	
0	

The solutions are -8 and -6.

3. $(x - 3)(x + 5) = 0$
$x - 3 = 0$ or $x + 5 = 0$
$x = 3$ or $x = -5$

Check:

For 3

$(x - 3)(x + 5) = 0$	
$(3 - 3)(3 + 5)$	0
$0\cdot 8$	
0	

For -5

$(x - 3)(x + 5) = 0$	
$(-5 - 3)(-5 + 5)$	0
$-8\cdot 0$	
0	

The solutions are 3 and -5.

5. $(x - 12)(x - 11) = 0$
$x - 12 = 0$ or $x - 11 = 0$
$x = 12$ or $x = 11$

The solutions are 12 and 11.

7. $y(y - 13) = 0$
$y = 0$ or $y - 13 = 0$
$y = 0$ or $y = 13$

The solutions are 0 and 13.

9. $0 = x(x + 21)$
$x = 0$ or $x + 21 = 0$
$x = 0$ or $x = -21$

The solutions are 0 and -21.

11. $(2x + 5)(x + 4) = 0$
$2x + 5 = 0$ or $x + 4 = 0$
$2x = -5$ or $x = -4$
$x = -\frac{5}{2}$ or $x = -4.$
The solutions are $-\frac{5}{2}$ and -4.

13. $(3x - 1)(x + 2) = 0$
$3x - 1 = 0$ or $x + 2 = 0$
$3x = 1$ or $x = -2$
$x = \frac{1}{3}$ or $x = -2$
The solutions are $\frac{1}{3}$ and -2.

15. $2x(3x - 2) = 0$
$2x = 0$ or $3x - 2 = 0$
$x = 0$ or $3x = 2$
$x = 0$ or $x = \frac{2}{3}$
The solutions are 0 and $\frac{2}{3}$.

17. $(\frac{1}{3}y - \frac{2}{3})(\frac{1}{4}y - \frac{3}{2}) = 0$
$\frac{1}{3}y - \frac{2}{3} = 0$ or $\frac{1}{4}y - \frac{3}{2} = 0$
$\frac{1}{3}y = \frac{2}{3}$ or $\frac{1}{4}y = \frac{3}{2}$
$y = 2$ or $y = 6$
The solutions are 2 and 6.

19. $(0.03x - 0.01)(0.05x - 1) = 0$
$0.03x - 0.01 = 0$ or $0.05x - 1 = 0$
$0.03x = 0.01$ or $0.05x = 1$
$x = \frac{0.01}{0.03}$ or $x = \frac{1}{0.05}$
$x = \frac{1}{3}$ or $x = 20$
The solutions are $\frac{1}{3}$ and 20.

21. $x^2 + 6x + 5 = 0$
$(x + 5)(x + 1) = 0$
$x + 5 = 0$ or $x + 1 = 0$
$x = -5$ or $x = -1$
The solutions are -5 and -1.

23. $x^2 - 5x = 0$
$x(x - 5) = 0$
$x = 0$ or $x - 5 = 0$
$x = 0$ or $x = 5$
The solutions are 0 and 5.

25. $x^2 + 6x + 9 = 0$
$(x + 3)(x + 3) = 0$
$x + 3 = 0$ or $x + 3 = 0$
$x = -3$ or $x = -3$
The solution is -3.

27. $6y^2 - 4y - 10 = 0$
$2(3y^2 - 2y - 5) = 0$
$2(3y - 5)(y + 1) = 0$
$3y - 5 = 0$ or $y + 1 = 0$
$3y = 5$ or $y = -1$
$y = \frac{5}{3}$ or $y = -1$
The solutions are $\frac{5}{3}$ and -1.

29. $3x^2 = 7x + 20$
$3x^2 - 7x - 20 = 0$
$(3x + 5)(x - 4) = 0$
$3x + 5 = 0$ or $x - 4 = 0$
$3x = -5$ or $x = 4$
$x = -\frac{5}{3}$ or $x = 4$
The solutions are $-\frac{5}{3}$ and 4.

31. $-5x = -12x^2 + 2$
$12x^2 - 5x - 2 = 0$
$(4x + 1)(3x - 2) = 0$
$4x + 1 = 0$ or $3x - 2 = 0$
$4x = -1$ or $3x = 2$
$x = -\frac{1}{4}$ or $x = \frac{2}{3}$
The solutions are $-\frac{1}{4}$ and $\frac{2}{3}$.

33. $0 = -3x + 5x^2$
$0 = x(-3 + 5x)$
$x = 0$ or $-3 + 5x = 0$
$x = 0$ or $5x = 3$
$x = 0$ or $x = \frac{3}{5}$
The solutions are 0 and $\frac{3}{5}$.

35. $x(x - 5) = 14$
$x^2 - 5x = 14$
$x^2 - 5x - 14 = 0$
$(x - 7)(x + 2) = 0$
$x - 7 = 0$ or $x + 2 = 0$
$x = 7$ or $x = -2$
The solutions are 7 and -2.

37.

$$64m^2 = 81$$
$$64m^2 - 81 = 0$$
$$(8m - 9)(8m + 9) = 0$$
$$8m - 9 = 0 \text{ or } 8m + 9 = 0$$
$$8m = 9 \text{ or } 8m = -9$$
$$m = \frac{9}{8} \text{ or } m = -\frac{9}{8}$$

The solutions are $\frac{9}{8}$ and $-\frac{9}{8}$.

39. $(-9)(16) = -144$

41.

$$b(b + 9) = 4(5 + 2b)$$
$$b^2 + 9b = 20 + 8b$$
$$b^2 + 9b - 8b - 20 = 0$$
$$b^2 + b - 20 = 0$$
$$(b + 5)(b - 4) = 0$$
$$b + 5 = 0 \text{ or } b - 4 = 0$$
$$b = -5 \text{ or } b = 4$$

The solutions are -5 and 4.

43.

$$(m - 5)^2 = 2(5 - m)$$
$$m^2 - 10m + 25 = 10 - 2m$$
$$m^2 - 8m + 15 = 0$$
$$(m - 5)(m - 3) = 0$$
$$m - 5 = 0 \text{ or } m - 3 = 0$$
$$m = 5 \text{ or } m = 3$$

The solutions are 5 and 3.

45.

$$(0.00005x + 0.1)(0.0097x + 0.5) = 0$$
$$0.00005x + 0.1 = 0 \text{ or } 0.0097x + 0.5 = 0$$
$$0.00005x = -0.1 \text{ or } 0.0097x = -0.5$$
$$x = -\frac{0.1}{0.00005} \text{ or } x = -\frac{0.5}{0.0097}$$
$$x = -2000 \text{ or } x \approx -51.546392$$

The solutions are -2000 and -51.546392.

47.

$$(x + 3)(4x - 5)(x - 7) = 0$$
$$x + 3 = 0 \text{ or } 4x - 5 = 0 \text{ or } x - 7 = 0$$
$$x = -3 \text{ or } x = \frac{5}{4} \text{ or } x = 7$$

The solutions are -3, $\frac{5}{4}$, and 7.

Exercise Set 5.8

1. Four times the square of a number minus the number is 3.

$$4x^2 \quad - \quad x \quad = 3$$

We solve the equation.

$$4x^2 - x = 3$$
$$4x^2 - x - 3 = 0$$
$$(4x + 3)(x - 1) = 0$$
$$4x + 3 = 0 \text{ or } x - 1 = 0$$
$$4x = -3 \text{ or } x = 1$$
$$x = -\frac{3}{4} \text{ or } x = 1$$

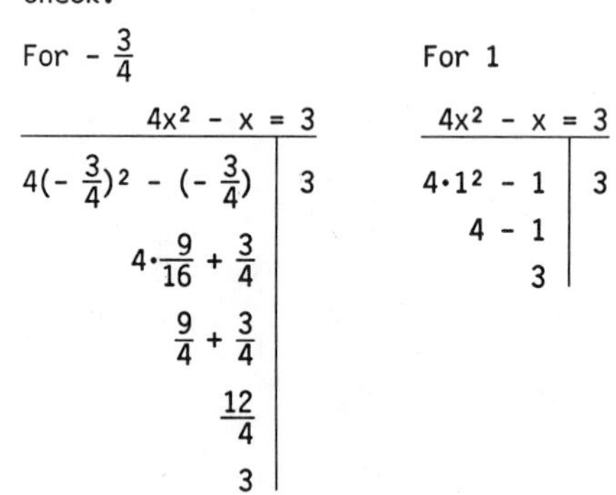

Check:

For $-\frac{3}{4}$

$4x^2 - x = 3$	
$4(-\frac{3}{4})^2 - (-\frac{3}{4})$	3
$4\cdot\frac{9}{16} + \frac{3}{4}$	
$\frac{9}{4} + \frac{3}{4}$	
$\frac{12}{4}$	
3	

For 1

$4x^2 - x = 3$	
$4\cdot1^2 - 1$	3
$4 - 1$	
3	

Both numbers check. There are two such numbers, $-\frac{3}{4}$ and 1.

3. Consecutive integers are next to each other. Let x represent the smaller integer, then x + 1 represents the larger integer.

Smaller integer times larger integer is 182.

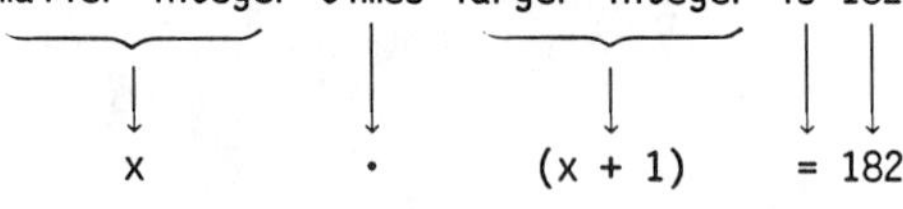

$$x(x + 1) = 182$$
$$x^2 + x = 182$$
$$x^2 + x - 182 = 0$$
$$(x + 14)(x - 13) = 0$$
$$x + 14 = 0 \text{ or } x - 13 = 0$$
$$x = -14 \text{ or } x = 13$$

The solutions of the equation are -14 and 13. When x is -14, then x + 1 is -13 and $-14\cdot(-13) = 182$. The numbers -14 and -13 are consecutive integers which are solutions to the problem. When x is 13, then x + 1 is 14 and $13\cdot14 = 182$. The numbers 13 and 14 are also consecutive integers which are solutions to the problem.

We have two solutions each of which consists of a pair of numbers: -14 and -13, and 13 and 14.

5. Consecutive odd integers are next to each other such as 11 and 13, or -9 and -7. Let x represent the smaller odd integer, then x + 2 represents the next odd integer.

Smaller odd integer times next odd integer is 195.

$x \quad \cdot \quad (x + 2) \quad = 195$

$$x(x + 2) = 195$$

$$x^2 + 2x = 195$$

$$x^2 + 2x - 195 = 0$$

$$(x + 15)(x - 13) = 0$$

$$x + 15 = 0 \quad \text{or} \quad x - 13 = 0$$

$$x = -15 \quad \text{or} \quad x = 13$$

The solutions of the equation are -15 and 13. When x is -15, then x + 2 is -13 and $(-15)\cdot(-13) = 195$. The numbers -15 and -13 are consecutive odd integers which are solutions to the problem. When x is 13, then x + 2 is 15 and $13\cdot15 = 195$. The numbers 13 and 15 are also consecutive odd integers which are solutions to the problem.

We have two solutions each of which consists of a pair of numbers: -15 and -13, and 13 and 15.

7. First draw a picture.

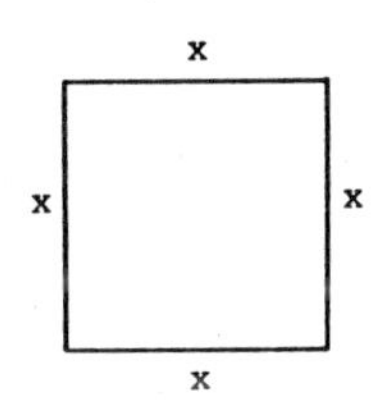

The area of the square is $x\cdot x$, or x^2. The perimeter of the square is x + x + x + x, or 4x.

Area of square is 5 more than perimeter of the square.

$x^2 \quad = 5 \quad + \quad 4x$

$$x^2 = 5 + 4x$$

$$x^2 - 4x - 5 = 0$$

$$(x - 5)(x + 1) = 0$$

$$x - 5 = 0 \quad \text{or} \quad x + 1 = 0$$

$$x = 5 \quad \text{or} \quad x = -1$$

The solutions of the equation are 5 and -1. The length of a side cannot be negative, so we only check 5. The area is $5\cdot5$, or 25. The perimeter is 5 + 5 + 5 + 5, or 20. The area, 25, is 5 more than the perimeter, 20. This checks.

The length of a side is 5.

9. First make a drawing.

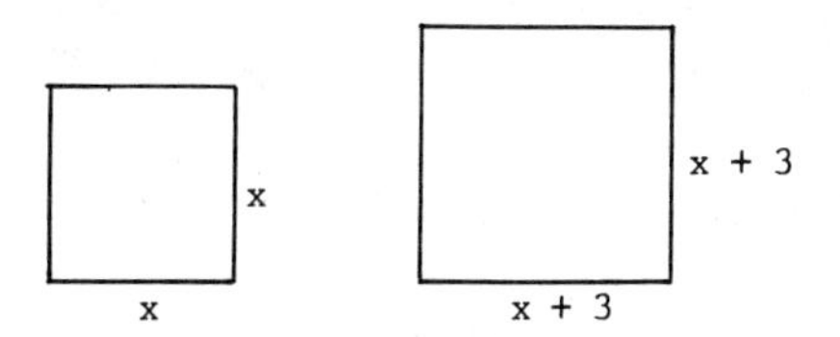

Let x represent the length of a side of the original square. When the length is increased by 3, the length of a side of the new square is x + 3. The area of the new square is 81 m^2. The area of a square is found by squaring the length of a side.

Area of new square is the square of the lengthened side.

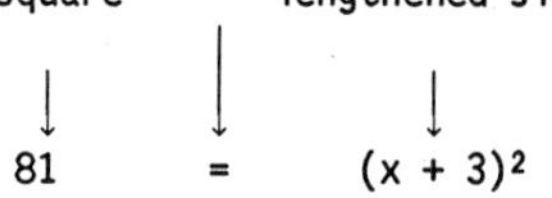

$81 \quad = \quad (x + 3)^2$

$$81 = (x + 3)^2$$

$$81 = x^2 + 6x + 9$$

$$0 = x^2 + 6x - 72$$

$$0 = (x + 12)(x - 6)$$

$$x + 12 = 0 \quad \text{or} \quad x - 6 = 0$$

$$x = -12 \quad \text{or} \quad x = 6$$

The solutions of the equation are -12 and 6. The length of a side cannot be negative, so -12 cannot be a solution. Suppose the length of a side of the original square is 6 m. Then the length of a side of the new square is 6 + 3, or 9 m. Its area is 9^2, or 81 m^2. The numbers check.

The length of a side of the original square is 6 m.

11. We substitute 23 for n in order to solve for N.

$$N = n^2 - n$$

$$N = 23^2 - 23$$

$$= 529 - 23$$

$$= 506$$

The total number of games to be played is 506.

13. We substitute 132 for N and solve for n.

$$N = n^2 - n$$

$$132 = n^2 - n$$

$$0 = n^2 - n - 132$$

$$0 = (n - 12)(n + 11)$$

$$n - 12 = 0 \quad \text{or} \quad n + 11 = 0$$

$$n = 12 \quad \text{or} \quad n = -11$$

The solutions of the equation are 12 and -11. Since the number of teams cannot be negative, -11 cannot be a solution. But 12 checks since $12^2 - 12 = 144 - 12 = 132$. There are 12 teams in the league.

15. We substitute 16 for n and solve for N.

$N = \frac{1}{2} n(n - 1)$

$N = \frac{1}{2}\cdot 16(16 - 1)$

$= 8\cdot 15$

$= 120$

Thus, 120 double teams are possible.

17. We substitute 190 for N and solve for n.

$N = \frac{1}{2} n(n - 1)$

$190 = \frac{1}{2} n(n - 1)$

$380 = n(n - 1)$

$380 = n^2 - n$

$0 = n^2 - n - 380$

$0 = (n - 20)(n + 19)$

$n - 20 = 0$ or $n + 19 = 0$

$n = 20$ or $n = -19$

The solutions of the equation are 20 and -19. Since the number of people cannot be negative, -19 cannot be a solution. But 20 checks since $\frac{1}{2}\cdot 20\cdot(20 - 1) = 10\cdot 19 = 190$.

There were 20 people at the tournament.

19. $S = 2\pi rh$

$\frac{1}{2\pi r}\cdot S = \frac{1}{2\pi r}\cdot 2\pi rh$

$\frac{S}{2\pi r} = h$

21. $-67.3 + (-32.8) = -100.1$

23. A number less than 100 is a two-digit number. Let x represent the tens digit. Then x + 4 represents the ones digit.

The number is $10x + (x + 4)$, or $11x + 4$.

The product of the digits is $x(x + 4)$, or $x^2 + 4x$.

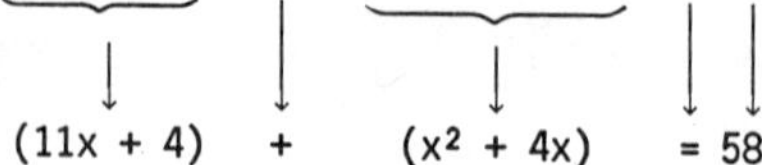

$11x + 4 + x^2 + 4x = 58$

$x^2 + 15x + 4 = 58$

$x^2 + 15x - 54 = 0$

$(x + 18)(x - 3) = 0$

$x + 18 = 0$ or $x - 3 = 0$

$x = -18$ or $x = 3$

23. (continued)

The solutions of the equation are -18 and 3. We only check 3 since a digit cannot be -18. If the tens digit is 3, the ones digit is 3 + 4, or 7, and the resulting number is 37. The product of the digits is $3\cdot 7$, or 21. The sum of the number and the product of the digits is 37 + 21, or 58. The value checks.

The number is 37.

25.

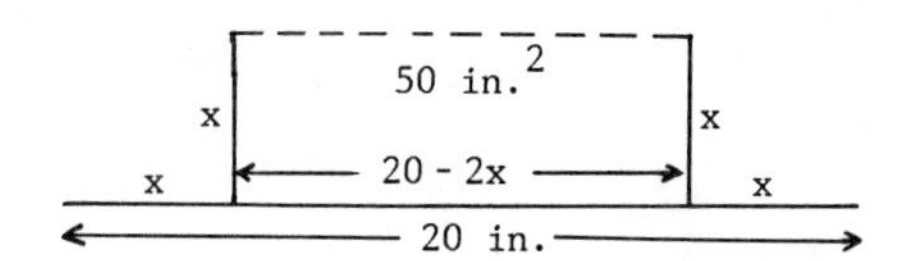

We let x represent the depth of the gutter. Then the width of the gutter is 20 - 2x.

Area of cross-section of gutter = Depth · Width

$50 = x \cdot (20 - 2x)$

$50 = x(20 - 2x)$

$50 = 20x - 2x^2$

$2x^2 - 20x + 50 = 0$

$x^2 - 10x + 25 = 0$

$(x - 5)(x - 5) = 0$

$x - 5 = 0$ or $x - 5 = 0$

$x = 5$ or $x = 5$

When x = 5, the depth is 5 in. and the width is $20 - 2\cdot 5$, or 10 in. The area is $5\cdot 10$, or 50 in^2. This checks.

The depth of the gutter is 5 in.

Exercise Set 5.9

1. We replace each x by 3 and each y by -2.

$x^2 - y^2 + xy$

$= 3^2 - (-2)^2 + 3(-2)$

$= 9 - 4 - 6$

$= -1$

3. We replace each x by 2, each y by -3, and each z by -1.

$xyz^2 + z$

$= 2(-3)(-1)^2 + (-1)$

$= -6 - 1$

$= -7$

5. We replace each P by 10,000 and each r by 0.08 ($r = 8\% = 0.08$).

$P + 3rP + 3r^2P + r^3P$

$= 10,000 + 3(0.08)(10,000) + 3(0.08)^2(10,000) + (0.08)^3(10,000)$

$= 10,000 + 3(0.08)(10,000) + 3(0.0064)(10,000) + 0.000512(10,000)$

$= 10,000 + 2400 + 192 + 5.12$

$= 12,597.12$

The amount to which \$10,000 will grow at 8% interest for 3 years is \$12,597.12.

7. We replace each h by 4.7, each r by 1.2, and π by 3.14.

$2\pi rh + 2\pi r^2$

$= 2(3.14)(1.2)(4.7) + 2(3.14)(1.2)^2$

$= 2(3.14)(1.2)(4.7) + 2(3.14)(1.44)$

$= 35.4192 + 9.0432$

$= 44.4624$

The area is 44.4624 in^2.

9. The coefficient is the number factor of the term.

The degree of a term is the sum of the exponents of the variables.

$x^3y - 2xy + 3x^2 - 5$

Term	Coefficient	Degree	
x^3y	1	4	(Think: $x^3y = x^3y^1$)
$-2xy$	-2	2	(Think: $-2xy = -2x^1y^1$)
$3x^2$	3	2	
-5	-5	0	(Think: $-5 = -5x^0$)

The degree of a polynomial is the degree of the term of highest degree.

The term of highest degree is x^3y. Its degree is 4. The degree of the polynomial is 4.

11. $17x^2y^3 - 3x^3yz - 7$

Term	Coefficient	Degree
$17x^2y^3$	17	5
$-3x^3yz$	-3	5 (Think: $-3x^3yz=-3x^3y^1z^1$)
-7	-7	0 (Think: $-7 = -7x^0$)

The terms of highest degree are $17x^2y^3$ and $-3x^3yz$. Each has degree 5. The degree of the polynomial is 5.

13. $a + b - 2a - 3b$

$= (1 - 2)a + (1 - 3)b$

$= -a - 2b$

15. $3x^2y - 2xy^2 + x^2$

There are no like terms, so none of the terms can be collected.

17. $2u^2v - 3uv^2 + 6u^2v - 2uv^2$

$= (2 + 6)u^2v + (-3 - 2)uv^2$

$= 8u^2v - 5uv^2$

19. $6au + 3av - 14au + 7av$

$= (6 - 14)au + (3 + 7)av$

$= -8au + 10av$

21. $(2x^2 - xy + y^2) + (-x^2 - 3xy + 2y^2)$

$= (2 - 1)x^2 + (-1 - 3)xy + (1 + 2)y^2$

$= x^2 - 4xy + 3y^2$

23. $(r - 2s + 3) + (2r + s) + (s + 4)$

$= (1 + 2)r + (-2 + 1 + 1)s + (3 + 4)$

$= 3r + 0s + 7$

$= 3r + 7$

25. $(2x^2 - 3xy + y^2) + (-4x^2 - 6xy - y^2) + (x^2 + xy - y^2)$

$= (2 - 4 + 1)x^2 + (-3 - 6 + 1)xy + (1 - 1 - 1)y^2$

$= -x^2 - 8xy - y^2$

27. $(xy - ab) - (xy - 3ab)$

$= xy - ab - xy + 3ab$

$= (1 - 1)xy + (-1 + 3)ab$

$= 0xy + 2ab$

$= 2ab$

29. $(-2a + 7b - c) - (-3b + 4c - 8d)$

$= -2a + 7b - c + 3b - 4c + 8d$

$= -2a + (7 + 3)b + (-1 - 4)c + 8d$

$= -2a + 10b - 5c + 8d$

31. $(3z - u)(2z + 3u)$

F O I L

$= 6z^2 + 9zu - 2uz - 3u^2$

$= 6z^2 + 7zu - 3u^2$

33. $(a^2b - 2)(a^2b - 5)$

F O I L

$= a^4b^2 - 5a^2b - 2a^2b + 10$

$= a^4b^2 - 7a^2b + 10$

35. $a^2 + a - 1$
$\underline{a^2 - y + 1}$
$a^4 + a^3 - a^2$ ①
$- a^2y - ay + y$ ②
$\underline{a^2 + a - 1}$ ③
$a^4 + a^3 - a^2y - ay + y + a - 1$ ④

① Multiplying by a^2
② Multiplying by $-y$
③ Multiplying by 1
④ Adding

37. $(a^3 + bc)(a^3 - bc)$
$= (a^3)^2 - (bc)^2$ $[(A + B)(A - B) = A^2 - B^2]$
$= a^6 - b^2c^2$

39. $y^4x + y^2 + 1$
$\underline{y^2 + 1}$
$y^6x + y^4 + y^2$ (Multiplying by y^2)
$\underline{y^2 + y^4x + 1}$ (Multiplying by 1)
$y^6x + y^4 + 2y^2 + y^4x + 1$ (Adding)

41. $(3xy - 1)(4xy + 2)$
F O I L
$= 12x^2y^2 + 6xy - 4xy - 2$
$= 12x^2y^2 + 2xy - 2$

43. $(3 - c^2d^2)(4 + c^2d^2)$
F O I L
$= 12 + 3c^2d^2 - 4c^2d^2 - c^4d^4$
$= 12 - c^2d^2 - c^4d^4$

45. $(m^2 - n^2)(m + n)$
F O I L
$= m^3 + m^2n - mn^2 - n^3$

47. $(x^5y^5 + xy)(x^4y^4 - xy)$
F O I L
$= x^9y^9 - x^6y^6 + x^5y^5 - x^2y^2$

49. $(x + h)^2$
$= x^2 + 2xh + h^2$ $[(A + B)^2 = A^2 + 2AB + B^2]$

51. $(r^3t^2 - 4)^2$
$= (r^3t^2)^2 - 2 \cdot r^3t^2 \cdot 4 + 4^2$
$[(A - B)^2 = A^2 - 2AB + B^2]$
$= r^6t^4 - 8r^3t^2 + 16$

53. $(p^4 + m^2n^2)^2$
$= (p^4)^2 + 2 \cdot p^4 \cdot m^2n^2 + (m^2n^2)^2$
$[(A + B)^2 = A^2 + 2AB + B^2]$
$= p^8 + 2p^4m^2n^2 + m^4n^4$

55. $(2a^3 - \frac{1}{2}b^3)^2$
$= (2a^3)^2 - 2 \cdot 2a^3 \cdot \frac{1}{2}b^3 + (\frac{1}{2}b^3)^2$
$[(A - B)^2 = A^2 - 2AB + B^2]$
$= 4a^6 - 2a^3b^3 + \frac{1}{4}b^6$

57. $3a(a - 2b)^2$
$= 3a(a^2 - 4ab + 4b^2)$
$[(A - B)^2 = A^2 - 2AB + B^2]$
$= 3a^3 - 12a^2b + 12ab^2$

59. $(2a - b)(2a + b)$
$= (2a)^2 - b^2$ $[(A - B)(A + B) = A^2 - B^2]$
$= 4a^2 - b^2$

61. $(c^2 - d)(c^2 + d)$
$= (c^2)^2 - d^2$ $[(A - B)(A + B) = A^2 - B^2]$
$= c^4 - d^2$

63. $(ab + cd^2)(ab - cd^2)$
$= (ab)^2 - (cd^2)^2$ $[(A - B)(A + B) = A^2 - B^2]$
$= a^2b^2 - c^2d^4$

65. The largest common factor is $2\pi r$.
$2\pi rh + 2\pi r^2$
$= 2\pi r \cdot h + 2\pi r \cdot r$
$= 2\pi r(h + r)$

67. The two terms have a common factor, $a + b$.
$(a + b)(x - 3) + (a + b)(x + 4)$
$= (a + b)[(x - 3) + (x + 4)]$
$= (a + b)(2x + 1)$

69. The two terms have a common factor, $x + 1$.
$(x - 1)(x + 1) - y(x + 1)$
$= [(x - 1) - y](x + 1)$
$= (x - 1 - y)(x + 1)$

71. Factor by grouping.
$n^2 + 2n + np + 2p$
$= n(n + 2) + p(n + 2)$
$= (n + p)(n + 2)$

73. Factor by grouping.
$2x^2 - 4x + xz - 2z$
$= 2x(x - 2) + z(x - 2)$
$= (2x + z)(x - 2)$

75. $x^2 - 2xy + y^2$
$= x^2 - 2xy + y^2$
$= (x - y)^2$ $[A^2 - 2AB + B^2 = (A - B)^2]$

77. $9c^2 + 6cd + d^2$
$= (3c)^2 + 2 \cdot 3c \cdot d + d^2$
$= (3c + d)^2$ $[A^2 + 2Ab + B^2 = (A + B)^2]$

79. $49m^4 - 112m^2n + 64n^2$
$= (7m^2)^2 - 2 \cdot 7m^2 \cdot 8n + (8n)^2$
$= (7m^2 - 8n)^2$ $[A^2 - 2AB + B^2 = (A - B)^2]$

81. $y^4 + 10y^2z^2 + 25z^4$
$= (y^2)^2 + 2 \cdot y^2 \cdot 5z^2 + (5z^2)^2$
$= (y^2 + 5z^2)^2$ $[A^2 + 2AB + B^2 = (A + B)^2]$

83. $\frac{1}{4} a^2 + \frac{1}{3} ab + \frac{1}{9} b^2$
$= (\frac{1}{2} a)^2 + 2 \cdot \frac{1}{2} a + \frac{1}{3} b + (\frac{1}{3} b)^2$
$= (\frac{1}{2} a + \frac{1}{3} b)^2$ $[A^2 + 2AB + B^2 = (A + B)^2]$

85. $a^2 - ab - 2b^2$
We look for two numbers whose product is -2 and whose sum is -1. They are -2 and 1.
$a^2 - ab - 2b^2 = (a - 2b)(a + b)$

87. $m^2 + 2mn - 360n^2$
We look for two numbers whose product is -360 and whose sum is 2. They are -18 and 20: $-18 \cdot 20 = -360$ and $-18 + 20 = 2$.
$m^2 + 2mn - 360n^2 = (m - 18n)(m + 20n)$

89. $m^2n^2 - 4mn - 32$
We look for two numbers whose product is -32 and whose sum is -4. They are -8 and 4: $-8 \cdot 4 = -32$ and $-8 + 4 = -4$.
$m^2n^2 - 4mn - 32 = (mn - 8)(mn + 4)$

91. $a^5b^2 + 3a^4b - 10a^3$
$= a^3(a^2b^2 + 3ab - 10)$
$= a^3(ab + 5)(ab - 2)$

93. $a^5 + 4a^4b - 5a^3b^2$
$= a^3(a^2 + 4ab - 5b^2)$
$= a^3(a + 5b)(a - b)$

95. $x^6 + x^3y - 2y^2$
$= (x^3)^2 + x^3y - 2y^2$
We look for two numbers whose product is -2 and whose sum is 1. They are 2 and -1: $2(-1) = -2$ and $2 + (-1) = 1$.
$x^6 + x^3y - 2y^2 = (x^3 + 2y)(x^3 - y)$

97. $x^2 - y^2$
$= (x - y)(x + y)$ $[A^2 - B^2 = (A - B)(A + B)]$

99. $a^2b^2 - 9$
$= (ab)^2 - 3^2$
$= (ab - 3)(ab + 3)$ $[A^2 - B^2 = (A - B)(A + B)]$

101. $9x^4y^2 - b^2$
$= (3x^2y)^2 - b^2$
$= (3x^2y - b)(3x^2y + b)$

103. $3x^2 - 48y^2$
$= 3(x^2 - 16y^2)$
$= 3[x^2 - (4y)^2]$
$= 3(x - 4y)(x + 4y)$

105. $64z^2 - 25c^2d^2$
$= (8z)^2 - (5cd)^2$
$= (8z - 5cd)(8z + 5cd)$

107. $7p^4 - 7q^4$
$= 7(p^4 - q^4)$
$= 7(p^2 + q^2)(p^2 - q^2)$
$= 7(p^2 + q^2)(p + q)(p - q)$
We factored a difference of squares twice.

109. $81a^4 - b^4$
$= (9a^2 + b^2)(9a^2 - b^2)$
$= (9a^2 + b^2)(3a + b)(3a - b)$

111. $18m^4 + 12m^3 + 2m^2$
$= 2m^2(9m^2 + 6m + 1)$
$= 2m^2(3m + 1)^2$

113. $xy^2 + 3y^2 - 4x - 12$
$= y^2(x + 3) - 4(x + 3)$
$= (y^2 - 4)(x + 3)$
$= (y + 2)(y - 2)(x + 3)$

115. $p^3 - p^2t - 2pt^2$
$= p(p^2 - pt - 2t^2)$
$= p(p - 2t)(p + t)$

117. $-a^2 - ab + 6b^2$
$= -1(a^2 + ab - 6b^2)$
$= -(a + 3b)(a - 2b)$

119. $ab^3 - ab^2 - ab$
$= ab(b^2 - b - 1)$

121. It is helpful to add additional labels to the figure.

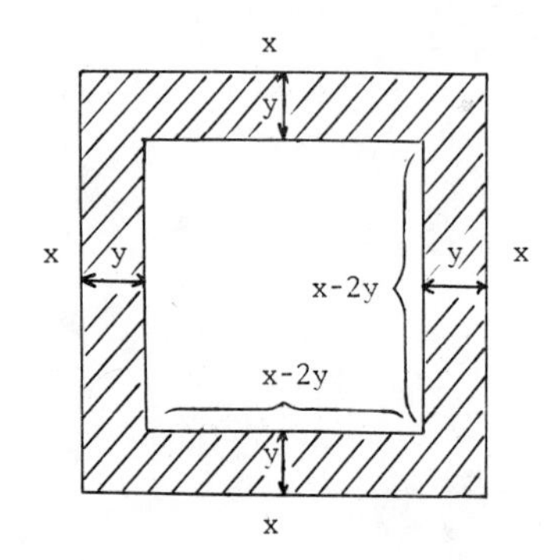

The area of the large square is $x \cdot x$, or x^2.

The area of the small square is $(x - 2y)(x - 2y)$, or $(x - 2y)^2$.

Area of shaded region = Area of large square - Area of small square

$$\text{Area of shaded region} = x^2 - (x - 2y)^2$$

$$= x^2 - (x^2 - 4xy + 4y^2)$$

$$= x^2 - x^2 + 4xy - 4y^2$$

$$= 4xy - 4y^2$$

$$= 4y(x - y)$$

123. It is helpful to add additional labels to the figure.

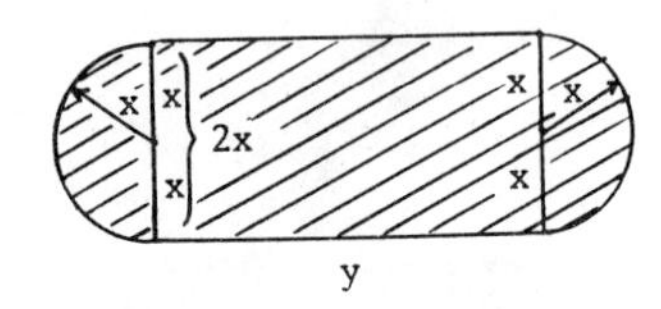

The two semicircles make a circle with radius x. The area of that circle is πx^2. The area of the rectangle is $2x \cdot y$. The sum of the two regions, $\pi x^2 + 2xy$, is the area of the shaded region.

$\pi x^2 + 2xy = x(\pi x + 2y)$

125. $(A + B)^3$

$= (A + B)(A + B)^2$

$= (A + B)(A^2 + 2AB + B^2)$

Now we can use columns.

$A^2 + 2AB + B^2$

$A + B$

$A^3 + 2A^2B + AB^2$ (Multiplying by A)

$A^2B + 2AB^2 + B^3$ (Multiplying by B)

$A^3 + 3A^2B + 3AB^2 + B^3$ (Adding)

$(A + B)^3 = A^3 + 3A^2B + 3AB^2 + B^3$

Exercise Set 6.1

1.

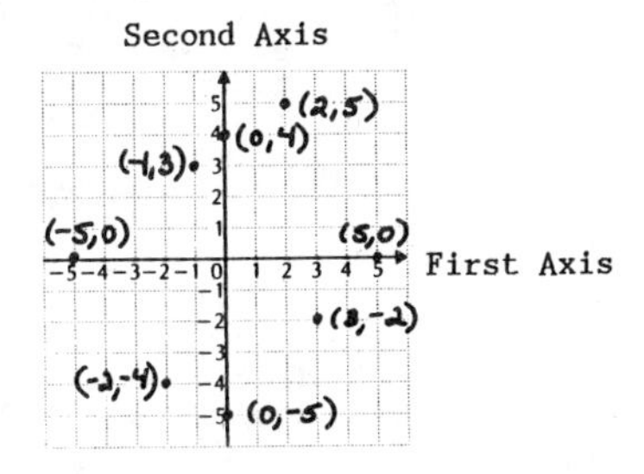

(2,5) is 2 units right and 5 units up.
(-1,3) is 1 unit left and 3 units up.
(3,-2) is 3 units right and 2 units down.
(-2,-4) is 2 units left and 4 units down.
(0,4) is 0 units left or right and 4 units up.
(0,-5) is 0 units left or right and 5 units down.
(5,0) is 5 units right and 0 units up or down.
(-5,0) is 5 units left and 0 units up or down.

3. Since the first coordinate is negative and the second coordinate positive, the point (-5,3) is located in the <u>second</u> quadrant.

5. Since the first coordinate is positive and the second coordinate negative, the point (100,-1) is in the <u>fourth</u> quadrant.

7. Since both coordinates are negative, the point (-6,-29) is in the <u>third</u> quadrant.

9. Since both coordinates are positive, the point (3.8,9.2) is in the <u>first</u> quadrant.

11. In quadrant III, first coordinates are always <u>negative</u> and second coordinates are always <u>negative</u>.

13.

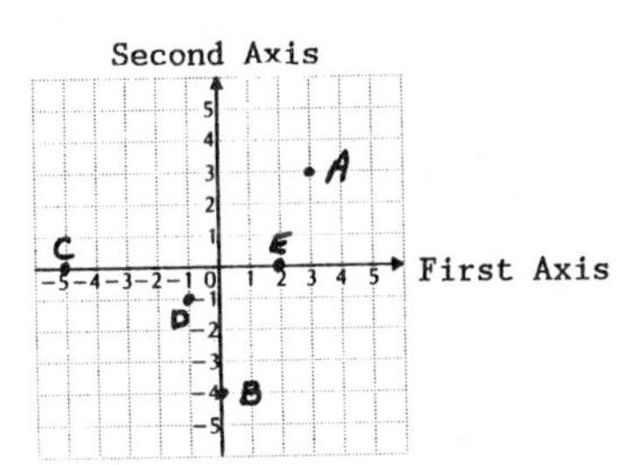

Point A is 3 units right and 3 units up.
The coordinates of A are (3,3).

Point B is 0 units left or right and 4 units down.
The coordinates of B are (0,-4).

Point C is 5 units left and 0 units up or down.
The coordinates of C are (-5,0).

Point D is 1 unit left and 1 unit down.
The coordinates of D are (-1,-1).

Point E is 2 units right and 0 units up or down.
The coordinates of E are (2,0).

15.

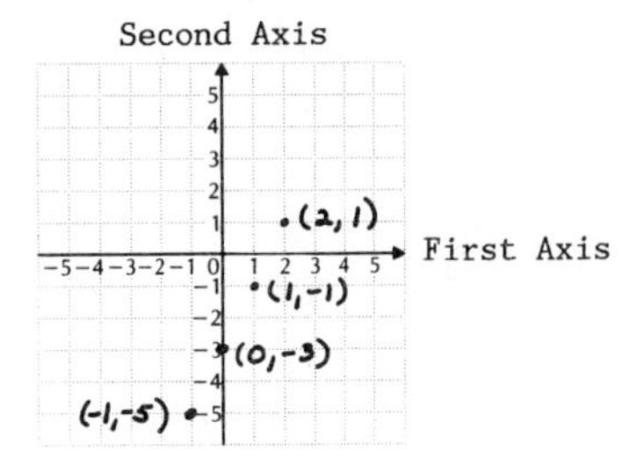

(0,-3) is 0 units left or right and 3 units down.
(-1,-5) is 1 unit left and 5 units down.
(1,-1) is 1 unit right and 1 unit left.
(2,1) is 2 units right and 1 unit up.

17. $\frac{5}{2} + y = \frac{1}{3}$

$y = \frac{1}{3} - \frac{5}{2}$

$y = \frac{2}{6} - \frac{15}{6}$

$y = -\frac{13}{6}$

19. $-8x + 3x = 25$

$-5x = 25$

$x = -5$

21.

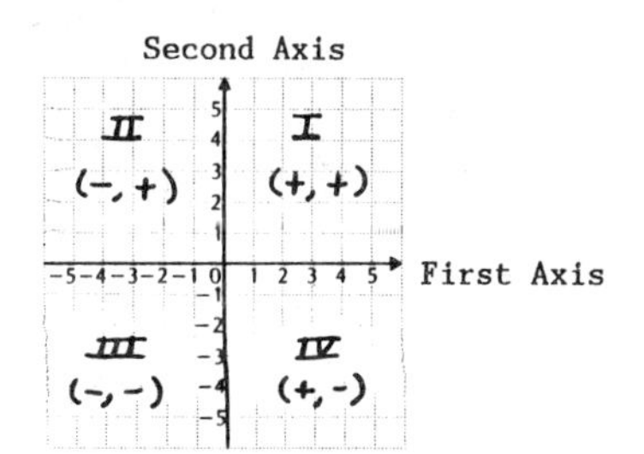

If the first coordinate is positive, then the point must be in either I or IV.

23. If the first and second coordinates are the same, they must either be both positive or both negative. The point must be in either I (both positive) or III (both negative).

Exercise Set 6.2

1. (2,5); $y = 3x - 1$

We take the variables in alphabetical order. We replace x by the first coordinate and y by the second coordinate.

$y = 3x - 1$		
5	$3 \cdot 2 - 1$	(Substituting 2 for x and 5 for y)
	$6 - 1$	
	5	

The equation becomes true, (2,5) is a solution of $y = 3x - 1$.

3. (2,-3); $3x - y = 4$

We take the variables in alphabetical order. We replace x by the first coordinate and y by the second coordinate.

$3x - y = 4$		
$3 \cdot 2 - (-3)$	4	(Substituting 2 for x and -3 for y)
$6 + 3$		
9		

Since $9 \neq 4$, (2,-3) is not a solution of $3x - y = 4$.

5. (-2,-1); $2a + 2b = -7$

We take the variables in alphabetical order. We replace a by the first coordinate and b by the second coordinate.

$2a + 2b = -7$		
$2(-2) + 2(-1)$	-7	(Substituting -2 for a and -1 for b)
$-4 - 2$		
6		

Since $-6 \neq 7$, (-2,-1) is not a solution of $2a + 2b = -7$.

7. $y = 4x$

We first make a table of values. We choose any number for x and then determine y by substitution.

When $x = 0$, $y = 4 \cdot 0 = 0$.
When $x = -1$, $y = 4(-1) = -4$.
When $x = 1$, $y = 4 \cdot 1 = 4$.

x	y
0	0
-1	-4
1	4

Since two points determine a line, that is all we really need to graph a line, but you should always plot a third point as a check.

Plot these points, draw the line they determine, and label the graph $y = 4x$.

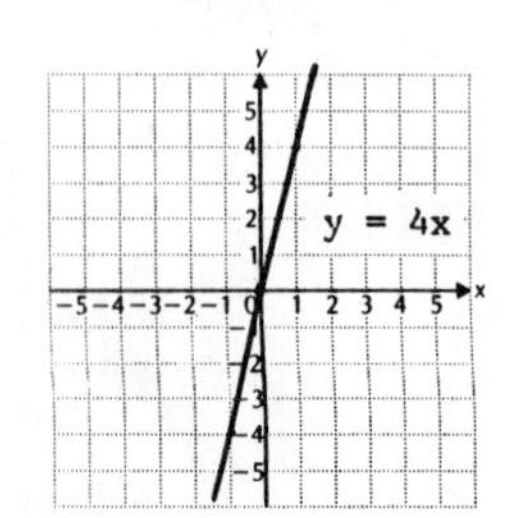

7. (continued)

You may have chosen different x values and thus have different ordered pairs in your table, but your graph will be the same.

In the equation $y = 4x$ the number 4, called the slope, tells us that the line slants up from left to right.

9. $y = -2x$

We first make a table of values. We choose any number for x and then determine y by substitution.

When $x = 0$, $y = -2 \cdot 0 = 0$.
When $x = 2$, $y = -2 \cdot 2 = -4$.
When $x = -1$, $y = -2(-1) = 2$.

x	y
0	0
2	-4
-1	2

Plot these points, draw the line they determine, and label the graph $y = -2x$.

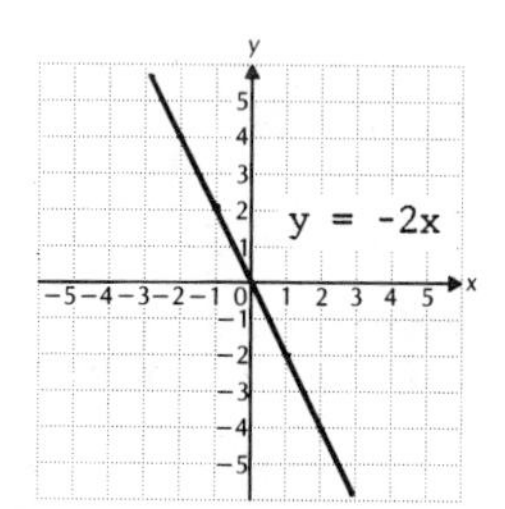

In the equation $y = -2x$ the number -2, called the slope, tells us that the line slants down from left to right.

11. $y = \frac{1}{3}x$

We first make a table of values. We choose any number for x and then determine y by substitution. Using multiples of 3 avoids fractions.

When $x = 0$, $y = \frac{1}{3} \cdot 0 = 0$.
When $x = 6$, $y = \frac{1}{3} \cdot 6 = 2$.
When $x = -3$, $y = \frac{1}{3}(-3) = -1$.

x	y
0	0
6	2
-3	-1

Plot these points, draw the line they determine, and label the graph $y = \frac{1}{3}x$.

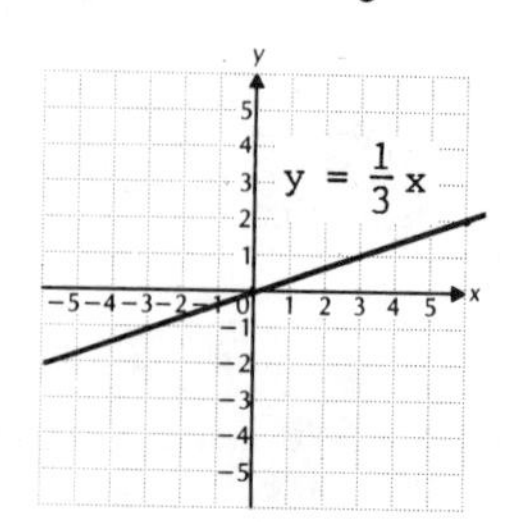

The number $\frac{1}{3}$, called the slope, tells us how the line slants. For a positive slope, a line slants up from left to right.

13. $y = -\frac{3}{2}x$

We first make a table of values. We choose any number for x and then determine y by substitution. Using multiples of 2 avoids fractions.

When $x = 0$, $y = -\frac{3}{2}\cdot 0 = 0$.

When $x = 2$, $y = -\frac{3}{2}\cdot 2 = -3$.

When $x = -2$, $y = -\frac{3}{2}(-2) = 3$.

x	y
0	0
2	-3
-2	3

Plot these points, draw the line they determine, and label the graph $y = -\frac{3}{2}x$.

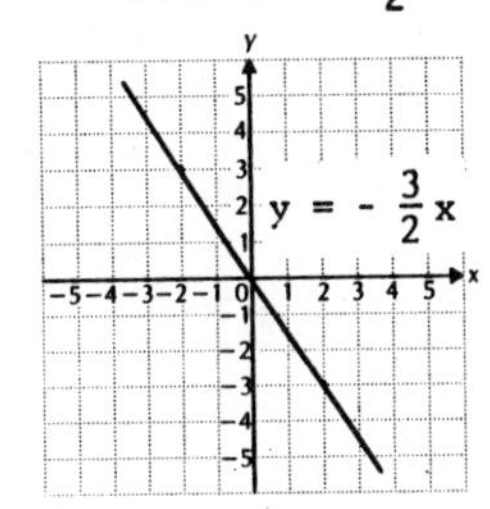

The number $-\frac{3}{2}$, called the slope, tells us how the line slants. For a negative slope, a line slants down from left to right.

15. $y = x + 1$

We first make a table of values. We choose any number for x and then determine y by substitution.

When $x = 0$, $y = 0 + 1 = 1$.
When $x = 3$, $y = 3 + 1 = 4$.
When $x = -5$, $y = -5 + 1 = -4$.

x	y
0	1
3	4
-5	-4

Plot these points, draw the line they determine, and label the graph $y = x + 1$.

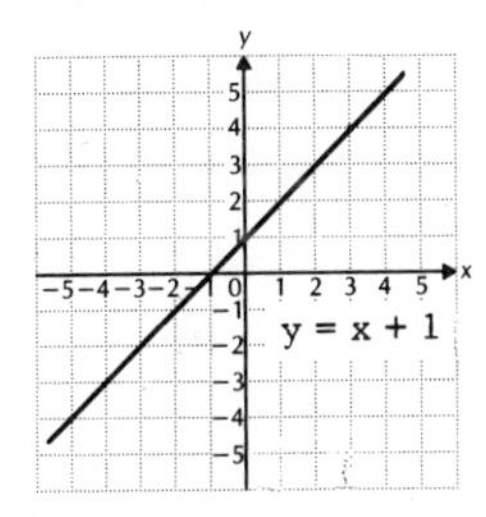

The graph of the equation $y = x + 1$ is a straight line that goes through the point (0,1), the y-intercept, and has slope 1.

17. $y = 2x + 2$

We first make a table of values. We choose any number for x and then determine y by substitution.

When $x = 0$, $y = 2\cdot 0 + 2 = 0 + 2 = 2$.
When $x = -3$, $y = 2(-3) + 2 = -6 + 2 = -4$.
When $x = 1$, $y = 2\cdot 1 + 2 = 2 + 2 = 4$.

x	y
0	2
-3	-4
1	4

Plot these points, draw the line they determine, and label the graph $y = 2x + 2$.

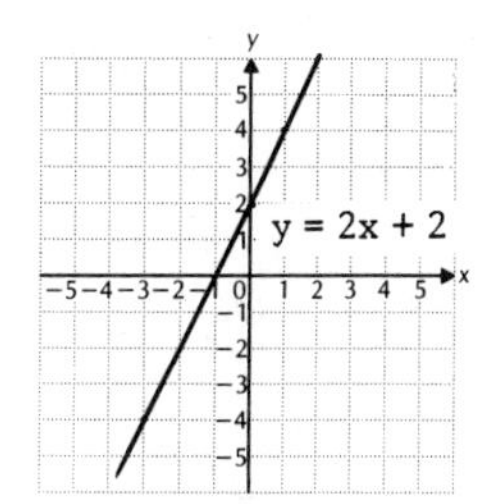

The graph of the equation $y = 2x + 2$ is a straight line that goes through the point (0,2), the y-intercept, and has slope 2.

19. $y = \frac{1}{3}x - 1$

We first make a table of values. We choose any number for x and then determine y by substitution. Using multiples of 3 avoids fractions.

When $x = 0$, $y = \frac{1}{3}\cdot 0 - 1 = 0 - 1 = -1$.

When $x = -6$, $y = \frac{1}{3}(-6) - 1 = -2 - 1 = -3$.

When $x = 3$, $y = \frac{1}{3}\cdot 3 - 1 = 1 - 1 = 0$.

x	y
0	-1
-6	-3
3	0

Plot these points, draw the line they determine, and label the graph $y = \frac{1}{3}x - 1$.

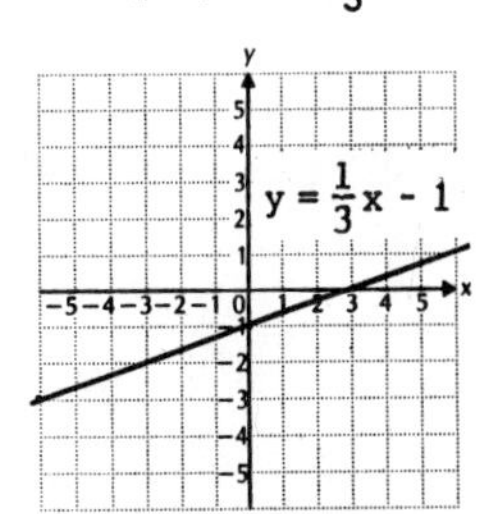

The graph of the equation $y = \frac{1}{3}x - 1$ is a straight line that goes through the point (0,-1), the y-intercept, and has slope $\frac{1}{3}$.

21. $y = -x - 3$ (Think: $y = -1 \cdot x - 3$)

We first make a table of values.

When $x = 0$, $y = -1 \cdot 0 - 3 = 0 - 3 = -3$.
When $x = 1$, $y = -1 \cdot 1 - 3 = -1 - 3 = -4$.
When $x = -5$, $y = -1(-5) - 3 = 5 - 3 = 2$.

x	y
0	-3
1	-4
-5	2

Plot these points, draw the line they determine, and label the graph.

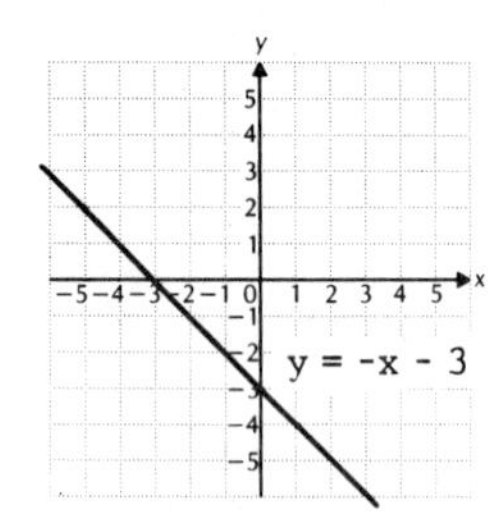

The graph of the equation $y = -x - 3$ is a straight line that goes through the point (0,-3), the y-intercept, and has slope -1.

23. $y = \frac{5}{2}x + 3$

We first make a table of values. Using multiples of 2 avoids fractions.

When $x = 0$, $y = \frac{5}{2} \cdot 0 + 3 = 0 + 3 = 3$.

When $x = -2$, $y = \frac{5}{2}(-2) + 3 = -5 + 3 = -2$.

When $x = -4$, $y = \frac{5}{2}(-4) + 3 = -10 + 3 = -7$.

x	y
0	3
-2	-2
-4	-7

Plot these points, draw the line they determine, and label the graph.

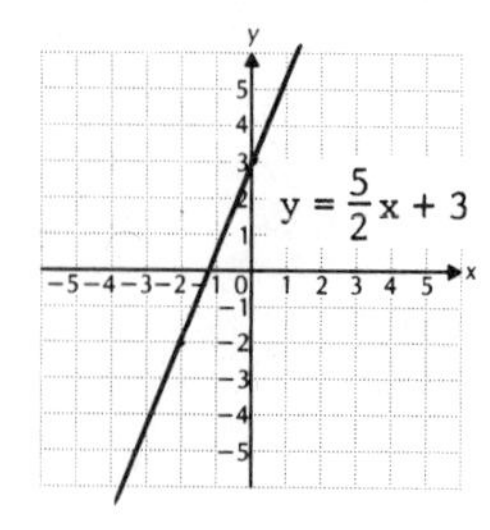

The graph of the equation $y = \frac{5}{2}x + 3$ is a straight line that goes through the point (0,3), the y-intercept, and has slope $\frac{5}{2}$.

25. $y = -\frac{5}{2}x - 2$

We first make a table of values. Using multiples of 2 avoids fractions.

When $x = 0$, $y = -\frac{5}{2} \cdot 0 - 2 = 0 - 2 = -2$.

When $x = -2$, $y = -\frac{5}{2}(-2) - 2 = 5 - 2 = 3$.

When $x = -4$, $y = -\frac{5}{2}(-4) - 2 = 10 - 2 = 8$.

x	y
0	-2
-2	3
-4	8

Plot these points, draw the line they determine, and label the graph.

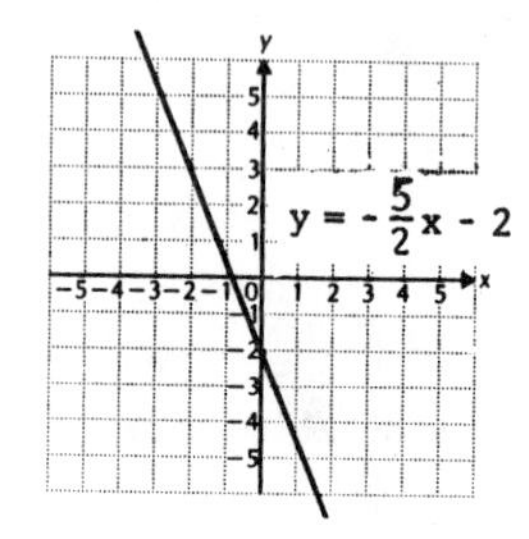

The graph of the equation $y = -\frac{5}{2}x - 2$ is a straight line that goes through the point (0,-2), the y-intercept, and has slope $-\frac{5}{2}$.

27. $16 - t^4$
$= 4^2 - (t^2)^2$
$= (4 + t^2)(4 - t^2)$
$= (4 + t^2)(2 + t)(2 - t)$

29. $x^5 - 2x^4 - 35x^3$
$= x^3(x^2 - 2x - 35)$
$= x^3(x - 7)(x + 5)$

31. $x + y = 6$

When $x = 0$, $0 + y = 6$
$y = 6$

When $x = 1$, $1 + y = 6$
$y = 5$

When $x = 2$, $2 + y = 6$
$y = 4$

When $x = 3$, $3 + y = 6$
$y = 3$

When $x = 4$, $4 + y = 6$
$y = 2$

When $x = 5$, $5 + y = 6$
$y = 1$

When $x = 6$, $6 + y = 6$
$y = 0$

The whole-number solutions are (0,6), (1,5), (2,4), (3,3), (4,2), (5,1), and (6,0).

Exercise Set 6.3

1. $5x - 3y = 15$

To find the x-intercept, let $y = 0$.

$5x - 3y = 15$
$5x - 3\cdot 0 = 15$
$5x = 15$
$x = 3$

Thus, (3,0) is the x-intercept.

To find the y-intercept, let $x = 0$.

$5x - 3y = 15$
$5\cdot 0 - 3y = 15$
$-3y = 15$
$y = -5$

Thus, (0,-5) is the y-intercept.

Plot these points and draw the line.

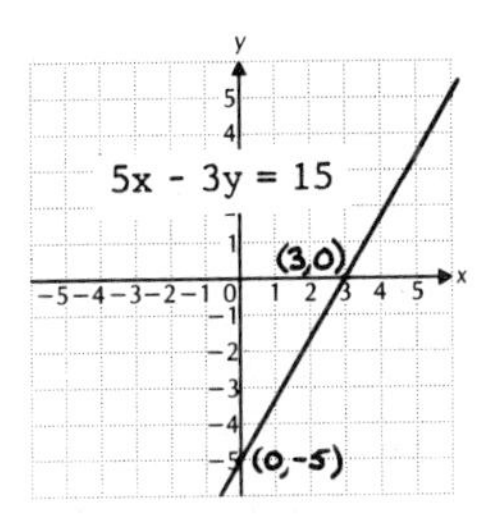

A third point should be used as a check. We substitute any value for x and solve for y.

We let $x = 2$. Then

$5x - 3y = 15$
$5\cdot 2 - 3y = 15$
$10 - 3y = 15$
$-3y = 5$
$y = -\frac{5}{3}$

The point $(2,-\frac{5}{3})$ is on the graph, so the graph is probably correct.

3. $4x + 2y = 8$

To find the x-intercept, let $y = 0$.

$4x + 2y = 8$
$4x + 2\cdot 0 = 8$
$4x = 8$
$x = 2$

Thus, (2,0) is the x-intercept.

3. (continued)

To find the y-intercept, let $x = 0$.

$4x + 2y = 8$
$4\cdot 0 + 2y = 8$
$2y = 8$
$y = 4$

Thus, (0,4) is the y-intercept.

Plot these points and draw the line.

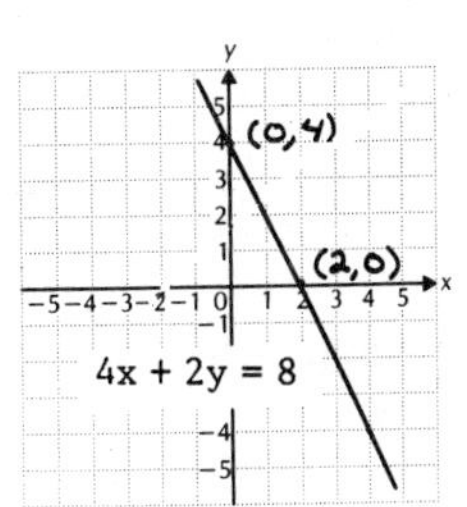

A third point should be used as a check. We substitute any value for x and solve for y.

We let $x = 1$. Then

$4x + 2y = 8$
$4\cdot 1 + 2y = 8$
$4 + 2y = 8$
$2y = 4$
$y = 2$

The point (1,2) is on the graph, so the graph is probably correct.

5. $x - 1 = y$

To find the x-intercept, let $y = 0$.

$x - 1 = y$
$x - 1 = 0$
$x = 1$

Thus, (1,0) is the x-intercept.

To find the y-intercept, let $x = 0$.

$x - 1 = y$
$0 - 1 = y$
$-1 = y$

Thus, (0,-1) is the y-intercept.

It is helpful to plot another point since the intercepts are so close together. This point can also serve as a check.

We let $x = 4$. Then

$x - 1 = y$
$4 - 1 = y$
$3 = y$

5. (continued)

Plot the point (4,3) and the intercepts and draw the line.

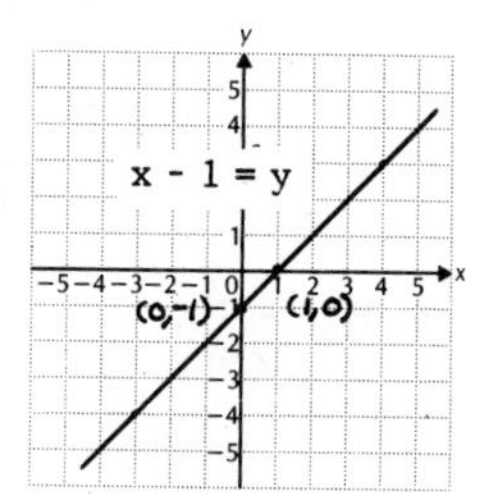

7. $2x - 1 = y$

To find the x-intercept, let $y = 0$.

$$2x - 1 = y$$
$$2x - 1 = 0$$
$$2x = 1$$
$$x = \frac{1}{2}$$

Thus, $(\frac{1}{2},0)$ is the x-intercept.

To find the y-intercept, let $x = 0$.

$$2x - 1 = y$$
$$2 \cdot 0 - 1 = y$$
$$-1 = y$$

Thus, (0,-1) is the y-intercept.

It is helpful to plot another point since the intercepts are so close together. This point can also serve as a check.

We let $x = 3$. Then

$$2x - 1 = y$$
$$2 \cdot 3 - 1 = y$$
$$6 - 1 = y$$
$$5 = y$$

Plot the point (3,5) and the intercepts and draw the line.

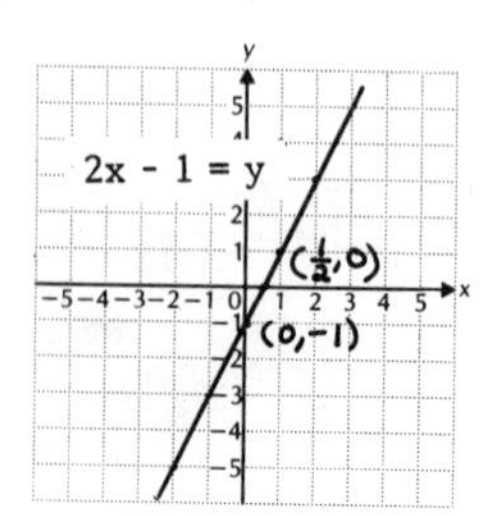

9. $4x - 3y = 12$

To find the x-intercept, let $y = 0$.

$$4x - 3y = 12$$
$$4x - 3 \cdot 0 = 12$$
$$4x = 12$$
$$x = 3$$

Thus, (3,0) is the x-intercept.

To find the y-intercept, let $x = 0$.

$$4x - 3y = 12$$
$$4 \cdot 0 - 3y = 12$$
$$-3y = 12$$
$$y = -4$$

Thus, (0,-4) is the y-intercept.

Plot these points and draw the line.

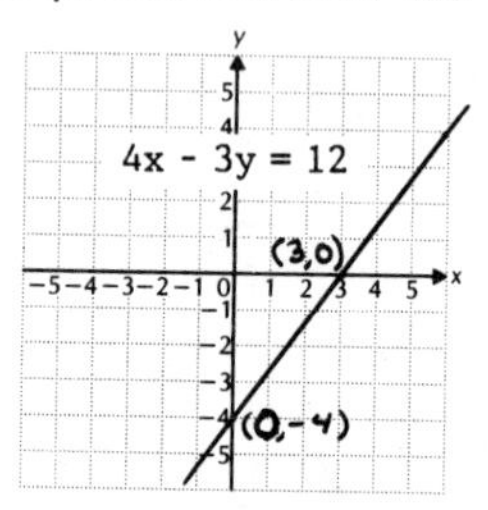

A third point should be used as a check. We substitute any value for x and solve for y.

We let $x = 6$. Then

$$4x - 3y = 12$$
$$4 \cdot 6 - 3y = 12$$
$$24 - 3y = 12$$
$$-3y = -12$$
$$y = 4$$

The point (6,4) is on the graph, so the graph is probably correct.

11. $7x + 2y = 6$

To find the x-intercept, let $y = 0$.

$$7x + 2y = 6$$
$$7x + 2 \cdot 0 = 6$$
$$7x = 6$$
$$x = \frac{6}{7}$$

Thus, $(\frac{6}{7},0)$ is the x-intercept.

To find the y-intercept, let $x = 0$.

$$7x + 2y = 6$$
$$7 \cdot 0 + 2y = 6$$
$$2y = 6$$
$$y = 3$$

Thus, (0,3) is the y-intercept.

Plot these points and draw the line.

11. (continued)

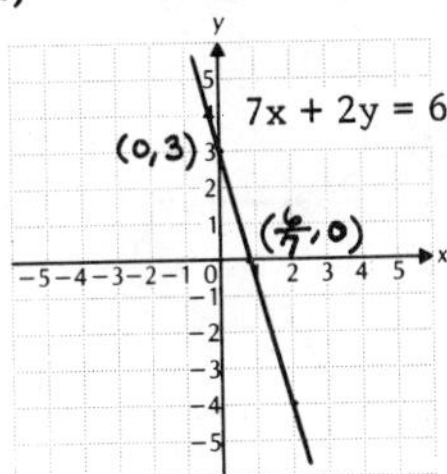

A third point should be used as a check. We substitute any value for x and solve for y.

We let x = 2. Then

$7x + 2y = 6$

$7 \cdot 2 + 2y = 6$

$14 + 2y = 6$

$2y = -8$

$y = -4$

The point (2,-4) is on the graph, so the graph is probably correct.

13. $y = -4 - 4x$

To find the x-intercept, let y = 0.

$y = -4 - 4x$

$0 = -4 - 4x$

$4x = -4$

$x = -1$

Thus, (-1,0) is the x-intercept.

To find the y-intercept, let x = 0.

$y = -4 - 4x$

$y = -4 - 4 \cdot 0$

$y = -4$

Thus, (0,-4) is the y-intercept.

Plot these points and draw the line.

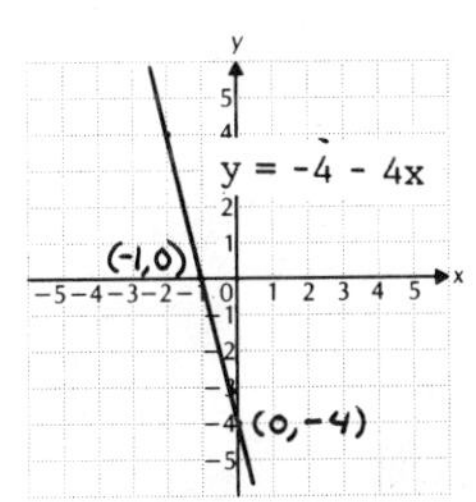

A third point should be used as a check. We substitute any value for x and solve for y.

We let x = -2. Then

$y = -4 - 4x$

$y = -4 - 4(-2)$

$y = -4 + 8$

$y = 4$

The point (-2,4) is on the graph, so the graph is probably correct.

15. $x = -2$

Any ordered pair (-2,y) is a solution. The variable x must be -2, but the y variable can be any number we choose. A few solutions are listed below. Plot these points and draw the line.

x	y
-2	-3
-2	0
-2	4

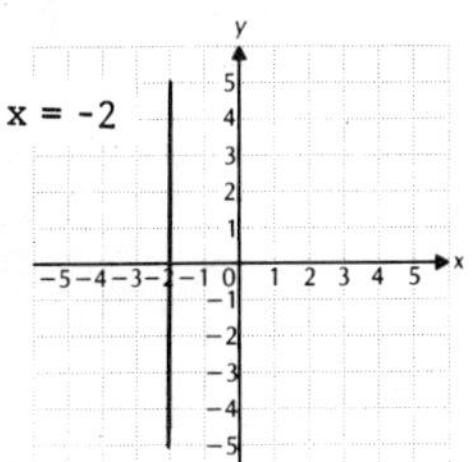

17. $y = 2$

Any ordered pair (x,2) is a solution. The variable y must be 2, but the x variable can be any number we choose. A few solutions are listed below. Plot these points and draw the line.

x	y
-4	2
-1	2
5	2

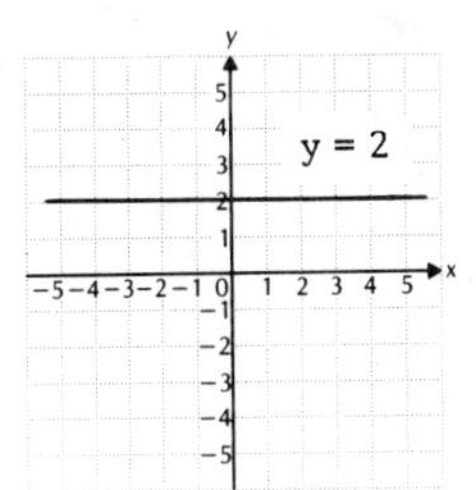

19. $x = 2$

Any ordered pair (2,y) is a solution. The variable x must be 2, but the y variable can be any number we choose. A few solutions are listed below. Plot these points and draw the line.

x	y
2	-2
2	1
2	4

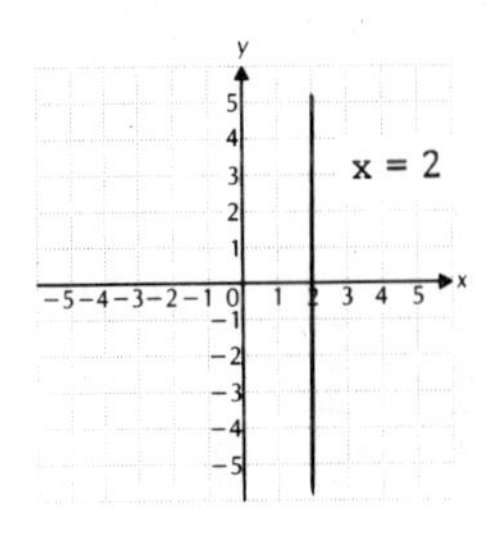

21. $y = 0$

Any ordered pair $(x,0)$ is a solution. The variable y must be 0, but the x variable can be any number we choose. A few solutions are listed below. Plot these points and draw the line.

x	y
-3	0
0	0
3	0

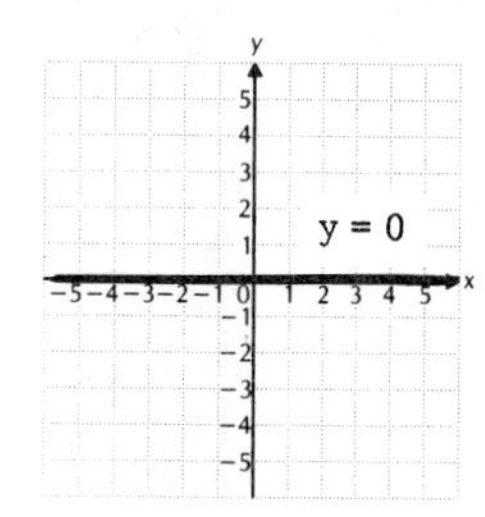

23. $x = \frac{3}{2}$

Any ordered pair $(\frac{3}{2},y)$ is a solution. The variable x must be $\frac{3}{2}$, but the y variable can be any number we choose. A few solutions are listed below. Plot these points and draw the line.

x	y
$\frac{3}{2}$	-5
$\frac{3}{2}$	-1
$\frac{3}{2}$	4

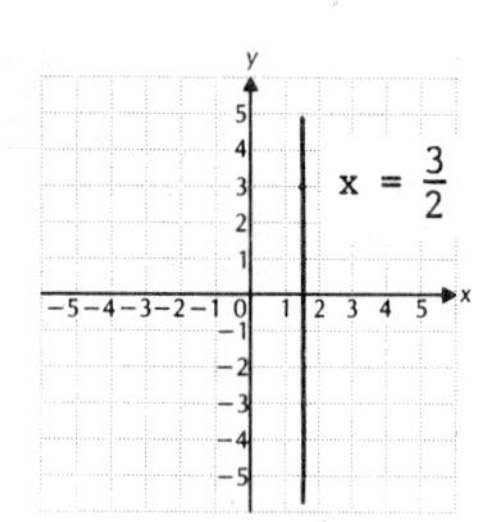

25. $(6x^2 + 7) - (4x^2 - 9)$
$= 6x^2 + 7 - 4x^2 + 9$
$= 2x^2 + 16$

27. $(\frac{1}{4}x^2 - \frac{3}{4}x + 3) - (\frac{3}{4}x^2 + \frac{1}{4}x - 3)$
$= \frac{1}{4}x^2 - \frac{3}{4}x + 3 - \frac{3}{4}x^2 - \frac{1}{4}x + 3$
$= -\frac{2}{4}x^2 - \frac{4}{4}x + 6$
$= -\frac{1}{2}x^2 - x + 6$

29. The x-axis is a horizontal line with y-intercept $(0,0)$. The equation of the x-axis is $y = 0$.

31. First graph the lines $x = -3$ and $y = 6$.

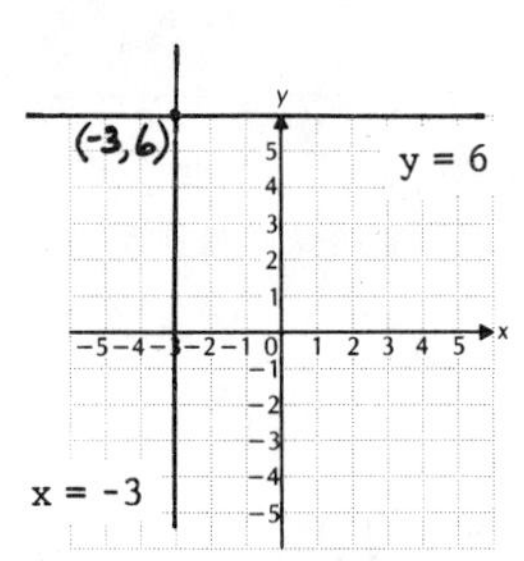

The coordinates of the point of intersection are $x = -3$ and $y = 6$. The ordered pair is $(-3,6)$.

Exercise Set 6.4

1. We substitute to find k.
$y = kx$
$28 = k \cdot 7$ (Substituting 28 for y and 7 for x)
$\frac{28}{7} = k$
$4 = k$ (k is the constant of variation.)

The equation of variation is $y = 4x$.

3. We substitute to find k.
$y = kx$
$0.7 = k \cdot 0.4$ (Substituting 0.7 for y and 0.4 for x)
$\frac{0.7}{0.4} = k$
$\frac{7}{4} = k$, or $k = 1.75$

The equation of variation is $y = 1.75x$.

5. We substitute to find k.
$y = kx$
$400 = k \cdot 125$ (Substituting 400 for y and 125 for x)
$\frac{400}{125} = k$
$\frac{16}{5} = k$, or $k = 3.2$

The equation of variation is $y = 3.2x$.

7. We substitute to find k.

$y = kx$

$200 = k \cdot 300$ (Substituting 200 for y and 300 for x)

$\frac{200}{300} = k$

$\frac{2}{3} = k$

The equation of variation is $y = \frac{2}{3}x$.

9. The problem states that we have direct variation between the variables P and H. Thus, an equation $P = kH$, $k > 0$, applies. As the number of hours increases, the paycheck increases.

a) First find an equation of variation.

$P = kH$

$78.75 = k \cdot 15$ (Substituting 78.75 for P and 15 for H)

$\frac{78.75}{15} = k$

$5.25 = k$

The equation of variation is $P = 5.25H$.

b) Use the equation to find the pay for 35 hours work.

$P = 5.25H$

$P = 5.25(35)$ (Substituting 35 for H)

$P = 183.75$

For 35 hours work, the paycheck is $183.75.

11. This problem states that we have direct variation between S and W. Thus, an equation $S = kW$, $k > 0$, applies. As the weight increases, the number of servings increases.

a) First find an equation of variation.

$S = kW$

$40 = k \cdot 14$ (Substituting 40 for S and 14 for W)

$\frac{40}{14} = k$

$\frac{20}{7} = k$

The equation of variation is $S = \frac{20}{7}W$.

b) Use the equation to find the number of servings from an 8-kg turkey.

$S = \frac{20}{7}W$

$S = \frac{20}{7} \cdot 8$ (Substituting 8 for W)

$S = \frac{160}{7}$, or $22\frac{6}{7}$

Thus, $22\frac{6}{7}$ servings can be obtained from an 8-kg turkey.

13. The problem states that we have direct variation between the variables M and E. Thus, an equation $M = kE$, $k > 0$, applies. As the weight on earth increases, the weight on moon increases.

a) First find an equation of variation.

$M = kE$

$13 = k \cdot 78$ (Substituting 13 for M and 78 for E)

$\frac{13}{78} = k$

$\frac{1}{6} = k$

The equation of variation is $M = \frac{1}{6}E$.

b) Use the equation to find how much a 100 kg-person would weigh on the moon.

$M = \frac{1}{6}E$

$M = \frac{1}{6} \cdot 100$ (Substituting 100 for E)

$M = \frac{100}{6}$, or $16\frac{2}{3}$

A 100-kg person would weigh $16\frac{2}{3}$ kg on the moon.

15. This problem states that we have direct variation between the variables W and B. Thus, an equation $W = kB$, $k > 0$, applies. As the body weight increases, the water weight increases.

a) First find an equation of variation.

$W = kB$

$54 = k \cdot 75$ (Substituting 54 for W and 75 for B)

$\frac{54}{75} = k$

$\frac{18}{25} = k$, or $k = 0.72$

The equation of variation is $W = 0.72B$.

b) Use the equation to find how many kilograms of water are in a person weighing 95 kg.

$W = 0.72B$

$W = 0.72(95)$ (Substituting 95 for B)

$W = 68.4$

There are 68.4 kg of water in a person weighing 95 kg.

17. $x^2 - 5x + 7$

Substitute -3 for x.

$(-3)^2 - 5(-3) + 7$

$= 9 + 15 + 7$

$= 31$

19. $121 = x^2$

$0 = x^2 - 121$

$0 = (x - 11)(x + 11)$

$x - 11 = 0$ or $x + 11 = 0$

$x = 11$ or $x = -11$

The solutions are 11 and -11.

21. $C = \pi d$ (C varies directly as the diameter)

$C = \pi(2r)$ (Substituting 2r for d)

$C = (2\pi)r$ (C varies directly as the radius)

In the equation of variation, $C = 2\pi r$, $k = 2\pi$.

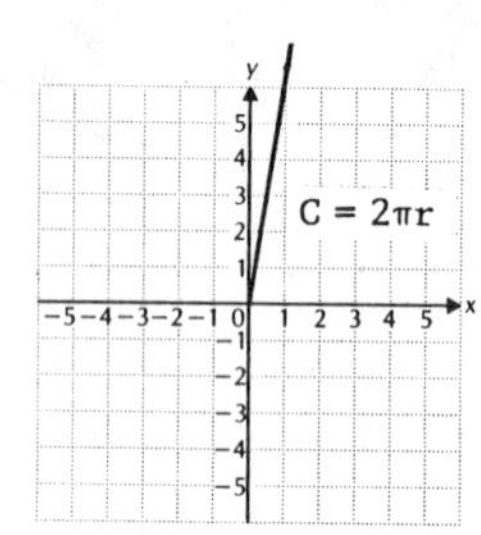

23. The formula for the area of a circle is $A = \pi r^2$, where A varies directly as r^2.

The variation constant is π.

Exercise Set 6.5

1. (3,2) and (-1,2)

$$m = \frac{2 - 2}{3 - (-1)} = \frac{2 - 2}{3 + 1} = \frac{0}{4} = 0$$

3. (-2,4) and (3,0)

$$m = \frac{4 - 0}{-2 - 3} = \frac{4}{-5} = -\frac{4}{5}$$

5. (4,0) and (5,7)

$$m = \frac{0 - 7}{4 - 5} = \frac{-7}{-1} = 7$$

7. (-3,-2) and (-5,-6)

$$m = \frac{-2 - (-6)}{-3 - (-5)} = \frac{-2 + 6}{-3 + 5} = \frac{4}{2} = 2$$

9. $(-2,\frac{1}{2})$ and $(-5,\frac{1}{2})$

$$m = \frac{\frac{1}{2} - \frac{1}{2}}{-2 - (-5)} = \frac{\frac{1}{2} - \frac{1}{2}}{-2 + 5} = \frac{0}{3} = 0$$

11. (9,-4) and (9,-7)

$$m = \frac{-4 - (-7)}{9 - 9} = \frac{-4 + 7}{9 - 9} = \frac{3}{0}$$

Since division by 0 is not defined, this line has no slope.

13. The line $x = -8$ is a vertical line.
A vertical line has no slope.

15. The line $y = 2$ is a horizontal line.
A horizontal line has slope 0.

17. The line $x = 9$ is a vertical line.
A vertical line has no slope.

19. The line $y = -4$ is a horizontal line.
A horizontal line has slope 0.

21. We solve for y.

$3x + 2y = 6$

$2y = -3x + 6$

$y = \frac{1}{2}(-3x + 6)$

$y = -\frac{3}{2}x + 3$

The slope is $-\frac{3}{2}$.

23. We solve for y.

$x + 4y = 8$

$4y = -x + 8$

$y = \frac{1}{4}(-x + 8)$

$y = -\frac{1}{4}x + 2$

The slope is $-\frac{1}{4}$.

25. We solve for y.

$-2x + y = 4$

$y = 2x + 4$

The slope is 2.

27. $y = -4x - 9$

The slope is -4 and the y-intercept is (0,-9).

29. $y = 1.8x$ (Think: $y = 1.8x + 0$)

The slope is 1.8 and the y-intercept is (0,0).

31. We solve for y.

$-8x - 7y = 21$

$-7y = 8x + 21$

$y = -\frac{1}{7}(8x + 21)$

$y = -\frac{8}{7}x - 3$

The slope is $-\frac{8}{7}$ and the y-intercept is (0,-3).

33. We solve for y.

$$9x = 3y + 5$$
$$9x - 5 = 3y$$
$$\frac{1}{3}(9x - 5) = y$$
$$3x - \frac{5}{3} = y$$

The slope is 3 and the y-intercept is $(0,-\frac{5}{3})$

35. We solve for y.

$$-6x = 4y + 2$$
$$-6x - 2 = 4y$$
$$\frac{1}{4}(-6x - 2) = y$$
$$-\frac{3}{2}x - \frac{1}{2} = y$$

The slope is $-\frac{3}{2}$ and the y-intercept is $(0,-\frac{1}{2})$.

37. The equation $y = -17$ could also be written as $y = 0 \cdot x - 17$. The slope is 0 and the y-intercept is $(0,-17)$.

39. $y - y_1 = m(x - x_1)$

We substitute 5 for m, 2 for x_1, and 5 for y_1.

$$y - 5 = 5(x - 2)$$
$$y - 5 = 5x - 10$$
$$y = 5x - 5$$

41. $y - y_1 = m(x - x_1)$

We substitute $\frac{3}{4}$ for m, 2 for x_1, and 4 for y_1.

$$y - 4 = \frac{3}{4}(x - 2)$$
$$y - 4 = \frac{3}{4}x - \frac{3}{2}$$
$$y = \frac{3}{4}x - \frac{3}{2} + 4$$
$$y = \frac{3}{4}x + \frac{5}{2}$$

43. $y - y_1 = m(x \cdot x_1)$

We substitute 1 for m, 2 for x_1, and -6 for y_1.

$$y - (-6) = 1(x - 2)$$
$$y + 6 = x - 2$$
$$y = x - 8$$

45. $y - y_1 = m(x - x_1)$

We substitute -3 for m, -3 for x_1, and 0 for y_1.

$$y - 0 = -3[x - (-3)]$$
$$y = -3(x + 3)$$
$$y = -3x - 9$$

47. (-6,1) and (2,3)

First we find the slope.

$$m = \frac{1 - 3}{-6 - 2} = \frac{-2}{-8} = \frac{1}{4}$$

Then we use the point-slope equation.

$$y - y_1 = m(x - x_1)$$

We substitute $\frac{1}{4}$ for m, -6 for x_1, and 1 for y_1.

$$y - 1 = \frac{1}{4}[x - (-6)]$$
$$y - 1 = \frac{1}{4}(x + 6)$$
$$y - 1 = \frac{1}{4}x + \frac{3}{2}$$
$$y = \frac{1}{4}x + \frac{3}{2} + 1$$
$$y = \frac{1}{4}x + \frac{5}{2}$$

We also could substitute $\frac{1}{4}$ for m, 2 for x_1, and 3 for y_1.

$$y - 3 = \frac{1}{4}(x - 2)$$
$$y - 3 = \frac{1}{4}x - \frac{1}{2}$$
$$y = \frac{1}{4}x - \frac{1}{2} + 3$$
$$y = \frac{1}{4}x + \frac{5}{2}$$

49. (0,4) and (4,2)

First we find the slope.

$$m = \frac{4 - 2}{0 - 4} = \frac{2}{-4} = -\frac{1}{2}$$

Then we use the point-slope equation.

$$y - y_1 = m(x - x_1)$$

We substitute $-\frac{1}{2}$ for m, 0 for x_1, and 4 for y_1.

$$y - 4 = -\frac{1}{2}(x - 0)$$
$$y - 4 = -\frac{1}{2}x$$
$$y = -\frac{1}{2}x + 4$$

51. (3,2) and (1,5)

First we find the slope.

$m = \frac{2 - 5}{3 - 1} = \frac{-3}{2} = -\frac{3}{2}$

Then we use the point-slope equation.

$y - y_1 = m(x - x_1)$

We substitute $-\frac{3}{2}$ for m, 3 for x_1, and 2 for y_1.

$y - 2 = -\frac{3}{2}(x - 3)$

$y - 2 = -\frac{3}{2}x + \frac{9}{2}$

$y = -\frac{3}{2}x + \frac{9}{2} + 2$

$y = -\frac{3}{2}x + \frac{13}{2}$

53. (-2,-4) and (2,-1)

We first find the slope.

$m = \frac{-4 - (-1)}{-2 - 2} = \frac{-4 + 1}{-2 - 2} = \frac{-3}{-4} = \frac{3}{4}$

Then we use the point-slope equation.

$y - y_1 = m(x - x_1)$

We substitute $\frac{3}{4}$ for m, -2 for x_1, and -4 for y_1.

$y - (-4) = \frac{3}{4}[x - (-2)]$

$y + 4 = \frac{3}{4}(x + 2)$

$y + 4 = \frac{3}{4}x + \frac{3}{2}$

$y = \frac{3}{4}x + \frac{3}{2} - 4$

$y = \frac{3}{4}x - \frac{5}{2}$

55.

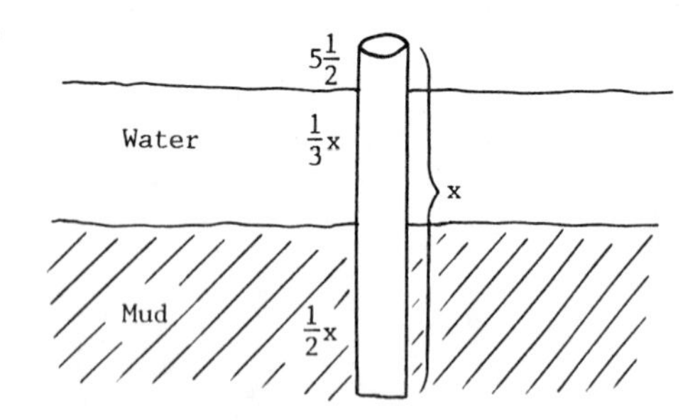

Let x represent the length of the post. Then $\frac{1}{3}x$ represents the length of the part in the water, and $\frac{1}{2}x$ represents the length of the part in the mud. The part above the water is $5\frac{1}{2}$ ft long.

55. (continued)

Above water + In water + In mud = Total length of post

$5\frac{1}{2} + \frac{1}{3}x + \frac{1}{2}x = x$

We solve:

$5\frac{1}{2} + \frac{1}{3}x + \frac{1}{2}x = x$

$5\frac{1}{2} = x - \frac{1}{3}x - \frac{1}{2}x$

$5\frac{1}{2} = \frac{6}{6}x - \frac{2}{6}x - \frac{3}{6}x$

$\frac{11}{2} = \frac{1}{6}x$

$\frac{66}{2} = x$

$33 = x$

If the post is 33 ft long, then the part in the mud measures $\frac{1}{2}\cdot 33$, or $16\frac{1}{2}$ ft, and the part in the water measures $\frac{1}{3}\cdot 33$, or 11 ft. This leaves $33 - 16\frac{1}{2} - 11$, or $5\frac{1}{2}$ ft above the water. The value checks.

The post is 33 ft long.

57. $[10 - 3(7 - 2)]$

$= [10 - 3\cdot 5]$

$= 10 - 15$

$= -5$

59. First find the slope of $3x - y + 4 = 0$

$3x - y + 4 = 0$

$3x + 4 = y$

The slope is 3.

Then find an equation of the line containing (2,-3) and having slope 3.

$y - y_1 = m(x - x_1)$

We substitute 3 for m, 2 for x_1, and -3 for y_1.

$y - (-3) = 3(x - 2)$

$y + 3 = 3x - 6$

$y = 3x - 9$

61. Solve each equation for y.

$3x - 2y = 8$ $\quad$ $2x + 3x = -4$

$3x - 8 = 2y$ $\quad$ $2y = -3x - 4$

$\frac{3}{2}x - 4 = y$ $\quad$ $y = -\frac{3}{2}x - 2$

The slope is $\frac{3}{2}$. The y-intercept is (0,-2).

Then find an equation of the line with slope $\frac{3}{2}$ and y-intercept (0,-2).

$y = mx + b$

$y = \frac{3}{2}x - 2$ (Substituting $\frac{3}{2}$ for m and -2 for b)

Exercise Set 7.1

1. We translate the first statement:

The sum of two numbers is 58.

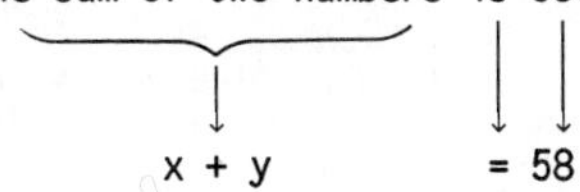

$x + y \quad = 58$

We have used x and y for the numbers.

Now we translate the second statement:

The difference between two numbers is 16.

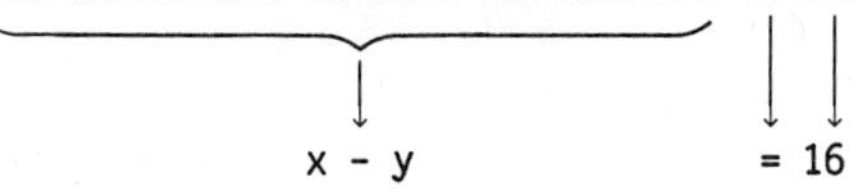

$x - y \quad = 16$

The system of equations is

$x + y = 58$

$x - y = 16$

3. First we make a drawing.

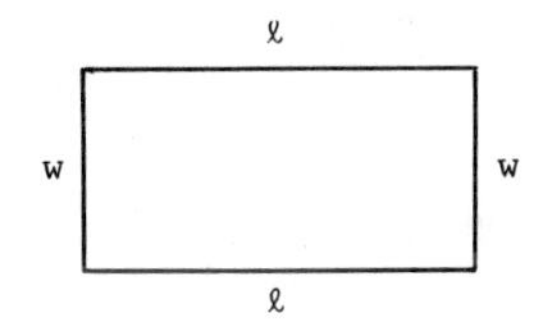

We have let ℓ = length and w = width. The perimeter of a rectangle is 2ℓ + 2w.

We translate the first statement:

The perimeter is 400 m.

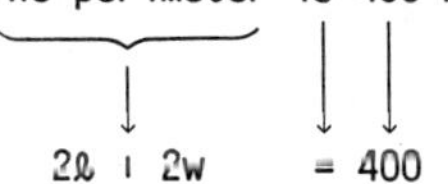

$2\ell + 2w \quad = 400$

We translate the second statement:

The width is 40 m less than the length.

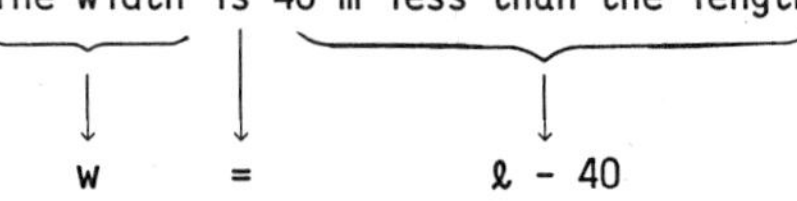

$w \quad = \quad \ell - 40$

The system of equations is

$2\ell + 2w = 400$

$w = \ell - 40$

5. We translate the first statement, using $0.30 for 30 ¢:

$53.95 plus 30¢ times the number of miles driven is cost.

$53.95 \quad + \quad 0.30 \quad \cdot \quad m \quad = \quad c$

We have let m represent the mileage and c the cost.

5. (continued)

We translate the second statement using $0.20 for 20¢:

$54.95 plus 20¢ times the number of miles driven is cost.

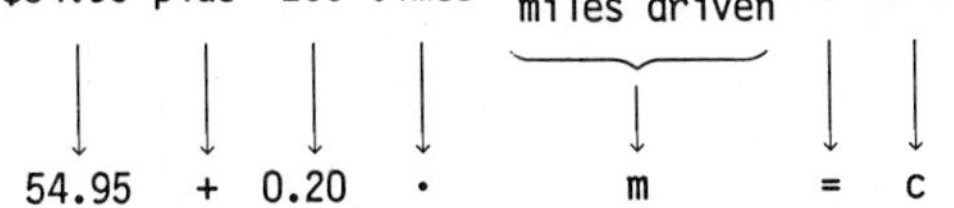

$54.95 \quad + \quad 0.20 \quad \cdot \quad m \quad = \quad c$

The system of equations is

$53.95 + 0.30m = c$

$54.95 + 0.20m = c$

7. We translate the first statement:

The difference between two numbers is 16.

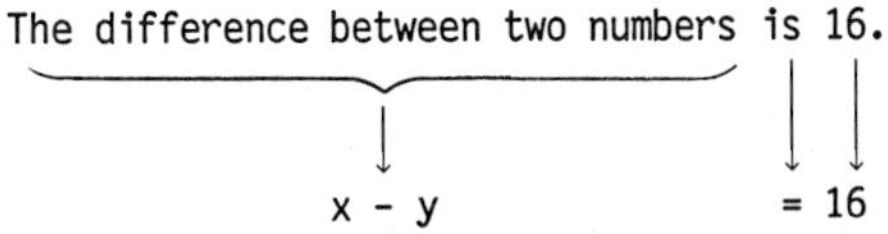

$x - y \quad = 16$

We have used x for the larger number and y for the smaller number.

Now we translate the second statement:

Three times the larger number is seven times the smaller number.

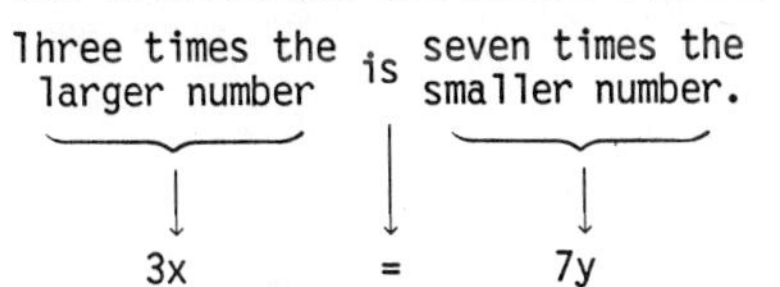

$3x \quad = \quad 7y$

The system of equations is

$x - y = 16$

$3x = 7y$

9. Supplementary angles are angles whose sum is 180°.

We translate the first statement:

Two angles are supplementary.

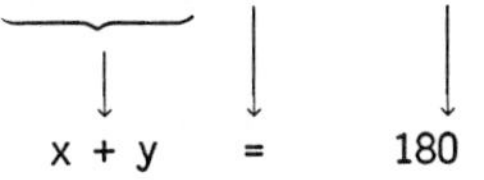

$x + y \quad = \quad 180$

We have used x and y for the angles.

Now we translate the second statement:

One is 8° more than 3 times the other.

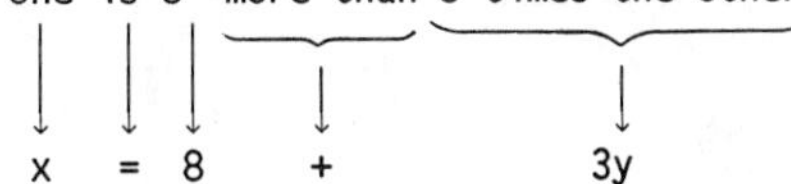

$x \quad = \quad 8 \quad + \quad 3y$

Here we have used x for the larger angle.

The system of equations is

$x + y = 180$

$x = 8 + 3y$

11. Complementary angles are angles whose sum is 90°.

We translate the first statement:

Two angles are complementary.

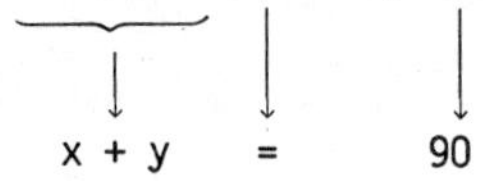

$x + y \quad = \quad 90$

We have used x and y for the angles.

Now we translate the second statement:

The difference between the angles is 34°.

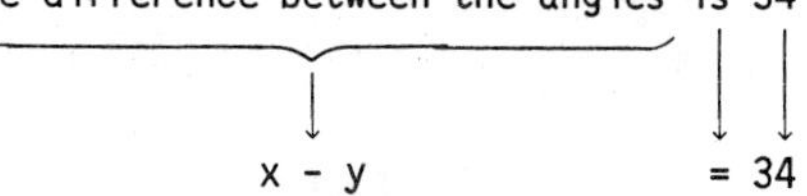

$x - y \quad = 34$

We are letting x represent the larger angle.

The system of equations is

$x + y = 90$

$x - y = 34$

13. We translate the first statement:

Hectares of Chardonnay grapes | plus | hectares of Riesling grapes | total | 820 hectares.

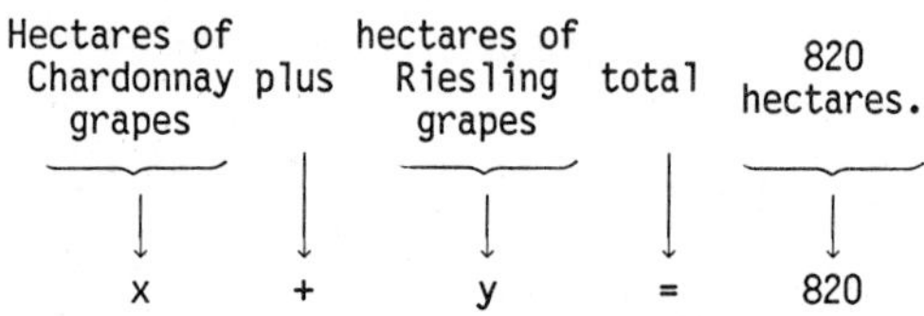

$x \quad + \quad y \quad = \quad 820$

We have used x for the number of hectares of Chardonnay grapes and y the number of hectares of Riesling grapes.

Now we translate the second statement:

Hectares of Chardonnay grapes | is | 140 hectares | more than | hectares of Riesling grapes.

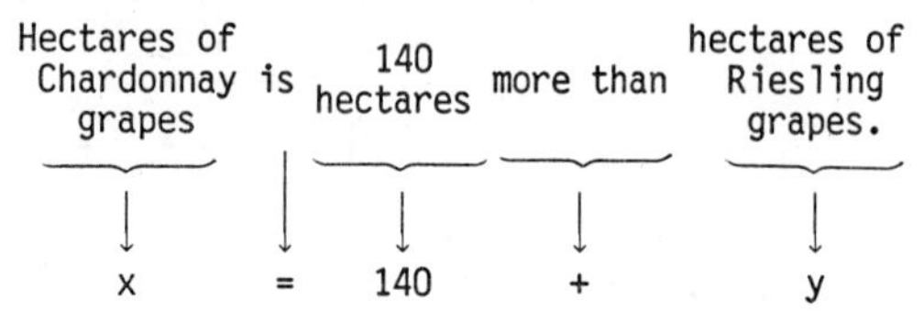

$x \quad = \quad 140 \quad + \quad y$

The system of equations is

$x + y = 820$

$x = 140 + y$

15. $3x^2 - x + 8$

We substitute 4 for x.

$3(4^2) - 4 + 8$

$= 3 \cdot 16 - 4 + 8$

$= 48 - 4 + 8$

$= 52$

17. $(9x^{-5})(12x^{-8})$

$= 9 \cdot 12 \cdot x^{-5} \cdot x^{-8}$

$= 108x^{-5+(-8)}$

$= 108x^{-13}$

19. We first list the information in a chart.

	Ages now	Ages 20 years from now
Patrick	x	x + 20
Father	y	y + 20

We have used x for Patrick's age now and y for the father's age now. In 20 years they will be x + 20 and y + 20.

We translate the first statement:

Patrick's age now is 20% of his father's age now.

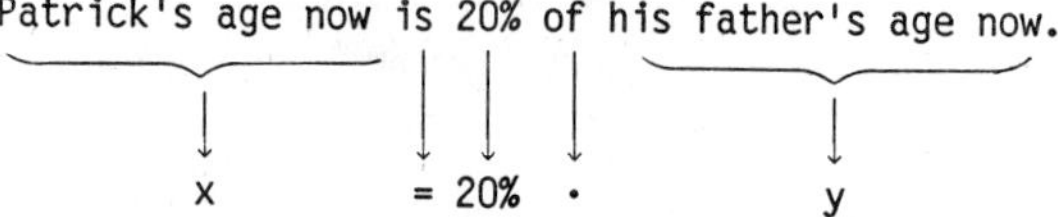

$x \quad = 20\% \quad \cdot \quad y$

Now we translate the second statement:

Patrick's age twenty years from now | will be | 52% | of | his father's age twenty years from now.

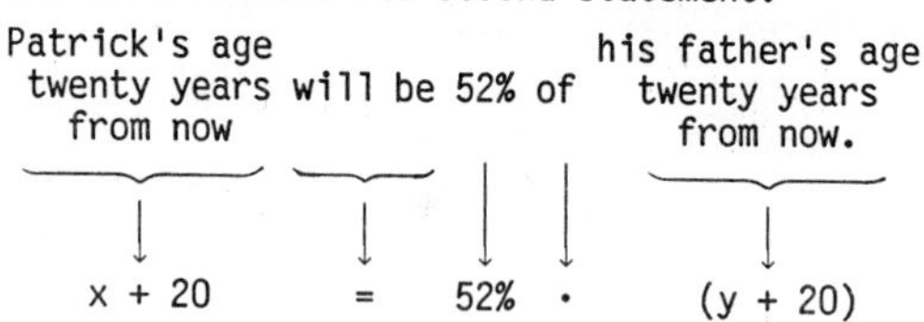

$x + 20 \quad = \quad 52\% \quad \cdot \quad (y + 20)$

The system of equations is

$x = 0.2y$

$x + 20 = 0.52(y + 20)$

21. We first make a drawing.

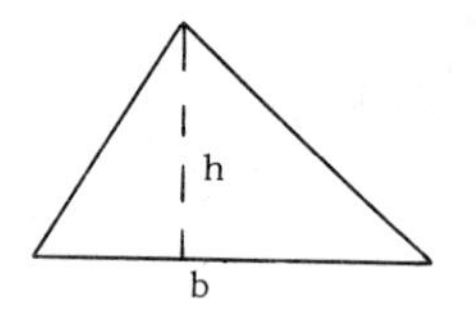

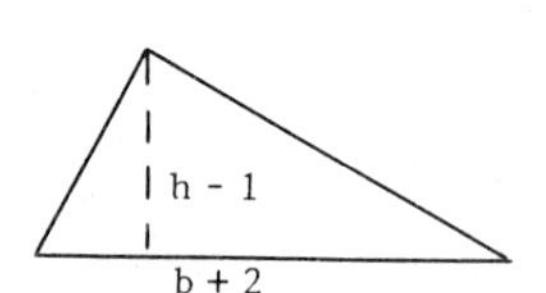

We have used h for height and b for base in the original triangle. Then h - 1 and b + 2 represent the height and base in the new triangle. The area of a triangle is $\frac{1}{2} \cdot$base$\cdot$height.

We translate the first statement:

The height of the new triangle | is | $\frac{1}{3}$ | of | the base of the new triangle.

$h - 1 \quad = \frac{1}{3} \quad \cdot \quad (b + 2)$

Now we translate the second statement:

The area of the new triangle | is | $\frac{1}{2}$ | $\cdot$ | base | $\cdot$ | height

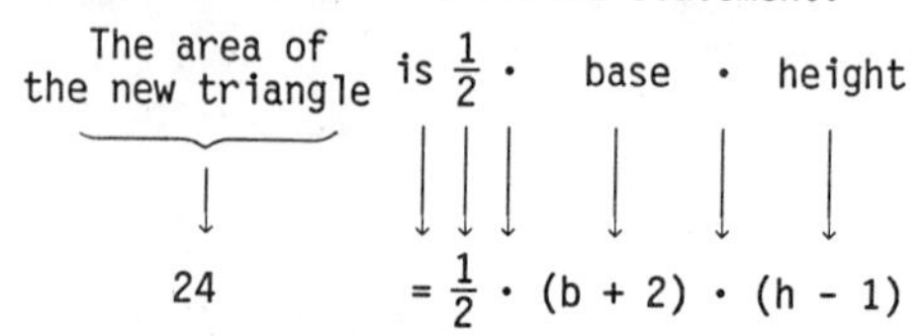

$24 \quad = \frac{1}{2} \cdot (b + 2) \cdot (h - 1)$

The system of equations is

$h - 1 = \frac{1}{3}(b + 2)$

$24 = \frac{1}{2}(b + 2)(h - 1)$

Exercise Set 7.2

1. Use alphabetical order of the variables. We substitute 3 for x and 2 for y.

$2x + 3y = 12$	
$2\cdot3 + 3\cdot2$	12
$6 + 6$	
12	

$x - 4y = -5$	
$3 - 4\cdot2$	-5
$3 - 8$	
-5	

The ordered pair (3,2) is a solution of each equation. Therefore it <u>is</u> a solution of the system of equations.

3. Use alphabetical order of the variables. We substitute -3 for x and 4 for y.

$2x = -y - 2$	
$2(-3)$	$-4 - 2$
-6	-6

$y + 7x = 9$	
$4 + 7(-3)$	9
$4 - 21$	
-17	

The ordered pair (-3,4) is not a solution of $y + 7x = 9$. Therefore it <u>is not</u> a solution of the system of equations.

5. Use alphabetical order of the variables. We substitute 2 for a and -2 for b.

$b + 2a = 2$	
$-2 + 2\cdot2$	2
$-2 + 4$	
2	

$b - a = -4$	
$-2 - 2$	-4
-4	

The ordered pair (2,-2) is a solution of each equation. Therefore it <u>is</u> a solution of the system of equations.

7. First make a table of values for each equation.

$x + 2y = 10$

x	y
0	5
-2	6
4	3

$3x + 4y = 8$

x	y
0	2
-4	5
4	-1

Plot these points and draw the line each set of points determines.

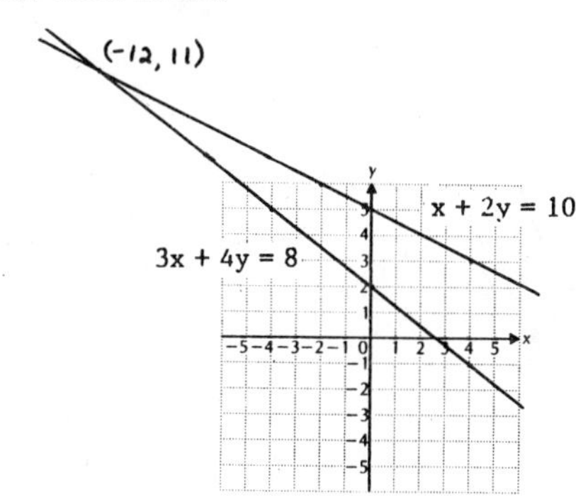

The point of intersection looks as if it has coordinates (-12,11).

7. (continued)

Check:

$x + 2y = 10$	
$-12 + 2\cdot11$	10
$-12 + 22$	
10	

$3x + 4y = 8$	
$3(-12) + 4(11)$	8
$-36 + 44$	
8	

The solution is (-12,11).

9. First make a table of values for each equation.

$8x - y = 29$

x	y
4	3
3	-5
5	11

$2x + y = 11$

x	y
3	5
5	1
4	3

Plot these points and draw the line each set of points determines.

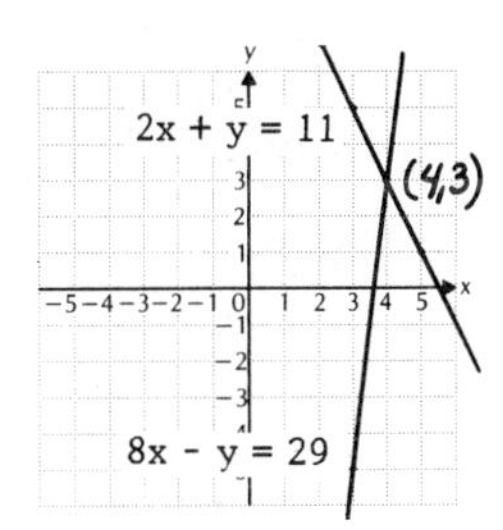

The point of intersection looks as if it has coordinates (4,3).

Check:

$8x - y = 29$	
$8\cdot4 - 3$	29
$32 - 3$	
29	

$2x + y = 11$	
$2\cdot4 + 3$	11
$8 + 3$	
11	

The solution is (4,3).

11. First make a table of values for each equation.

$a - b = 6$

a	b
2	-4
4	-2
1	-5

$a - 7 = b$

a	b
2	-5
4	-3
6	-1

11. (continued)

Plot these points and draw the line each set of points determines.

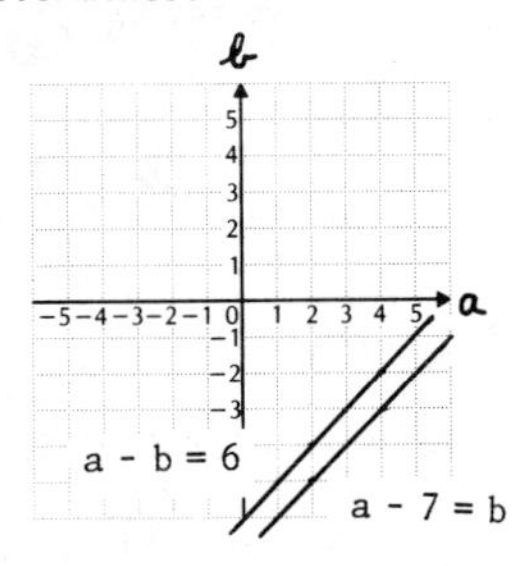

The lines are parallel. There is no solution.

13. $y = 3$ is parallel to the x-axis (horizontal) with y-intercept (0,3).

$x = 5$ is parallel to the y-axis (vertical) with x-intercept (5,0).

Draw these lines.

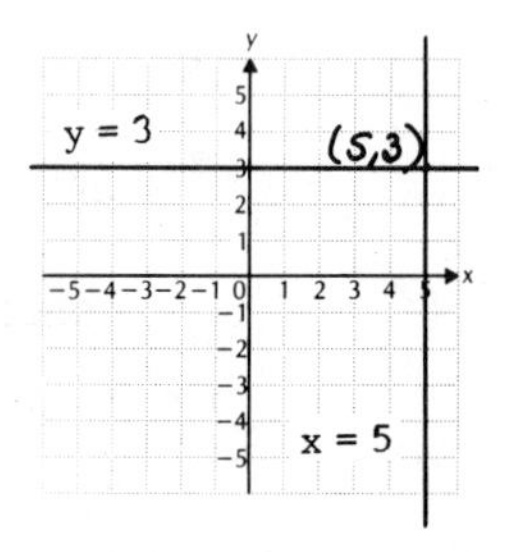

The point of intersection looks as if it has coordinates (5,3).

Check:

$y = 3$	
3	3

$x = 5$	
5	5

The solution is (5,3).

15. $(5x^{-2}y^5)^{-4}$

$= 5^{-4}(x^{-2})^{-4}(y^5)^{-4}$

$= \frac{1}{5^{-4}}\ x^8\ y^{-20}$

$= \frac{1}{625}\ x^8y^{-20}$

17. $(5x - 6)^2$

$= (5x)^2 - 2(5x)(6) + 6^2$

$= 25x^2 - 60x + 36$

19. (2,-3) is a solution of $Ax - 3y = 13$. Substitute 2 for x and -3 for y and solve for A.

$Ax - 3y = 13$

$A\cdot 2 - 3(-3) = 13$

$2A + 9 = 13$

$2A = 4$

$A = 2$

(2,-3) is a solution of $x - By = 8$. Substitute 2 for x and -3 for y and solve for B.

$x - By = 8$

$2 - B(-3) = 8$

$2 + 3B = 8$

$3B = 6$

$B = 2$

21. First graph each equation.

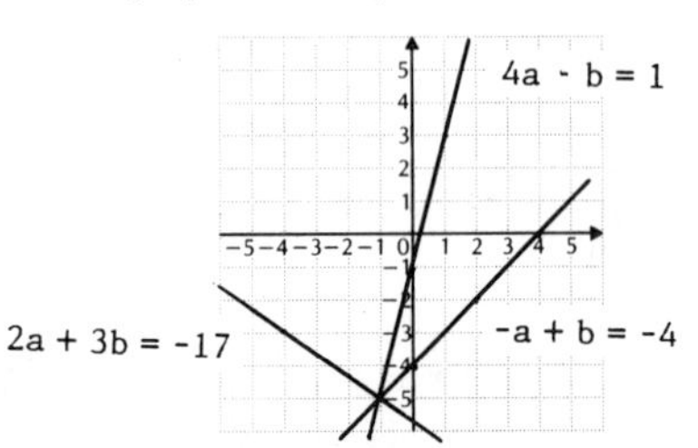

The graph of the system is three lines intersecting in the same point. The solution is one point, the ordered pair (-1,-5).

Exercise Set 7.3

1. $x + y = 4$ (1)

$y = 2x + 1$ (2)

We substitute $2x + 1$ for y in the first equation and solve for x.

$x + y = 4$ (1)

$x + (2x + 1) = 4$ (Substituting)

$3x + 1 = 4$

$3x = 3$

$x = 1$

Next we substitute 1 for x in either equation in the original system and solve for y.

$x + y = 4$ (1)

$1 + y = 4$ (Substituting)

$y = 3$

We check the ordered pair (1,3).

$x + y = 4$	
1 + 3	4
4	

$y = 2x + 1$	
3	$2\cdot 1 + 1$
	2 + 1
	3

Since (1,3) checks in both equations, it is the solution.

3. $y = x + 1$ (1)
$2x + y = 4$ (2)

We substitute $x + 1$ for y in the second equation and solve for x.

$2x + y = 4$ (2)
$2x + (x + 1) = 4$ (Substituting)
$3x + 1 = 4$
$3x = 3$
$x = 1$

Next we substitute 1 for x in either equation of the original system and solve for y.

$y = x + 1$ (1)
$y = 1 + 1$ (Substituting)
$y = 2$

We check the ordered pair (1,2).

$y = x + 1$	
2	$1 + 1$
	2

$2x + y = 4$	
$2\cdot1 + 2$	4
$2 + 2$	
4	

Since (1,2) checks in both equations, it is the solution.

5. $y = 2x - 5$ (1)
$3y - x = 5$ (2)

We substitute $2x - 5$ for y in the second equation and solve for x.

$3y - x = 5$ (2)
$3(2x - 5) - x = 5$ (Substituting)
$6x - 15 - x = 5$
$5x - 15 = 5$
$5x = 20$
$x = 4$

Next we substitute 4 for x in either equation of the original system and solve for y.

$y = 2x - 5$ (1)
$y = 2\cdot4 - 5$ (Substituting)
$y = 8 - 5$
$y = 3$

We check the ordered pair (4,3).

$y = 2x - 5$	
3	$2\cdot4 - 5$
	$8 - 5$
	3

$3y - x = 5$	
$3\cdot3 - 4$	5
$9 - 4$	
5	

Since (4,3) checks in both equations, it is the solution.

7. $x = -2y$ (1)
$x + 4y = 2$ (2)

We substitute $-2y$ for x in the second equation and solve for y.

$x + 4y = 2$ (2)
$-2y + 4y = 2$
$2y = 2$
$y = 1$

Next we substitute 1 for y in either equation of the original system and solve for x.

$x = -2y$ (1)
$x = -2\cdot1$
$x = -2$

We check the ordered pair (-2,1).

$x = -2y$	
-2	$-2\cdot1$
	-2

$3y - x = 5$	
$3\cdot1 - (-2)$	5
$3 + 2$	
5	

Since (-2,1) checks in both equations, it is the solution.

9. $s + t = -4$ (1)
$s - t = 2$ (2)

We solve the second equation for s.

$s - t = 2$ (2)
$s = t + 2$

We substitute $t + 2$ for s in the first equation and solve for t.

$s + t = -4$ (1)
$(t + 2) + t = -4$ (Substituting)
$2t + 2 = -4$
$2t = -6$
$t = -3$

Now we substitute -3 for t in either equation of the original system and solve for s.

$s + t = -4$ (1)
$s + (-3) = -4$ (Substituting)
$s = -1$

We check the ordered pair (-1,-3).

$s + t = -4$	
$-1 + (-3)$	-4
-4	

$s - t = 2$	
$-1 - (-3)$	2
$-1 + 3$	
2	

Since (-1,-3) checks in both equations, it is the solution.

11. $y - 2x = -6$ (1)

$2y - x = 5$ (2)

We solve the first equation for y.

$y - 2x = -6$ (1)

$y = 2x - 6$

We substitute $2x - 6$ for y in the second equation and solve for x.

$2y - x = 5$ (2)

$2(2x - 6) - x = 5$ (Substituting)

$4x - 12 - x = 5$

$3x - 12 = 5$

$3x = 17$

$x = \frac{17}{3}$

Now we substitute $\frac{17}{3}$ for x in either equation of the original system and solve for y.

$y - 2x = -6$ (1)

$y - 2\cdot\frac{17}{3} = -6$ (Substituting)

$y - \frac{34}{3} = -\frac{18}{3}$

$y = \frac{16}{3}$

The ordered pair $(\frac{17}{3},\frac{16}{3})$ checks in both equations. It is the solution.

13. $2x + 3y = -2$ (1)

$2x - y = 9$ (2)

We solve the second equation for y.

$2x - y = 9$ (2)

$2x - 9 = y$

We substitute $2x - 9$ for y in the first equation and solve for x.

$2x + 3y = -2$ (1)

$2x + 3(2x - 9) = -2$

$2x + 6x - 27 = -2$

$8x - 27 = -2$

$8x = 25$

$x = \frac{25}{8}$

Now we substitute $\frac{25}{8}$ for x in either equation of the original system and solve for y.

$2x - y = 9$ (2)

$2\cdot\frac{25}{8} - y = 9$ (Substituting)

$\frac{25}{4} - y = \frac{36}{4}$

$-y = \frac{11}{4}$

$y = -\frac{11}{4}$

13. (continued)

We check the ordered pair $(\frac{25}{8},-\frac{11}{4})$.

$2x + 3y = -2$	
$2\cdot\frac{25}{8} + 3(-\frac{11}{4})$	-2
$\frac{25}{4} - \frac{33}{4}$	
$-\frac{8}{4}$	
-2	

$2x - y = 9$	
$2\cdot\frac{25}{8} - (-\frac{11}{4})$	9
$\frac{25}{4} + \frac{11}{4}$	
$\frac{36}{4}$	
9	

Since $(\frac{25}{8},-\frac{11}{4})$ checks in both equations, it is the solution.

15. $x - y = -3$ (1)

$2x + 3y = -6$ (2)

We solve the first equation for x.

$x - y = -3$ (1)

$x = y - 3$

We substitute $y - 3$ for x in the second equation and solve for y.

$2x + 3y = -6$ (2)

$2(y - 3) + 3y = -6$ (Substituting)

$2y - 6 + 3y = -6$

$5y - 6 = -6$

$5y = 0$

$y = 0$

Now we substitute 0 for y in either equation of the original system and solve for x.

$x - y = -3$ (1)

$x - 0 = -3$ (Substituting)

$x = -3$

We check the ordered pair (-3,0).

$x - y = -3$	
$-3 - 0$	-3
-3	

$2x + 3y = -6$	
$2(-3) + 3\cdot 0$	-6
$-6 + 0$	
-6	

Since (-3,0) checks in both equations, it is the solution.

17. $r - 2s = 0$ (1)
$4r - 3s = 15$ (2)

We solve the first equation for r.

$r - 2s = 0$ (1)
$r = 2s$

We substitute 2s for r in the second equation and solve for s.

$4r - 3s = 15$ (2)
$4(2s) - 3s = 15$ (Substituting)
$8s - 3s = 15$
$5s = 15$
$s = 3$

Now we substitute 3 for s in either equation of the original system and solve for r.

$r - 2s = 0$ (1)
$r - 2\cdot 3 = 0$ (Substituting)
$r - 6 = 0$
$r = 6$

We check the ordered pair (6,3).

$r - 2s = 0$	
$6 - 2\cdot 3$	0
$6 - 6$	
0	

$4r - 3s = 15$	
$4\cdot 6 - 3\cdot 3$	15
$24 - 9$	
15	

Since (6,3) checks in both equations, it is the solution.

19. $x - 3y = 7$ (1)
$-4x + 12y = 28$ (2)

We solve the first equation for x.

$x - 3y = 7$ (1)
$x = 3y + 7$

We substitute 3y + 7 for x in the second equation and solve for y.

$-4x + 12y = 28$ (2)
$-4(3y + 7) + 12y = 28$ (Substituting)
$-12y - 28 + 12y = 28$
$-28 = 28$

We obtain a false equation, -28 = 28, so there is no solution. The graphs of the equations are parallel lines. They do not intersect.

21. $(7x^2 - 4)(7x^2 + 4)$
$= (7x^2)^2 - 4^2$
$= 49x^4 - 16$

23. $3x + 5x - 4 = 2(x - 7)$
$8x - 4 = 2x - 14$
$6x = -10$
$x = -\frac{10}{6}$
$x = -\frac{5}{3}$

25. $y - 2.35x = -5.97$ (1)
$2.14y - x = 4.88$ (2)

We solve the first equation for y.

$y - 2.35x = -5.97$
$y = 2.35x - 5.97$

We substitute 2.35x - 5.97 for y in the second equation and solve for x.

$2.14y - x = 4.88$ (2)
$2.14(2.35x - 5.97) - x = 4.88$ (Substituting)
$5.029x - 12.7758 - x = 4.88$
$4.029x = 17.6558$
$x = \frac{17.6558}{4.029}$
$x \approx 4.382179$

Now we substitute 4.382179 for x in either of the original equations and solve for y.

$y - 2.35x = -5.97$ (1)
$y - 2.35(4.382179) = -5.97$ (Substituting)
$y \approx -5.97 + 10.298121$
$y \approx 4.328121$

The ordered pair (4.382179,4.328121) checks in both equations. It is the solution.

27. $\frac{x}{2} + \frac{3y}{2} = 2$ (1)
$\frac{x}{5} - \frac{y}{2} = 3$ (2)

We first clear of fractions.

$x + 3y = 4$ [Multiplying (1) by 2]
$2x - 5y = 30$ [Multiplying (2) by 10]

We solve the first equation for x.

$x + 3y = 4$ (1)
$x = 4 - 3y$

We substitute 4 - 3y for x in the second equation and solve for y.

$2x - 5y = 30$ (2)
$2(4 - 3y) - 5y = 30$ (Substituting)
$8 - 6y - 5y = 30$
$-11y = 22$
$y = -2$

Now we substitute -2 for y in either equation of the original system and solve for x.

$x + 3y = 4$ (1)
$x + 3(-2) = 4$ (Substituting)
$x - 6 = 4$
$x = 10$

The ordered pair (10,-2) checks in both equations. It is the solution.

Exercise Set 7.4

1. $x + y = 10$
$x - y = 8$
$2x + 0 = 18$ (Adding)
$2x = 18$
$x = 9$

Substitute 9 for x in one of the original equations and solve for y.

$x + y = 10$
$9 + y = 10$ (Substituting)
$y = 1$

Check: For (9,1)

$x + y = 10$	
$9 + 1$	10
10	

$x - y = 8$	
$9 - 1$	8
8	

Since (9,1) checks, it is the solution.

3. $x + y = 8$
$2x - y = 7$
$3x + 0 = 15$ (Adding)
$3x = 15$
$x = 5$

Substitute 5 for x in one of the original equations and solve for y.

$x + y = 8$
$5 + y = 8$ (Substituting)
$y = 3$

Check: For (5,3)

$x + y = 8$	
$5 + 3$	8
8	

$2x - y = 7$	
$2\cdot5 - 3$	7
$10 - 3$	
7	

Since (5,3) checks, it is the solution.

5. $3a + 4b = 7$
$a - 4b = 5$
$4a + 0 = 12$ (Adding)
$4a = 12$
$a = 3$

Substitute 3 for a in one of the original equations and solve for b.

$3a + 4b = 7$
$3\cdot3 + 4b = 7$ (Substituting)
$9 + 4b = 7$
$4b = -2$
$b = -\frac{1}{2}$

5. (continued)

Check: For $(3,-\frac{1}{2})$

$3a + 4b = 7$	
$3(3) + 4(-\frac{1}{2})$	7
$9 - 2$	
7	

$a - 4b = 5$	
$3 - 4(-\frac{1}{2})$	5
$3 + 2$	
5	

Since $(3,-\frac{1}{2})$ checks, it is the solution.

7. $8x - 5y = -9$
$3x + 5y = -2$
$11x = -11$ (Adding)
$x = -1$

Substitute -1 for x in one of the original equations and solve for y.

$3x + 5y = -2$
$3(-1) + 5y = -2$ (Substituting)
$-3 + 5y = -2$
$5y = 1$
$y = \frac{1}{5}$

Check: For $(-1,\frac{1}{5})$

$8x - 5y = -9$	
$8(-1) - 5(\frac{1}{5})$	-9
$-8 - 1$	
-9	

$3x + 5y = -2$	
$3(-1) + 5(\frac{1}{5})$	-2
$-3 + 1$	
-2	

Since $(-1,\frac{1}{5})$ checks, it is the solution.

9. $-x - y = 8$
$2x - y = -1$

We multiply by -1 on both sides of the first equation and then add.

$x + y = -8$ (Multiplying by -1)
$2x - y = -1$
$3x = -9$ (Adding)
$x = -3$

Substitute -3 for x in one of the original equations and solve for y.

$2x - y = -1$
$2(-3) - y = -1$ (Substituting)
$-6 - y = -1$
$-y = 5$
$y = -5$

9. (continued)

Check: For (-3,-5)

$-x - y = 8$	
$-(-3) - (-5)$	8
$3 + 5$	
8	

$2x - y = -1$	
$2(-3) - (-5)$	-1
$-6 + 5$	
-1	

Since (-3,-5) checks, it is the solution.

11. $x + 3y = 19$

$x + 3y = -1$

We multiply by -1 on both sides of the second equation and then add.

$x + 3y = 19$

$-x - 3y = 1$ (Multiplying by -1)

$0 = 20$ (Adding)

We obtain a false equation, 0 = 20, so there is no solution. The graphs of the equations are parallel lines. They do not intersect.

13. $3x - 2y = 10$

$5x + 3y = 4$

We use the multiplication principle with both equations and add.

$9x - 6y = 30$ (Multiplying by 3)

$10x + 6y = 8$ (Multiplying by 2)

$19x = 38$ (Adding)

$x = 2$

Substitute 2 for x in one of the original equations and solve for y.

$5x + 3y = 4$

$5\cdot2 + 3y = 4$ (Substituting)

$10 + 3y = 4$

$3y = -6$

$y = -2$

Check: For (2,-2)

$3x - 2y = 10$	
$3\cdot2 - 2(-2)$	10
$6 + 4$	
10	

$5x + 3y = 4$	
$5\cdot2 + 3(-2)$	4
$10 - 6$	
4	

Since (2,-2) checks, it is the solution.

15. $2a + 3b = -1$

$3a + 5b = -2$

We use the multiplication principle with both equations and then add.

$-10a - 15b = 5$ (Multiplying by -5)

$9a + 15b = -6$ (Multiplying by 3)

$-a = -1$ (Adding)

$a = 1$

Substitute 1 for a in one of the original equations and solve for b.

$2a + 3b = -1$

$2\cdot1 + 3b = -1$ (Substituting)

$3b = -3$

$b = -1$

Check: For (1,-1)

$2a + 3b = -1$	
$2\cdot1 + 3(-1)$	-1
$2 - 3$	
-1	

$3a + 5b = -2$	
$3\cdot1 + 5(-1)$	-2
$3 - 5$	
-2	

Since (1,-1) checks, it is the solution.

17. $0.3x + 0.2y = 0$

$x + 0.5y = -0.5$

We first multiply each equation by 10 to clear decimals.

$3x + 2y = 0$

$10x + 5y = -5$ (Multiplying each equation by 10)

We use the multiplication principle with both equations of the resulting system.

$-15x - 10y = 0$ (Multiplying by -5)

$20x + 10y = -10$ (Multiplying by 2)

$5x = -10$ (Adding)

$x = -2$

Substitute -2 for x in either equation of the system in which decimals were cleared and solve for y.

$3x + 2y = 0$

$3(-2) + 2y = 0$

$-6 + 2y = 0$

$2y = 6$

$y = 3$

Check: For (-2,3)

$0.3x + 0.2y = 0$	
$0.3(-2) + 0.2(3)$	0
$-0.6 + 0.6$	
0	

$x + 0.5y = -0.5$	
$-2 + 0.5(3)$	-0.5
$-2 + 1.5$	
-0.5	

Since (-2,3) checks, it is the solution.

19. $\frac{3}{4}x + \frac{1}{3}y = 8$

$\frac{1}{2}x - \frac{5}{6}y = -1$

We first multiply each equation by a multiple of its denominators to clear fractions.

$12(\frac{3}{4}x + \frac{1}{3}y) = 12 \cdot 8$ (Multiplying by 12, a multiple of 4 and 3)

$6(\frac{1}{2}x - \frac{5}{6}y) = 6(-1)$ (Multiplying by 6, a multiple of 2 and 6)

The resulting system is:

$9x + 4y = 96$

$3x - 5y = -6$

We use the multiplication principle with both equations of the resulting system.

$45x + 20y = 480$ (Multiplying by 5)

$12x - 20y = -24$ (Multiplying by 4)

$57x = 456$ (Adding)

$x = 8$

Substitute 8 for x in either equation of the system in which fractions were cleared and solve for y.

$9x + 4y = 96$

$9 \cdot 8 + 4y = 96$ (Substituting)

$72 + 4y = 96$

$4y = 24$

$y = 6$

Check: For (8,6)

$\frac{3}{4}x + \frac{1}{3}y = 8$	
$\frac{3}{4} \cdot 8 + \frac{1}{3} \cdot 6$	8
$6 + 2$	
8	

$\frac{1}{2}x - \frac{5}{6}y = -1$	
$\frac{1}{2} \cdot 8 - \frac{5}{6} \cdot 6$	-1
$4 - 5$	
-1	

Since (8,6) checks, it is the solution.

21. $m - n = 32$

$3m - 8n - 6 = 0$

or

$m - n = 32$

$3m - 8n = 6$ (Adding 6)

We multiply by -8 on both sides of the first equation and then add.

$-8m + 8n = -256$

$3m - 8n = 6$

$-5m + 0 = -250$ (Adding)

$-5m = -250$

$m = 50$

21. (continued)

Substitute 50 for m in one of the original equations and solve for n.

$m - n = 32$

$50 - n = 32$ (Substituting)

$-n = -18$

$n = 18$

Check: For (50,18)

$m - n = 32$	
$50 - 18$	32
32	

$3m - 8n - 6 = 0$	
$3 \cdot 50 - 8 \cdot 18 - 6$	0
$150 - 144 - 6$	
0	

Since (50,18) checks, it is the solution.

23. $0.06x + 0.05y = 0.07$

$0.04x - 0.03y = 0.11$

We first multiply each equation by 100 to clear decimals.

$6x + 5y = 7$

$4x - 3y = 11$

We use the multiplication principle with both equations of the resulting system.

$18x + 15y = 21$ (Multiplying by 3)

$20x - 15y = 55$ (Multiplying by 5)

$38x = 76$ (Adding)

$x = 2$

Substitute 2 for x in one of the equations in which the decimals were cleared and solve for y.

$6x + 5y = 7$

$6 \cdot 2 + 5y = 7$

$12 + 5y = 7$

$5y = -5$

$y = -1$

Check: For (2,-1)

$0.06x + 0.05y = 0.07$	
$0.06(2) + 0.05(-1)$	0.07
$0.12 - 0.05$	
0.07	

$0.04x - 0.03y = 0.11$	
$0.04(2) - 0.03(-1)$	0.11
$0.08 + 0.03$	
0.11	

Since (2,-1) checks, it is the solution.

25. $5x^4 - 7x^3 + 9x^4 + 16x^3 - 19x^2$

$= (5 + 9)x^4 + (-7 + 16)x^3 - 19x^2$

$= 14x^4 + 9x^3 - 19x^2$

27. $15 = 3\cdot5$

$45 = 3\cdot3\cdot5$

$60 = 2\cdot2\cdot3\cdot5$

LCM = $2\cdot2\cdot3\cdot3\cdot5$, or 180

29. $3(x - y) = 9$ or $3x - 3y = 9$

$x + y = 7$ $\quad x + y = 7$

Multiply the second equation by 3 and then add.

$3x - 3y = 9$

$\underline{3x + 3y = 21}$ (Multiplying by 3)

$6x = 30$

$x = 5$

Substitute 5 for x in one of the original equations and solve for y.

$x + y = 7$

$5 + y = 7$ (Substituting)

$y = 2$

The ordered pair (5,2) checks and is the solution.

31. $\frac{x}{3} + \frac{y}{2} = 1\frac{1}{3}$ or $\frac{x}{3} + \frac{y}{2} = \frac{4}{3}$

$x + 0.05y = 4$

We first clear of decimals and fractions.

$2x + 3y = 8$ (Multiplying by 6)

$100x + 5y = 400$ (Multiplying by 100)

We multiply by -50 on both sides of the first equation and then add.

$-100x - 150y = -400$

$\underline{100x + 5y = 400}$

$-100y = 0$ (Adding)

$y = 0$

Substitute 0 for y in one of the original equations and solve for x.

$x + 0.05y = 4$

$x + 0.05(0) = 4$

$x = 4$

The ordered pair (4,0) checks and is the solution.

Exercise Set 7.5

1. The translation has been done in Exercise Set 7.1, Problem 5. The resulting system is

$53.95 + 0.30m = c$

$54.95 + 0.20m = c$

where m represents the mileage and c the cost.

We use substitution since there is a variable alone on one side of an equation - in fact, both.

$53.95 + 0.30m = 54.95 + 0.20m$

$5395 + 30m = 5495 + 20m$ (Multiplying by 100)

$10m = 100$

$m = 10$

For 10 mi, the cost of the Badger car is $53.95 + 0.30(10)$, or $53.95 + 3$, or \$56.95.

For 10 mi, the cost of the other car is $54.95 + 0.20(10)$, or $54.95 + 2$, or \$56.95.

The costs are the same when the mileage is 10.

3. We make a table to organize the information. We let x represent Sammy's age now and y his daughter's age now.

	Age now	Age in 4 years	Age 6 years ago
Sammy	x	x + 4	x - 6
Daughter	y	y + 4	y - 6

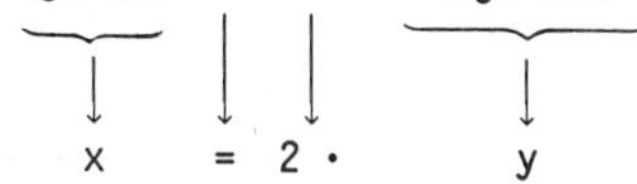

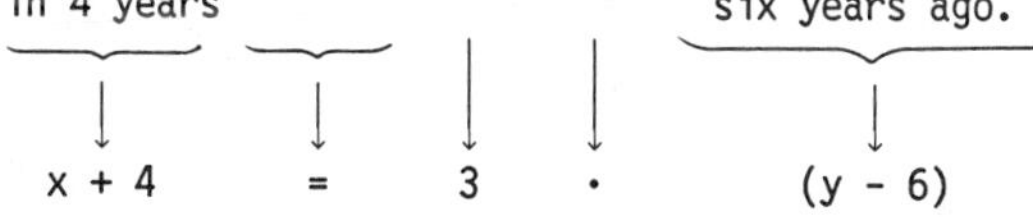

The resulting system is

$x = 2y$

$x + 4 = 3(y - 6)$

We use the substitution method since there is a variable alone on one side of an equation. We substitute 2y for x in the second equation and solve for y.

$x + 4 = 3(y - 6)$

$2y + 4 = 3(y - 6)$ (Substituting)

$2y + 4 = 3y - 18$

$22 = y$

Next we substitute 22 for y in one of the original equations and solve for x.

$x = 2y$

$x = 2\cdot22$ (Substituting)

$x = 44$

3. (continued)

Sammy's age is 44, which is twice his daughter's age, 22. In four years Sammy will be 44 + 4, or 48. Six years ago his daughter was 22 - 6, or 16. Sammy's age four years from now, 48, is three times his daughter's age six years ago, 16.

Sammy is now 44 and his daughter is 22.

5. The translation has been done in Exercise Set 7.1, Problem 7. The resulting system is

$x - y = 16$

$3x = 7y$

Here we use the substitution method. We solve the first equation for x.

$x - y = 16$

$x = y + 16$

We substitute y + 16 for x in the second equation and solve for y.

$3x = 7y$

$3(y + 16) = 7y$ (Substituting)

$3y + 48 = 7y$

$48 = 4y$

$12 = y$

Next we substitute 12 for y in either of the original equations and solve for x.

$x - y = 16$

$x - 12 = 16$ (Substituting)

$x = 28$

The difference between 28 and 12, 28 - 12, is 16. Three times the larger, 3·28 or 84, is seven times the smaller, 7·12 or 84. The numbers check.

The numbers are 28 and 12.

7. The translation has been done in Exercise Set 7.1, Problem 9. The resulting system is

$x + y = 180$ or $x + y = 180$

$x = 8 + 3y$ $\quad x - 3y = 8$ (Adding -3y)

We multiply the first equation by 3 and then add.

$3x + 3y = 540$ (Multiplying by 3)

$x - 3y = 8$

$4x = 548$ (Adding)

$x = 137$

We substitute 137 for x in one of the original equations and solve for y.

$x + y = 180$

$137 + y = 180$ (Substituting)

$y = 43$

The sum of 137° and 43° is 180°. Three times 43° plus 8°, 3·43 + 8 or 129 + 8, is 137°. The numbers check.

The angles measure 137° and 43°.

9. The translation has been done in Exercise Set 7.1, Problem 11.

We solve the system of equations using the addition method.

$x + y = 90$

$x - y = 34$

$2x = 124$ (Adding)

$x = 62$

Substitute 62 for x in one of the original equations and solve for y.

$x + y = 90$

$62 + y = 90$ (Substituting)

$y = 28$

The sum of 62° and 28° is 90°. The difference between 62° and 28°, 62 - 28, is 34°. The numbers check.

The angles measure 62° and 28°.

11. The translation has been done in Exercise Set 7.1, Problem 13.

We solve the system of equations using the addition method.

$x + y = 820$ or $x + y = 820$

$x = 140 + y$ $\quad x - y = 140$ (Adding -y)

$2x = 960$ (Adding)

$x = 480$

Substitute 480 for x in one of the original equations and solve for y.

$x + y = 820$

$480 + y = 820$

$y = 340$

The sum of 480 hectares and 340 hectares is 820 hectares. The difference between 480 hectares and 340 hectares is 140 hectares. The numbers check.

The vintner should plan 480 hectares of Chardonnay grapes and 340 hectares of Riesling grapes.

13. The translation has been done in Exercise Set 7.1, Problem 3. The resulting system is

$2\ell + 2w = 400$

$w = \ell - 40$

We use the substitution method since there is a variable on one side of an equation. We substitute $\ell - 40$ for w in the first equation and solve for ℓ.

$2\ell + 2w = 400$

$2\ell + 2(\ell - 40) = 400$ (Substituting)

$2\ell + 2\ell - 80 = 400$

$4\ell - 80 = 400$

$4\ell = 480$

$\ell = 120$

13. (continued)

Next we substitute 120 for ℓ in either of the original equations and solve for w.

$w = \ell - 40$

$w = 120 - 40$ (Substituting)

$w = 80$

A possible solution is a length of 120 m and a width of 80 m. The perimeter would be $2(120 + 2(80)$, or $240 + 160$, or 400 m. Also, the width, 80 m, is 40 m less than the length. These check.

The length is 120 m, and the width is 80 m.

15. Let d represent the number of dimes and q the number of quarters. Then, 10d represents the value of the dimes in cents, and 25q represents the value of the quarters in cents. The total value is \$15.25, or 1525¢. The total number of coins is 103.

Number of dimes plus number of quarters is 103.

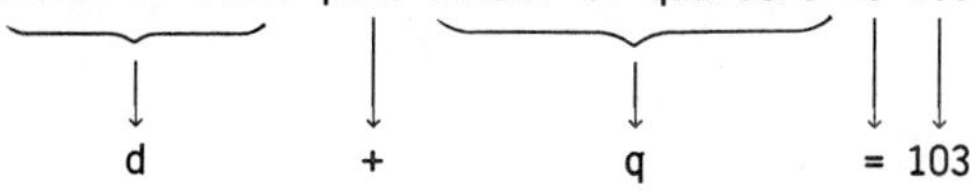

Value of dimes plus value of quarters is \$15.25

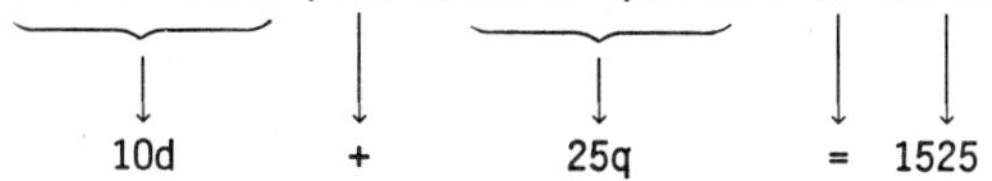

The resulting system is

$d + q = 103$

$10d + 25q = 1525$

We use the addition method. We multiply the first equation by -10 and then add.

$-10d - 10q = -1030$ (Multiplying by -10)

$10d + 25q = 1525$

$15q = 495$ (Adding)

$q = 33$

Next we substitute 33 for q in one of the original equations and solve for d.

$d + q = 103$

$d + 33 = 103$ (Substituting)

$d = 70$

The number of dimes plus the number of quarters is $70 + 33$, or 103. The total value in cents is $10 \cdot 70 + 25 \cdot 33$, or $700 + 825$, or 1525. This is equal to \$15.25. This checks.

There are 70 dimes and 33 quarters.

17. Let n represent the number of nickels and d the number of dimes. Then, 5n represents the value of the nickels in cents, and 10d represents the value of the dimes in cents. The total value is \$25, or 2500 cents.

Value of nickels plus value of dimes is \$25.

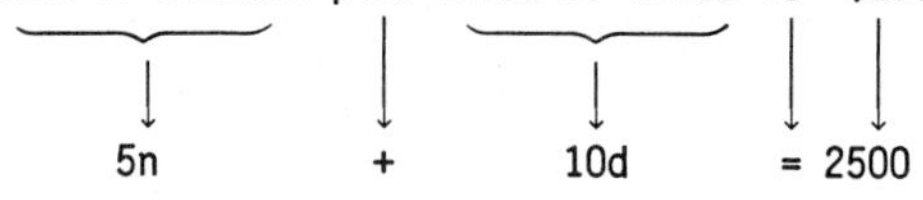

Number of nickels is three times the number of dimes

$n = 3 \cdot d$

The resulting system is

$5n + 10d = 2500$

$n = 3d$

We use the substitution method. We substitute 3d for n in the first equation and solve for d.

$5n + 10d = 2500$

$5 \cdot 3n + 10d = 2500$ (Substituting)

$15d + 10d = 2500$

$25d = 2500$

$d = 100$

We now substitute 100 for d in one of the original equations and solve for n.

$n = 3d$

$n = 3 \cdot 100$ (Substituting)

$n = 300$

The number of nickels is three times the number of dimes ($300 = 3 \cdot 100$). The total value in cents is $5 \cdot 300 + 10 \cdot 100$, or $1500 + 1000$, or 2500. This is equal to \$25. This checks.

There are 300 nickels and 100 dimes.

19. It helps to list the information in a table. We let x = the number of adults and y = the number of children.

People	Paid	Number attending	Money taken in
Adults	\$1.00	x	1.00x
Children	\$0.75	y	0.75y
	Totals	429	\$372.50

19. (continued)

The total number of people attending was 429, so

$$x + y = 429.$$

The total amount taken in was \$372.50, so

$$1.00x + 0.75y = 372.50.$$

or $100x + 75y = 37,250$ (Multiplying by 100)

We use the addition method.

$x + y = 429$

$100x + 75y = 37,250$

We multiply on both sides of the first equation by -75 and then add.

$-75x - 75y = -32,175$ (Multiplying by -75)

$100x + 75y = 37,250$

$25x = 5075$ (Adding)

$x = 203$

Next we substitute 203 for x in one of the original equations and solve for y.

$x + y = 429$

$203 + y = 429$ (Substituting)

$y = 226$

The total attending was 203 adults plus 226 children, or 429. The total receipts were 1.00(203) + 0.75(226). This is \$203 + \$169.50, or \$372.50. The numbers check.

Thus, 203 adults and 226 children attended the play.

21. It helps to list the information in a table. We let x = the number of student tickets sold and y = the number of adult tickets sold.

Ticket	Paid	Number sold	Money taken in
Student	\$0.50	x	0.50x
Adult	\$0.75	y	0.75y
	Totals	200	\$132.50

The total number of tickets sold was 200, so

$$x + y = 200$$

The total amount collected was \$132.50, so

$$0.50x + 0.75y = 132.50$$

or $50x + 75y = 13,250$ (Multiplying by 100)

We use the addition method.

$x + y = 200$

$50x + 75y = 13,250$

21. (continued)

We multiply on both sides of the first equation by -50 and then add.

$-50x - 50y = -10,000$ (Multiplying by -50)

$50x + 75y = 13,250$

$25y = 3250$

$y = 130$

Next we substitute 130 for y in one of the original equations and solve for x.

$x + y = 200$

$x + 130 = 200$ (Substituting)

$x = 70$

The total number of tickets sold was 70 students plus 130 adults, or 200. The total receipts were 0.50(70) + 0.75(130). This amount is \$35 + \$97.50, or \$132.50. The numbers check.

Thus, 70 student tickets and 130 adult tickets were sold.

23. We can arrange the information in a table.

Type of solution	Amount of solution	Percent of acid	Amount of acid in solution
A	x	50%	50%x
B	y	80%	80%y
Mixture	100 grams	68%	68% × 100 or 68 grams

We let x represent the number of milliliters of solution A and y represent the number of milliliters of solution B.

Since the total is 100 mL, we have

$$x + y = 100.$$

The amount of acid in the mixture is to be 68% of 100, or 68 mL. The amounts of acid from the two solutions are 50%x and 80%y. Thus

$50\%x + 80\%y = 68$

or $0.5x + 0.8y = 68$

or $5x + 8y = 680$

We use the addition method.

$x + y = 100$

$5x + 8y = 680$

We multiply the first equation by -5 and then add.

$-5x - 5y = -500$ (Multiplying by -5)

$5x + 8y = 680$

$3y = 180$

$y = 60$

23. (continued)

Next we substitute 60 for y in one of the original equations and solve for x.

$x + y = 100$

$x + 60 = 100$ (Substituting)

$x = 40$

We consider $x = 40$ and $y = 60$. The sum is 100. Now 50% of 40 is 20 and 80% of 60 is 48. These add up to 68. The numbers check.

Thus, 40 mL of solution A and 60 mL of solution B should be used.

25. We can arrange the information in a table. We let x represent the number of gallons of Milk A and y represent the number of gallons of Milk B.

Type of milk	Amt. of milk	Percent of butterfat	Amt. of butterfat in milk
Milk A	x	1%	0.01x
Milk B	y	5%	0.05y
Mixture	400 gallons	2%	0.02 × 400, or 8 gallons

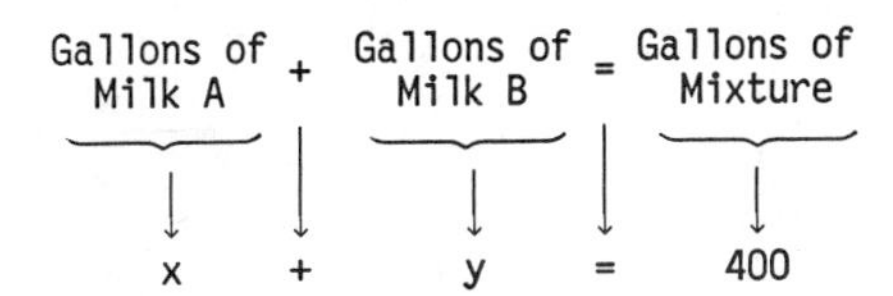

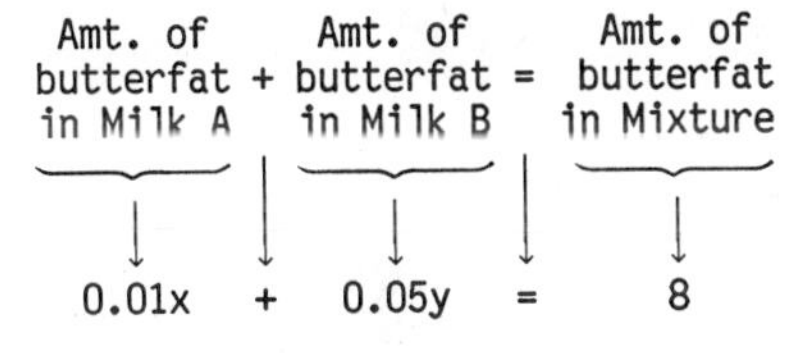

The resulting system is

$x + y = 400$ or $x + y = 400$

$0.01x + 0.05y = 8$ $\quad x + 5y = 800$

We multiply the first equation by -1 and then add.

$-x - y = -400$ (Multiplying by -1)

$x + 5y = 800$

$4y = 400$ (Adding)

$y = 100$

We substitute 100 for y in the first equation and solve for x.

$x + y = 400$

$x + 100 = 400$ (Substituting)

$x = 300$

We consider 300 gallons of Milk A and 100 gallons of Milk B. The sum is 400 gallons. The amount of butterfat in Milk A is 0.01×300, or 3 gallons. The amount of butterfat in Milk B is 0.05×100, or 5 gallons. The total amount of butterfat in the mixture is 8 gallons. These values check.

Thus, 300 gallons of Milk A and 100 gallons of Milk B are needed.

27. We can arrange the information in a table. We let x represent the amount of 30% solution and y represent the amount of 50% solution.

Type of insecticide	Amount of solution	Percent of insec- ticide	Amount of insecticide in the solution
30% solution	x	30%	30%x
50% solution	y	50%	50%x
Mixture	200 liters	42%	42% × 200 or 84 liters

Since the total is 200 liters, we have

$x + y = 200.$

The amount of insecticide in the mixture is to be 42% of 200, or 84 liters. The amounts of insecticide from the two solutions are 30%x and 50%y. Thus

$30\%x + 50\%y = 84$

or $0.3x + 0.5y = 84$

or $3x + 5y = 840$

We use the addition method.

$x + y = 200$

$3x + 5y = 840$

We mutliply the first equation by -3 and then add.

$-3x - 3y = -600$ (Multiplying by -3)

$3x + 5y = 840$

$2y = 240$

$y = 120$

Next we substitute 120 for y in one of the original equations and solve for x.

$x + y = 200$

$x + 120 = 200$ (Substituting)

$x = 80$

We consider $x = 80$ and $y = 120$. The sum is 200. Now 30% of 80 is 24 and 50% of 120 is 60. These add up to 84. The numbers check.

Thus, 80 L of the 30% solution and 120 L of the 50% solution should be used.

29. We organize the information in a table.

Nuts	Price per kilogram	Amount (kg)
Cashews	\$8.00	x
Pecans	\$9.00	y
Mixture	\$8.40	10 kg

We use x for the number of kilograms of cashews and y for the number of kilograms of pecans.

Since the total number of kilograms is 10, we have

$x + y = 10.$

The value of the cashews is 8x (x pounds at \$8 per pound). The value of pecans is 9x (x pounds at \$9 per pound). The value of the mixture is 8.40×10, or \$84. Thus we have

$8x + 9y = 84.$

We use the addition method.

$x + y = 10$

$8x + 9y = 84$

We multiply the first equation by -8 and then add.

$-8x - 8y = -80$ (Multiplying by -8)

$\underline{8x + 9y = \ 84}$

$y = 4$

Next we substitute 4 for y in one of the original equations and solve for x.

$x + y = 10$

$x + 4 = 10$ (Substituting)

$x = 6$

We consider x = 6 kg and y = 4 kg. The sum is 10 kg. The value of the mixture of nuts is \$8·6 + \$9·4, or \$48 + \$36, or \$84. These values check.

The mixture consists of 6 kg of cashews and 4 kg of pecans.

31. $y = -\frac{1}{4}x$

x	y
0	0
4	-1
-4	1

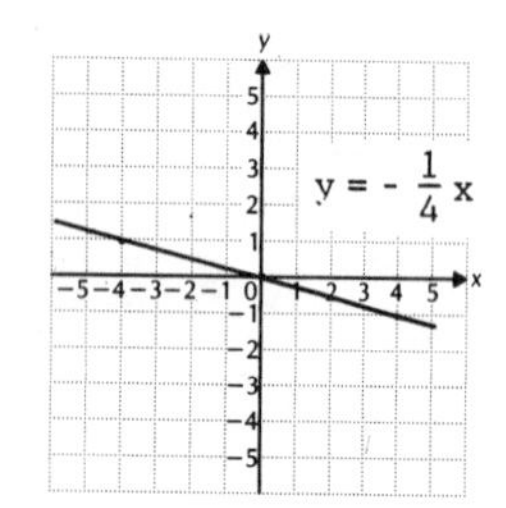

33. $2x - 3y = 6$

x	y	
0	-2	(y-intercept)
3	0	(x-intercept)
-3	-4	

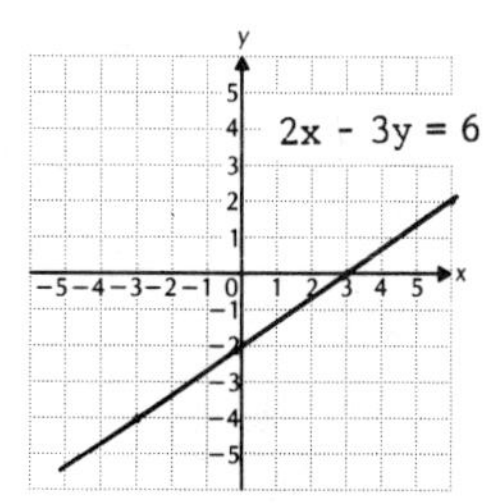

35. It helps to organize the information in a chart. We let x represent the number of liters of the alcohol-water solution added to the original wine.

	Amount	% of Alcohol
Original (5%)	1000	0.05×1000, or 50
Added solution (90%)	x	$0.90 \times x$, or $0.9x$
Mixture (12%)	1000 + x	$0.12 \times (1000 + x)$, or $120 + 0.12x$

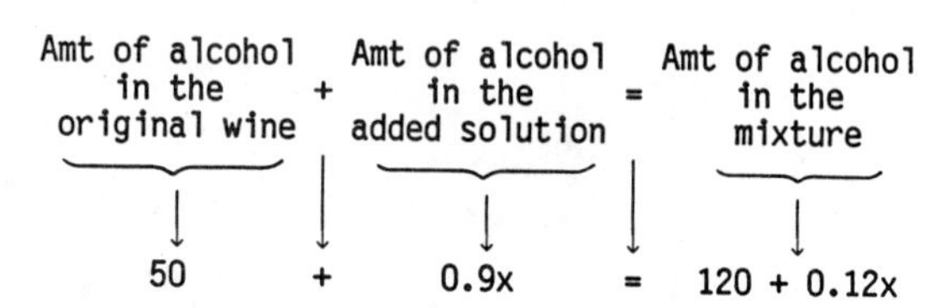

$50 + 0.9x = 120 + 0.12x$

$0.78x = 70$

$x = \frac{70}{0.78}$

$x \approx 89.7$

The amount of alcohol in the 1000 liters is 0.05×1000, or 50 liters. The amount of alcohol in the added solution is 0.9×89.7, or ≈ 80.7 liters. The amount of alcohol in the new mixture is $0.12 \times (1000 + 89.7)$, or ≈ 130.8 liters. The value checks, $50 + 80.7 = 130.7$.

Thus, 89.7 liters of the solution of alcohol and water should be added.

37. Let x represent the tens digit and y the units digit in a two-digit number.

Two-digit number: $10x + y$
Sum of digits: $x + y$

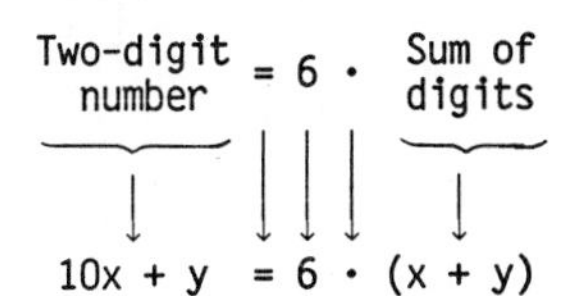

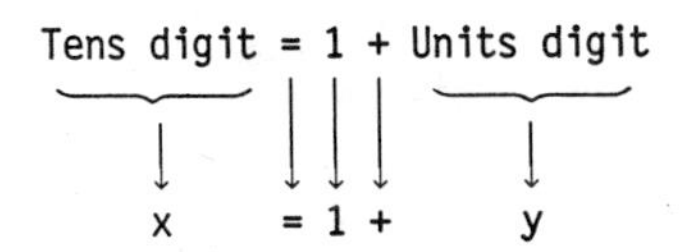

The resulting system is

$10x + y = 6(x + y)$ or $4x - 5y = 0$

$x = 1 + y$ $\qquad$ $x = 1 + y$

Here we use the substitution method. We substitute $1 + y$ for x in the first equation and solve for y.

$$4x - 5y = 0$$
$$4(1 + y) - 5y = 0$$
$$4 + 4y - 5y = 0$$
$$-y = -4$$
$$y = 4$$

Substitute 4 for y in $x = 1 + y$ and solve for x.

$$x = 1 + y$$
$$x = 1 + 4$$
$$x = 5$$

If the tens digit is 5 and the units digit is 4, the number is 54. The sum of the digits is 5 + 4, or 9. The number, 54, is six times the sum of the digits, 9. The tens digit, 5, is one more than the units digit, 4. The values check.

The number is 54.

39. Let x represent Tweedledee's weight and y represent Tweedledum's weight.

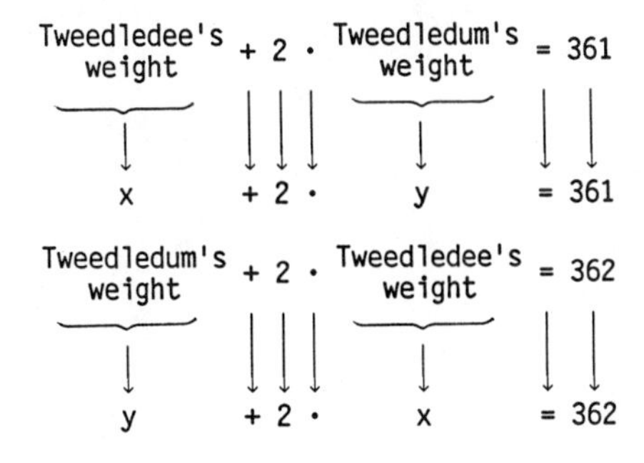

39. (continued)

The resulting system is

$x + 2y = 361$ or $(x = 361 - 2y)$

$y + 2x = 362$

Substitute $361 - 2y$ for x in the second equation and solve for y.

$$y + 2x = 362$$
$$y + 2(361 - 2y) = 362$$
$$y + 722 - 4y = 362$$
$$-3y = -360$$
$$y = 120$$

Next, substitute 120 for y in $x + 2y = 361$ and solve for x.

$$x + 2y = 361$$
$$x + 2 \cdot 120 = 361 \quad \text{(Substituting)}$$
$$x + 240 = 361$$
$$x = 121$$

We check x = 121 lb and y = 120 lb.

$121 + 2 \cdot 120 = 361$

$120 + 2 \cdot 121 = 362$

Tweedledee weighs 121 pounds, and Tweedledum weighs 120 pounds.

Exercise Set 7.6

1. First make a drawing.

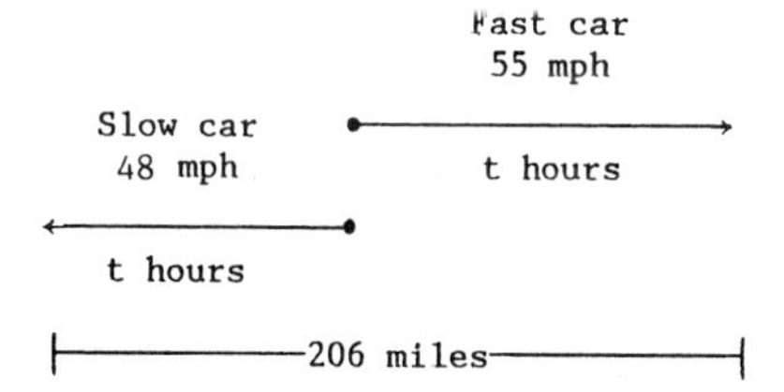

The sum of the distances is 206 miles. The times the cars travel are the same. We organize the information in a chart.

Car	Distance	Speed	Time
Slow car	Slow car distance	48	t
Fast car	Fast car distance	55	t
Total	206		

1. (continued)

From the drawing we see that

Slow car distance + Fast car distance = 206

Then using $d = rt$ in each row of the chart, we get

$48t + 55t = 206$

We solve this equation for t.

$$48t + 55t = 206$$
$$103t = 206$$
$$t = 2$$

If the time is 2 hr, then the distance the slow car travels is $48 \cdot 2$, or 96 mi. The fast car travels $55 \cdot 2$, or 110 mi. Since the sum of the distances, 96 + 110, is 206 mi, the problem checks

In 2 hours, the cars will be 206 miles apart.

3. First make a drawing.

Town — 30 mph
Slow car — t hours — d miles
Town — 46 mph
Fast car — t hours — (d + 72) miles

We see that the distances are not the same, but the times the cars travel are the same. We organize the information in a chart.

Car	Distance	Speed	Time
Slow car	d	30	t
Fast car	d + 72	46	t

Using $d = rt$ in each row of the chart, we get the following system of equations:

$d = 30t$

$d + 72 = 46t$

Substitute 30t for d in the second equation and solve for t.

$$d + 72 = 46t$$
$$30t + 72 = 46t \quad \text{(Substituting)}$$
$$72 = 16t$$
$$\frac{72}{16} = t$$
$$t = \frac{9}{2}, \text{ or } 4.5$$

In 4.5 hours the slow car will travel 30(4.5), or 135 mi. In 4.5 hours the fast car will travel 46(4.5), or 207 miles. The difference between 207 and 135 is 72.

In 4.5 hours the cars will be 72 miles apart.

5. First make a drawing.

Station — 72 km/h
Slow train — t + 3 hours — d kilometers
Station — 120 km/h
Fast train — t hours — d kilometers
Trains meet here

From the drawing we see that the distances are the same. Let's call the distance d. Let t represent the time for the faster train and t + 3 represent the time for the slower train. We organize the information in a chart.

Train	Distance	Speed	Time
Slow train	d	72	t + 3
Fast train	d	120	t

Using $d = rt$ in each row of the chart, we get the following system of equations:

$d = 72(t + 3)$

$d = 120t$

Substitute 120t for d in the first equation and solve for t.

$$d = 72(t + 3)$$
$$120t = 72(t + 3) \quad \text{(Substituting)}$$
$$120t = 72t + 216$$
$$48t = 216$$
$$t = \frac{216}{48}$$
$$t = 4.5$$

When t = 4.5 hours, the faster train will travel 120(4.5), or 540 km, and the slower train will travel 72(7.5), or 540 km. In both cases we get the distance 540 km.

In 4.5 hours after the second train leaves, the second train will overtake the first train. We can also state the answer as 7.5 hours after the first train leaves.

7. We first make a drawing.

With the current r + 6
4 hours d kilometers

Against the current r - 6
10 hours d kilometers

From the drawing we see that the distances are the same. Let d represent the distance. Let r represent the speed of the canoe in still water. Then, when the canoe is traveling with the current, its speed is r + 6. When it is traveling against the current, its speed is r - 6. We organize the information in a chart.

	Distance	Speed	Time
With current	d	r + 6	4
Against current	d	r - 6	10

Using d = rt in each row of the chart, we get the following system of equations:

$d = (r + 6)4$

$d = (r - 6)10$

Substitute (r + 6)4 for d in the second equation and solve for r.

$d = (r - 6)10$

$(r + 6)4 = (r - 6)10$ (Substituting)

$4r + 24 = 10r - 60$

$84 = 6r$

$14 = r$

When r = 14, r + 6 = 20 and 20·4 = 80, the distance. When r = 14, r - 6 = 8 and 8·10 = 80. In both cases, we get the same distance.

The speed of the canoe in still water is 14 km/h.

9. First make a drawing.

Passenger 96 km/h
t - 2 hours d kilometers

Freight 64 km/h
t hours d kilometers

Central City Clear Creek

From the drawing we see that the distances are the same. Let d represent the distance. Let t represent the time for the slower train (freight). Then the time for the faster train (passenger) is t - 2. We organize the information in a chart.

Train	Distance	Speed	Time
Passenger	d	96	t - 2
Freight	d	64	t

From each row of the chart we get an equation, d = rt.

$d = 96(t - 2)$

$d = 64t$

Substitute 64t for d in the first equation and solve for t.

$d = 96(t - 2)$

$64t = 96(t - 2)$ (Substituting)

$64t = 96t - 192$

$192 = 32t$

$6 = t$

Next we substitute 6 for t in one of the original equations and solve for d.

$d = 64t$

$d = 64 \cdot 6$ (Substituting)

$d = 384$

If the time is 6 hr, then the distance the passenger train travels is 96(6 - 2), or 384 km. The freight train travels 64(6), or 384 km. The distances are the same.

It is 384 km from Central City to Clear Creek.

11. We first make a drawing.

Against the wind r - w

2 hours 600 kilometers

With the wind r + w

$1\frac{2}{3}$ hours 600 kilometers

We let r represent the speed of the airplane in still air and w represent the speed of the wind. Then when flying against a head wind, the rate is r - w, and when flying with the wind, the rate is r + w. We organize the information in a chart.

	Distance	Speed	Time
Against	600	r - w	2
With	600	r + w	$1\frac{2}{3}$, or $\frac{5}{3}$

Using d = rt in each row of the chart, we get the following system of equations:

$600 = (r - w)2$ $\quad$ $300 = r - w$

or

$600 = (r + w)\frac{5}{3}$ $\quad$ $360 = r + w$

We use the addition method with the resulting system.

$300 = r - w$

$\underline{360 = r + w}$

$660 = 2r$

$330 = r$

Next we substitute 330 for r and solve for w.

$360 = r + w$

$360 = 330 + w$ (Substituting)

$30 = w$

If r = 330 and w = 30, then r - w = 300, and r + w = 360. If the plane flies 2 hours against the wind, it travels 300·2 or 600 km. If the plane flies $1\frac{2}{3}$ hours with the wind, it travels $360\cdot\frac{5}{3}$, or 600 km. All values check.

The speed of the plane in still air is 330 km/h.

13. First make a drawing.

Home t hr 45 mph | (2-t) hr 6 mph Work

Motorcycle distance | Walking distance

←25 miles→

Let t represent the time the motorcycle was driven, then 2 - t represents the time the rider walked. We organize the information in a chart.

	Distance	Speed	Time
Motorcycling	Motorcycle distance	45	t
Walking	Walking distance	6	2 - t

From the drawing we see that

Motorcycle distance + Walking distance = 25

Then using d = rt in each row of the chart we get

$45t + 6(2 - t) = 25$

We solve this equation for t.

$45t + 12 - 6t = 25$

$39t + 12 = 25$

$39t = 13$

$t = \frac{13}{39}$

$t = \frac{1}{3}$

The problem asks us to find how far the motorcycle went before it broke down. If $t = \frac{1}{3}$, then 45t (the distance the motorcycle traveled) = $45\cdot\frac{1}{3}$, or 15 and 6(2 - t) (the distance walked) = $6(2 - \frac{1}{3}) = 6\cdot\frac{5}{3}$, or 10. The total of these distances is 25, so $\frac{1}{3}$ checks.

The motorcycle went 15 miles before it broke down.

15. We first make a drawing.

With the wind $r + 25.5$

4.23 hours d kilometers

Against the wind $r - 25.5$

4.97 hours d kilometers

We let r represent the speed of the airplane in still air. When flying with a tail wind, the rate is $r + 25.5$. When flying against a head wind, the rate is $r - 25.5$. The distance, d, is the same for both directions. We organize the information in a chart.

	Distance	Speed	Time
With	d	r + 25.5	4.23
Against	d	r - 25.5	4.97

Using $d = rt$ in each row of the chart, we get the following system of equations:

$d = (r + 25.5)4.23$

$d = (r - 25.5)4.97$

Substitute $(r + 25.5)4.23$ for d in the second equation and solve for r.

$$d = (r - 25.5)4.97$$
$$(r + 25.5)4.23 = (r - 25.5)4.97$$
$$4.23r + 107.865 = 4.97r - 126.735$$
$$234.6 = 0.74r$$
$$317.03 \approx r$$

When $r = 317.03$, $r + 25.5 = 342.53$ and $342.53(4.23) \approx 1448.9$, the distance. When $r = 317.03$, $r - 25.5 = 291.53$ and $291.53(4.97) \approx 1448.9$. In both cases, we get the same distance.

The speed of the plane in still air is approximately 317.03 km/h.

17. First we make a drawing.

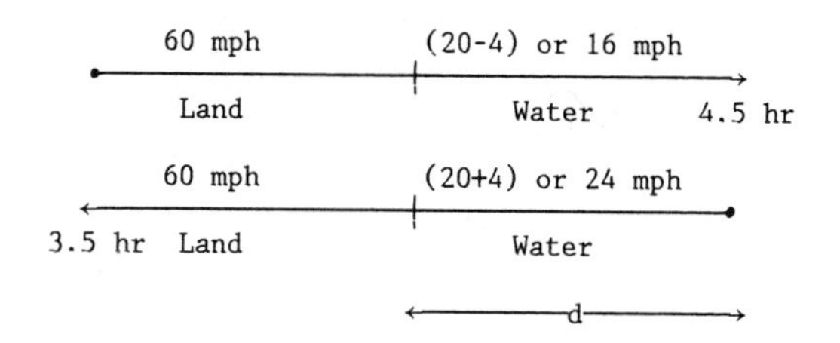

Let t = time spent on land. Then $4.5 - t$ represents the time spent in water before delivering the package and $3.5 - t$ represents the time spent in water after delivering the package. Let d = distance traveled in water. In a chart we organize the information concerning the part of the trip in water.

	Distance	Speed	Time
Before delivery	d	20 - 4, or 16	4.5 - t
After delivery	d	20 + 4, or 24	3.5 - t

Using $d = rt$ in each row of the chart, we get the following system of equations:

$d = 16(4.5 - t)$

$d = 24(3.5 - t)$

Substitute $16(4.5 - t)$ for d in the second equation and solve for t.

$$d = 24(3.5 - t)$$
$$16(4.5 - t) = 24(3.5 - t) \quad \text{(Substituting)}$$
$$72 - 16t = 84 - 24t$$
$$8t = 12$$
$$t = \frac{12}{8}$$
$$t = 1.5$$

Substitute 1.5 for t and solve for d.

$d = 16(4.5 - t)$

$d = 16(4.5 - 1.5)$

$d = 16(3)$

$d = 48$

If the time spent on land is 1.5 hours, then the distance traveled on land is $d = 60(1.5)$, or 90 miles. If $t = 1.5$, then $4.5 - t$ is 3 and the distance in water before delivery is 16(3), or 48, and the distance in water after delivery is 24(2), or 48. The distances are the same. The values check.

For the round trip, the messenger traveled 180 mi by land and 96 mi by water.

Exercise Set 8.1

1. $x > -4$

   a) Since $4 > -4$ is true, 4 is a solution.
   b) Since $0 > -4$ is true, 0 is a solution.
   c) Since $-4 > -4$ is false, -4 is not a solution.
   d) Since $6 > -4$ is true, 6 is a solution.

3. $x \geqslant 8$

   a) Since $-6 \geqslant 8$ is false, -6 is not a solution.
   b) Since $0 \geqslant 8$ is false, 0 is not a solution.
   c) Since $60 \geqslant 8$ is true, 60 is a solution.
   d) Since $8 \geqslant 8$ is true, 8 is a solution.

5. $t < -8$

   a) Since $0 < -8$ is false, 0 is not a solution.
   b) Since $-8 < -8$ is false, -8 is not a solution.
   c) Since $-9 < -8$ is true, -9 is a solution.
   d) Since $-7 < -8$ is false, -7 is not a solution.

7.
$$x + 7 > 2$$
$$x + 7 - 7 > 2 - 7$$
$$x > -5$$

The solution set is $\{x \mid x > -5\}$.

9.
$$y + 5 > 8$$
$$y + 5 - 5 > 8 - 5$$
$$y > 3$$

The solution set is $\{y \mid y > 3\}$.

11.
$$x + 8 \leqslant -10$$
$$x + 8 - 8 \leqslant -10 - 8$$
$$x \leqslant -18$$

The solution set is $\{x \mid x \leqslant -18\}$.

13.
$$x - 7 \leqslant 9$$
$$x - 7 + 7 \leqslant 9 + 7$$
$$x \leqslant 16$$

The solution set is $\{x \mid x \leqslant 16\}$.

15.
$$y - 7 > -12$$
$$y - 7 + 7 > -12 + 7$$
$$y > -5$$

The solution set is $\{y \mid y > -5\}$.

17.
$$2x + 4 > x + 7$$
$$2x + 4 - 4 > x + 7 - 4$$
$$2x > x + 3$$
$$2x - x > x + 3 - x$$
$$x > 3$$

The solution set is $\{x \mid x > 3\}$.

19.
$$3x - 6 \geqslant 2x + 7$$
$$3x - 6 + 6 \geqslant 2x + 7 + 6$$
$$3x \geqslant 2x + 13$$
$$3x - 2x \geqslant 2x + 13 - 2x$$
$$x \geqslant 13$$

The solution set is $\{x \mid x \geqslant 13\}$.

21.
$$5x - 6 < 4x - 2$$
$$5x - 6 + 6 < 4x - 2 + 6$$
$$5x < 4x + 4$$
$$5x - 4x < 4x + 4 - 4x$$
$$x < 4$$

The solution set is $\{x \mid x < 4\}$.

23.
$$-7 + c > 7$$
$$7 - 7 + c > 7 + 7$$
$$c > 14$$

The solution set is $\{c \mid c > 14\}$.

25.
$$y + \frac{1}{4} \leqslant \frac{1}{2}$$
$$y + \frac{1}{4} - \frac{1}{4} \leqslant \frac{1}{2} - \frac{1}{4}$$
$$y \leqslant \frac{2}{4} - \frac{1}{4}$$
$$y \leqslant \frac{1}{4}$$

The solution set is $\{y \mid y \leqslant \frac{1}{4}\}$.

27.
$$x - \frac{1}{3} > \frac{1}{4}$$
$$x - \frac{1}{3} + \frac{1}{3} > \frac{1}{4} + \frac{1}{3}$$
$$x > \frac{3}{12} + \frac{4}{12}$$
$$x > \frac{7}{12}$$

The solution set is $\{x \mid x > \frac{7}{12}\}$.

29. $-14x + 21 > 21 - 15x$

$-14x + 21 + 15x > 21 - 15x + 15x$

$x + 21 > 21$

$x + 21 - 21 > 21 - 21$

$x > 0$

The solution set is $\{x | x > 0\}$.

31. $3(r + 2) < 2r + 4$

$3r + 6 < 2r + 4$

$3r + 6 - 6 < 2r + 4 - 6$

$3r < 2r - 2$

$3r - 2r < 2r - 2 - 2r$

$r < -2$

The solution set is $\{r | r < -2\}$.

33. $0.8x + 5 \geqslant 6 - 0.2x$

$0.8x + 5 - 5 \geqslant 6 - 0.2x - 5$

$0.8x \geqslant 1 - 0.2x$

$0.8x + 0.2x \geqslant 1 - 0.2x + 0.2x$

$x \geqslant 1$

The solution set is $\{x | x \geqslant 1\}$.

35. $\frac{5}{3} = 1\frac{2}{3}$

The point is located two-thirds of the way from 1 to 2.

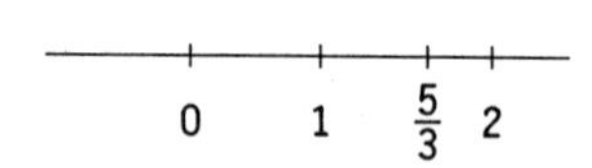

37. $-2y - 3 = 11$

$-2y = 14$

$y = -7$

39. $17x + 9{,}479{,}756 \leqslant 16x - 8{,}579{,}243$

$17x - 16x \leqslant -8{,}579{,}243 - 9{,}479{,}756$

$x \leqslant -18{,}058{,}999$

The solution set is $\{x | x \leqslant -18{,}058{,}999\}$.

41. If $a > b$ and $b > c$, then $a > c$.

Since $2x - 5 \geqslant 9$ and $9 \geqslant 8$, then $2x - 5 \geqslant 8$.

Exercise Set 8.2

1. $5x < 35$

$\frac{1}{5}\cdot 5x < \frac{1}{5}\cdot 35$

$x < 7$

The solution set is $\{x | x < 7\}$.

3. $9y \leqslant 81$

$\frac{1}{9}\cdot 9y \leqslant \frac{1}{9}\cdot 81$

$y \leqslant 9$

The solution set is $\{y | y \leqslant 9\}$.

5. $7x < 13$

$\frac{1}{7}\cdot 7x < \frac{1}{7}\cdot 13$

$x < \frac{13}{7}$

The solution set is $\{x | x < \frac{13}{7}\}$.

7. $12x > -36$

$\frac{1}{12}\cdot 12x > \frac{1}{12}\cdot(-36)$

$x > -3$

The solution set is $\{x | x > -3\}$.

9. $5y \geqslant -2$

$\frac{1}{5}\cdot 5y \geqslant \frac{1}{5}\cdot(-2)$

$y \geqslant -\frac{2}{5}$

The solution set is $\{y | y \geqslant -\frac{2}{5}\}$.

11. $-2x \leqslant 12$

$-\frac{1}{2}\cdot(-2x) \geqslant -\frac{1}{2}\cdot 12$

$x \geqslant -6$

The solution set is $\{x | x \geqslant -6\}$.

13. $-4y \geqslant -16$

$-\frac{1}{4}\cdot(-4y) \leqslant -\frac{1}{4}\cdot(-16)$

$y \leqslant 4$

The solution set is $\{y | y \leqslant 4\}$.

15. $-3x < -17$

$-\frac{1}{3}\cdot(-3x) > -\frac{1}{3}\cdot(-17)$

$x > \frac{17}{3}$

The solution set is $\{x | x > \frac{17}{3}\}$.

17. $$-2y > \frac{1}{7}$$
$$-\frac{1}{2}\cdot(-2y) < -\frac{1}{2}\cdot\frac{1}{7}$$
$$y < -\frac{1}{14}$$

The solution set is $\{y \mid y < -\frac{1}{14}\}$.

19. $$-\frac{6}{5} \leqslant -4x$$
$$-\frac{1}{4}\cdot(-\frac{6}{5}) \geqslant -\frac{1}{4}\cdot(-4x)$$
$$\frac{6}{20} \geqslant x$$
$$\frac{3}{10} \geqslant x, \text{ or } x \leqslant \frac{3}{10}$$

The solution set is $\{x \mid x \leqslant \frac{3}{10}\}$.

21. $$2x + 5 < 3$$
$$2x + 5 - 5 < 3 - 5$$
$$2x < -2$$
$$\frac{1}{2}\cdot 2x < \frac{1}{2}\cdot(-2)$$
$$x < -1$$

The solution set is $\{x \mid x < -1\}$.

23. $$-3x + 7 \geqslant -2$$
$$-3x + 7 - 7 \geqslant -2 - 7$$
$$-3x \geqslant -9$$
$$-\frac{1}{3}\cdot(-3x) \leqslant -\frac{1}{3}\cdot(-9)$$
$$x \leqslant 3$$

The solution set is $\{x \mid x \leqslant 3\}$.

25. $$6t - 8 \leqslant 4t + 1$$
$$6t - 8 + 8 \leqslant 4t + 1 + 8$$
$$6t \leqslant 4t + 9$$
$$-4t + 6t \leqslant -4t + 4t + 9$$
$$2t \leqslant 9$$
$$\frac{1}{2}\cdot 2t \leqslant \frac{1}{2}\cdot 9$$
$$t \leqslant \frac{9}{2}$$

The solution set is $\{t \mid t \leqslant \frac{9}{2}\}$.

27. $$3 - 6c > 15$$
$$-3 + 3 - 6c > -3 + 15$$
$$-6c > 12$$
$$-\frac{1}{6}\cdot(-6c) < -\frac{1}{6}\cdot 12$$
$$c < -2$$

The solution set is $\{c \mid c < -2\}$.

29. $$15x - 21 \geqslant 8x + 7$$
$$15x - 21 + 21 \geqslant 8x + 7 + 21$$
$$15x \geqslant 8x + 28$$
$$-8x + 15x \geqslant -8x + 8x + 28$$
$$7x \geqslant 28$$
$$\frac{1}{7}\cdot 7x \geqslant \frac{1}{7}\cdot 28$$
$$x \geqslant 4$$

The solution set is $\{x \mid x \geqslant 4\}$.

31. When $x = -17$, $-x = -(-17)$, or 17.

33. $$x - y = 7$$
$$x + y = 9$$
$$2x + 0 = 16 \quad \text{(Adding)}$$
$$2x = 16$$
$$x = 8$$

Substitute 8 for x in one of the original equations and solve for y.

$$x + y = 9$$
$$8 + y = 9 \quad \text{(Substituting)}$$
$$y = 1$$

The ordered pair (8,1) checks and is the solution.

35. $$5(12 - 3t) \geqslant 15(t + 4)$$
$$60 - 15t \geqslant 15t + 60$$
$$60 - 60 \geqslant 15t + 15t$$
$$0 \geqslant 30t$$
$$\frac{0}{30} \geqslant t$$
$$0 \geqslant t, \text{ or } t \leqslant 0$$

The solution set is $\{t \mid t \leqslant 0\}$.

37. $x^2 > 0$

The square of any nonzero real number is positive. All positive numbers are greater than 0. Thus the solution is the set of all nonzero real numbers.

39. Let m represent the mileage. Then $53.95 + 0.40m$ represents the cost of renting a car from Badger and $52.95 + 0.42m$ represents the cost of renting a car from Beaver.

We solve the following inequality:

Cost at Badger < Cost at Beaver

$$53.95 + 0.40m < 52.95 + 0.42m$$
$$53.95 - 52.95 < 0.42m - 0.40m$$
$$1 < 0.02m$$
$$\frac{1}{0.02}\cdot 1 < \frac{1}{0.02}\cdot 0.02m$$
$$50 < m$$

Thus, the cost at Badger is cheaper when the mileage is greater than 50 miles.

Exercise Set 8.3

1. $x < 5$

We shade all points to the left of 5. The open circle at 5 indicates that 5 is not part of the graph.

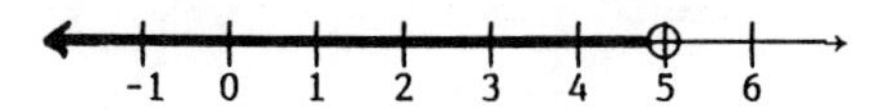

3. $y \geq -4$

We shade the point -4 and all points to the right of -4. The closed circle at -4 indicates that -4 is part of the graph.

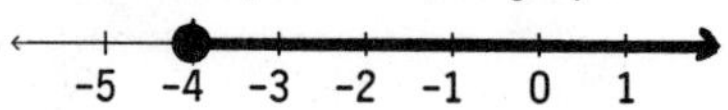

5. $t - 3 \leq -7$

$t \leq -4$

We shade the point for -4 and all points to the left of -4. The closed circle at -4 indicates that -4 is part of the graph.

7. $2x + 6 < 14$

$2x < 8$

$x < 4$

We shade all points to the left of 4. The open circle at 4 indicates that 4 is not part of the graph.

9. $4y + 9 > 11y - 12$

$9 + 12 > 11y - 4y$

$21 > 7y$

$\frac{1}{7}\cdot 21 > \frac{1}{7}\cdot 7y$

$3 > y$, or $y < 3$

We shade all points to the left of 3. The open circle at 3 indicates that 3 is not part of the graph.

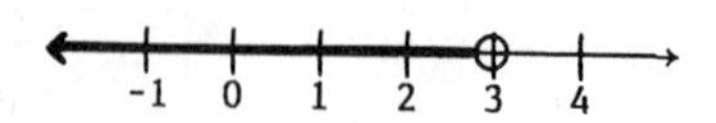

11. $|x| < 2$

The absolute value of a number is its distance from 0 on a number line. For the absolute value of a number to be less than 2 it must be between -2 and 2. Therefore, we use open circles at -2 and 2 and shade the points between these two numbers.

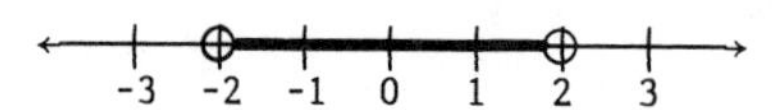

13. $|a| \geq 4$

For the absolute value of a number to be greater than or equal to 4 its distance from 0 must be 4 or more. Thus the number must be less than or equal to -4, or it must be 4 or greater. We shade the point for -4 and all points to its left. We also shade the point for 4 and all points to its right. The closed circles at -4 and 4 indicate that they are part of the graph.

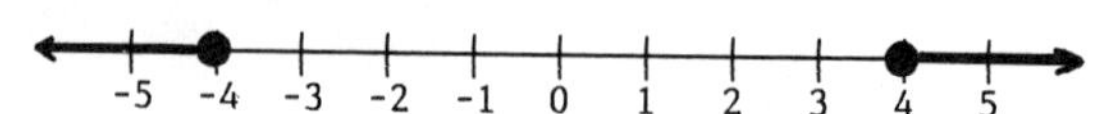

15. Graph $y > x - 2$.

First graph the line $y = x - 2$.

We make a table of values.

x	y	
0	-2	When $x = 0$, $y = 0 - 2 = -2$.
3	1	When $x = 3$, $y = 3 - 2 = 1$.
-2	-4	When $x = -2$, $y = -2 - 2 = -4$.

We plot these points and draw the line they determine. We use a dashed line for the graph since we have >.

Then pick a point which does not belong to the line. The origin, (0,0), is an easy one to use. (If the line goes through the origin, then we test some other point.) Substitute (0,0) into $y > x - 2$ to see if it is a solution.

$y > x - 2$

$0 > 0 - 2$ (Substituting)

$0 > -2$

$0 > -2$ is true, so the origin is a solution. This means we shade the upper half-plane.

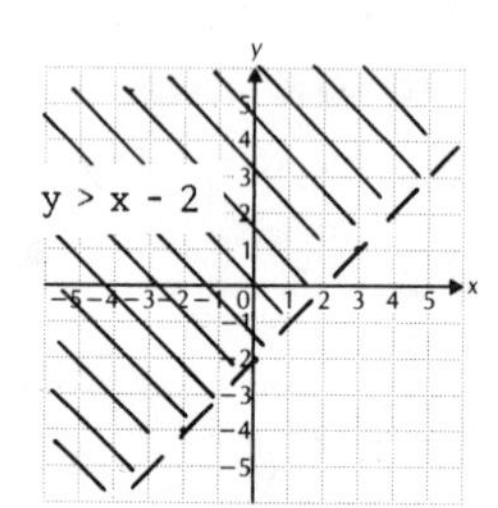

17. Graph $6x - 2y \leqslant 12$.

First graph the line $6x - 2y = 12$. The intercepts are (2,0) and (0,-6). We use a solid line since we have $\leqslant$.

Pick a point which does not belong to the line. The origin is easy to use. Substitute (0,0) into $6x - 2y \leqslant 12$ to see if it is a solution.

$$6x - 2y \leqslant 12$$
$$6\cdot 0 - 2\cdot 0 \leqslant 12 \quad \text{(Substituting)}$$
$$0 - 0 \leqslant 12$$
$$0 \leqslant 12$$

$0 \leqslant 12$ is true, so the origin is a solution. This means we shade the upper half-plane.

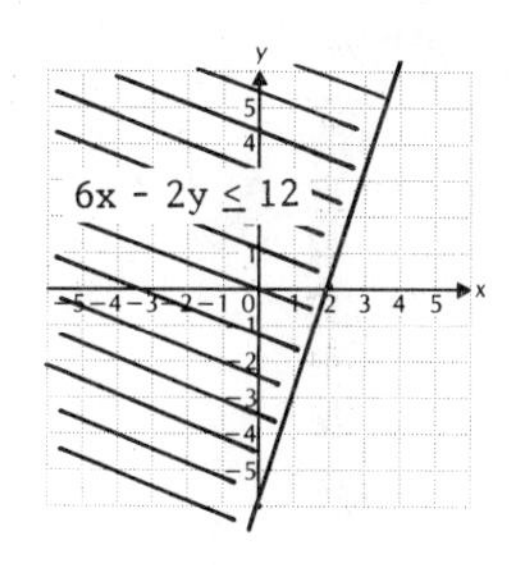

19. Graph $3x - 5y \geqslant 15$.

First graph the line $3x - 5y = 15$. The intercepts are (5,0) and (0, -3). We use a solid line since we have $\geqslant$.

Pick a point which does not belong to the line. The origin is easy to use. Substitute (0,0) into $3x - 5y \geqslant 15$ to see if it is a solution.

$$3x - 5y \geqslant 15$$
$$3\cdot 0 - 5\cdot 0 \geqslant 15 \quad \text{(Substituting)}$$
$$0 - 0 \geqslant 15$$
$$0 \geqslant 15$$

$0 \geqslant 15$ is false, so the origin is not a solution. This means we shade the lower half-plane.

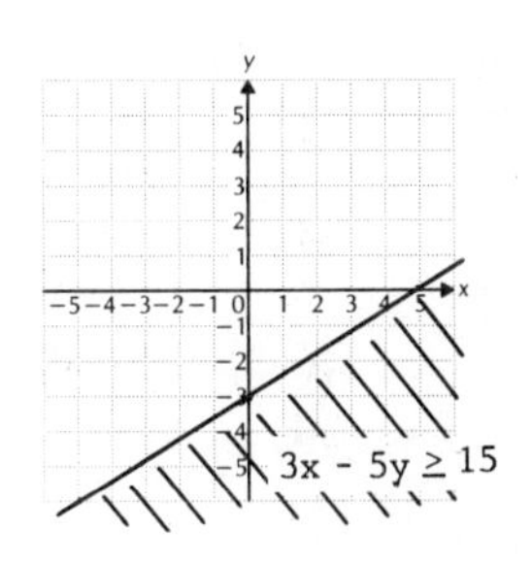

21. Graph $y - 2x < 4$.

First graph the line $y - 2x = 4$. The intercepts are (-2,0) and (0,4). We use a dashed line since we have <.

Pick a point which does not belong to the line. The origin is easy to use. Substitute (0,0) into $y - 2x < 4$ to see if it is a solution.

$$y - 2x < 4$$
$$0 - 2\cdot 0 < 4 \quad \text{(Substituting)}$$
$$0 - 0 < 4$$
$$0 < 4$$

$0 < 4$ is true, so the origin is a solution. This means we shade the lower half-plane.

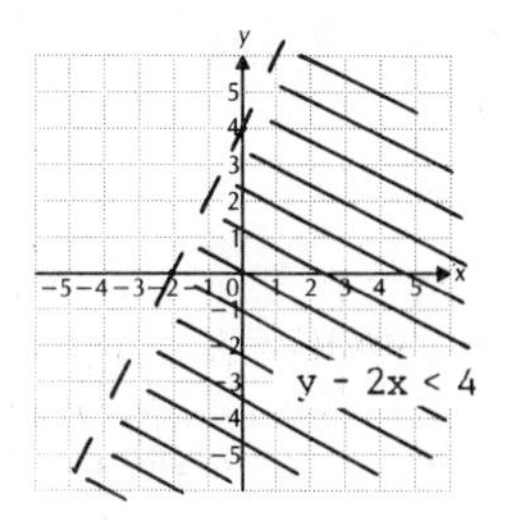

23.

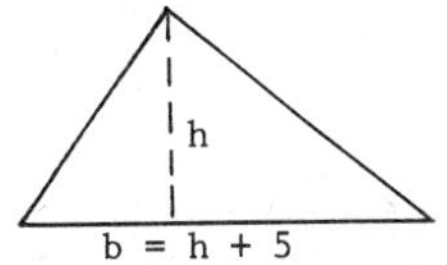

We let h represent the height and b the base.

The base is 5 m greater than the height.

$b = h + 5$

The area ($A = \frac{1}{2}bh$) is 7 m^2.

$\frac{1}{2}bh = 7$

We have the resulting system.

$b = h + 5$

$\frac{1}{2}bh = 7$

Substitute $h + 5$ for b in the second equation and solve for h.

$$\frac{1}{2}bh = 7$$
$$\frac{1}{2}(h + 5)h = 7 \quad \text{(Substituting)}$$
$$(h + 5)h = 14$$
$$h^2 + 5h = 14$$
$$h^2 + 5h - 14 = 0$$
$$(h + 7)(h - 2) = 0$$

$h + 7 = 0$ or $h - 2 = 0$

$h = -7$ or $h = 2$

23. (continued)

Since the height cannot be negative, we only check h = 2. If h = 2 m, then b = 2 + 5, or 7 m. The area is $\frac{1}{2}\cdot 7\cdot 2$, or 7 m^2. The values check.

The base is 7 m, and the height is 2 m.

25. Graph $x \geqslant 3$. Think of this as $x + 0\cdot y \geqslant 3$.

First graph the line x = 3. This is a vertical line with (3,0) as the x-intercept. We use a solid line since we have $\geqslant$.

Pick a point which does not belong to the line. The origin is easy to use. Substitute (0,0) into $x + 0\cdot y \geqslant 3$ to see if it is a solution.

$x + 0\cdot y \geqslant 3$

$0 + 0\cdot 0 \geqslant 3$ (Substituting)

$0 \geqslant 3$

$0 \geqslant 3$ is false, so the origin is not a solution. This means we shade the right half-plane.

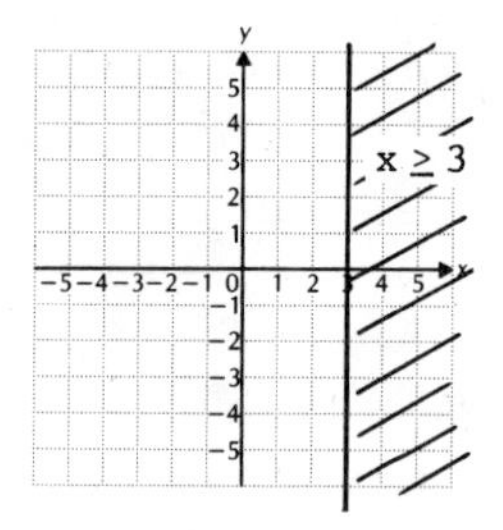

27. Graph $y \leqslant 0$. Think of this as $0\cdot x + y \leqslant 0$.

First graph the line y = 0. This is a horizontal line with (0,0) as the y-intercept (the x-axis). We use a solid line since we have $\leqslant$.

Pick a point which does not belong to the line. Substitute (4,-3) to see if it is a solution.

$0\cdot x + y \leqslant 0$

$0\cdot 4 + (-3) \leqslant 0$ (Substituting)

$-3 \leqslant 0$

$-3 \leqslant 0$ is true, so (4,-3) is a solution. This means we shade the lower half-plane.

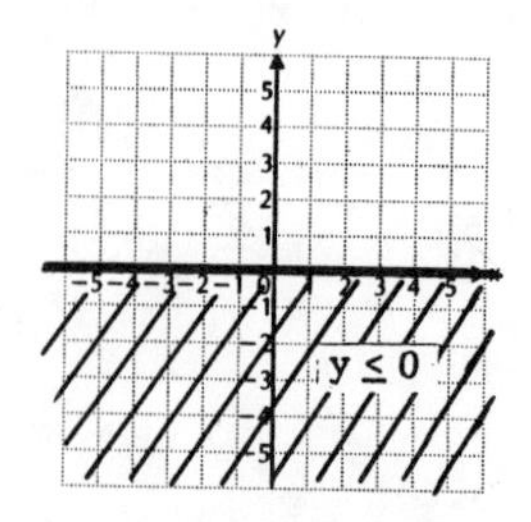

Exercise Set 8.4

1. The word "through" indicates that we are to include both 3 and 8.

{3, 4, 5, 6, 7, 8}

3. The word "between" indicates that we are not to include 40 and 50.

{41, 43, 45, 47, 49}

5. {3, -3}; $3\cdot 3 = 9$, $(-3)(-3) = 9$

7. False, 2 is not a member of the set of odd numbers.

9. True, Bruce Springsteen is a rock star.

11. True, -3 is a member of the set {-4, -3, 0, 1}.

13. {a, b, c, d, e} ∩ {c, d, e, f, g}

represents the set {c, d, e}.

The elements c, d, and e are common to both sets.

15. {1, 2, 5, 10} ∩ {0, 1, 7, 10}

represents the set {1, 10}.

The elements 1 and 10 are common to both sets.

17. {1, 2, 5, 10} ∩ {3, 4, 7, 8} represents the empty set, ∅.

There is no member common to both sets.

19. {a, e, i, o, u} ∪ {q, u, i, c, k}

represents the set {a, e, i, o, u, q, c, k}.

Note that both i and u are each listed only once in the union.

21. {0, 1, 7, 10} ∪ {0, 1, 2, 5}

represents the set {0, 1, 7, 10, 2, 5}.

Note that both 0 and 1 are each listed only once in the union.

23. {a, e, i, o, u} ∪ {m, n, f, g, h}

represents the set {a, e, i, o, u, m, n, f, g, h}.

25. $3x + 5y = 6$

$5x + 3y = 4$

We use the multiplication principle with both equations and then add.

$-9x - 15y = -18$ (Multiplying by -3)

$25x + 15y = 20$ (Multiplying by 5)

$16x = 2$ (Adding)

$x = \frac{2}{16}$

$x = \frac{1}{8}$

Substitute $\frac{1}{8}$ for x in one of the original equations and solve for y.

$5x + 3y = 4$

$5\cdot\frac{1}{8} + 3y = 4$

$3y = \frac{32}{8} - \frac{5}{8}$

$3y = \frac{27}{8}$

$y = \frac{9}{8}$

The ordered pair $(\frac{1}{8},\frac{9}{8})$ checks. It is the solution.

27. When $x = 9$, $-x = -9$ and $-(-x) = -(-9)$, or 9.

29. Original New

x x

x + 0.2 x + 0.2

We let x represent the length of a side of the original square. Then $x + 0.2$ represents the length of a side of the new square. Thus the area of the new square is

$(x + 0.2)(x + 0.2) = 0.64$

We solve this equation for x.

$x^2 + 0.4x + 0.04 = 0.64$

$x^2 + 0.4x - 0.60 = 0$

$(x + 1)(x - 0.6) = 0$

$x + 1 = 0$ or $x - 0.6 = 0$

$x = -1$ or $x = 0.6$

Since the length of a side cannot be negative, we only check $x = 0.6$. If $x = 0.6$, then the length of a side of the new square is $0.6 + 0.2$, or 0.8. The area is $(0.8)^2$, or 0.64 km^2. The value checks.

The length of a side of the original square is 0.6 km.

31. Set of even integers:

{..., -6, -4, -2, 0, 2, 4, 6, ...}

Set of positive rational numbers:

$\{x | x$ is rational and $x > 0\}$.

The only even integers that are also positive rationals are 2, 4, 6, ...

The intersection of the two sets is the set of positive even integers, {2, 4, 6, ...}.

Exercise Set 9.1

1. $\frac{x-2}{x-5} \cdot \frac{x-2}{x+5} = \frac{(x-2)(x-2)}{(x-5)(x+5)}$

3. $\frac{c-3d}{c+d} \cdot \frac{c+3d}{c-d} = \frac{(c-3d)(c+3d)}{(c+d)(c-d)}$

5. $\frac{2a-1}{2a-1} \cdot \frac{3a-1}{3a+2} = \frac{(2a-1)(3a-1)}{(2a-1)(3a+2)}$

7. $\frac{x(3x+2)(x+1)}{x(x+1)(3x-2)}$

$= \frac{x(x+1)}{x(x+1)} \cdot \frac{3x+2}{3x-2}$

$= \frac{3x+2}{3x-2}$

9. $\frac{8a+8b}{8a-8b}$

$= \frac{8(a+b)}{8(a-b)}$

$= \frac{8}{8} \cdot \frac{a+b}{a-b}$

$= \frac{a+b}{a-b}$

11. $\frac{t^2-25}{t^2+t-20}$

$= \frac{(t-5)(t+5)}{(t-4)(t+5)}$

$= \frac{t-5}{t-4} \cdot \frac{t+5}{t+5}$

$= \frac{t-5}{t-4}$

13. $\frac{2x^2+6x+4}{4x^2-12x-16}$

$= \frac{2(x^2+3x+2)}{4(x^2-3x-4)}$

$= \frac{2(x+2)(x+1)}{2\cdot 2(x-4)(x+1)}$

$= \frac{2(x+1)}{2(x+1)} \cdot \frac{x+2}{2(x-4)}$

$= \frac{x+2}{2(x-4)}$

15. $\frac{a^2-10a+21}{a^2-11a+28}$

$= \frac{(a-7)(a-3)}{(a-7)(a-4)}$

$= \frac{a-7}{a-7} \cdot \frac{a-3}{a-4}$

$= \frac{a-3}{a-4}$

17. $\frac{6x+12}{x^2-x-6}$

$= \frac{6(x+2)}{(x-3)(x+2)}$

$= \frac{6}{x-3} \cdot \frac{x+2}{x+2}$

$= \frac{6}{x-3}$

19. $\frac{a^2+1}{a+1}$ cannot be simplified.

Neither the numerator nor the denominator can be factored.

21. $\frac{t^2}{t^2-4} \cdot \frac{t^2-5t+6}{t^2-3t}$

$= \frac{t^2(t^2-5t+6)}{(t^2-4)(t^2-3t)}$

$= \frac{t\cdot t(t-3)(t-2)}{(t-2)(t+2)\cdot t\cdot(t-3)}$

$= \frac{t(t-3)(t-2)}{t(t-3)(t-2)} \cdot \frac{t}{t+2}$

$= \frac{t}{t+2}$

23. $\frac{24a^2}{3(a^2-4a+4)} \cdot \frac{3a-6}{2a}$

$= \frac{24a^2(3a-6)}{3(a^2-4a+4)2a}$

$= \frac{2\cdot 2\cdot 2\cdot 3\cdot a\cdot a\cdot 3(a-2)}{3(a-2)(a-2)2a}$

$= \frac{2\cdot 3\cdot a(a-2)}{2\cdot 3\cdot a(a-2)} \cdot \frac{2\cdot 2\cdot a\cdot 3}{a-2}$

$= \frac{12a}{a-2}$

25. $\frac{ab-b^2}{2a} \cdot \frac{2a+2b}{a^2b-b^3}$

$= \frac{(ab-b^2)(2a+2b)}{2a(a^2b-b^3)}$

$= \frac{b(a-b)2(a+b)}{2ab(a-b)(a+b)}$

$= \frac{2b(a-b)(a+b)}{2b(a-b)(a+b)} \cdot \frac{1}{a}$

$= \frac{1}{a}$

27. $\frac{1}{6} \cdot \frac{3}{8} = \frac{3}{48} = \frac{1\cdot 3}{16\cdot 3} = \frac{1}{16} \cdot \frac{3}{3} = \frac{1}{16}$

29. $\frac{1}{2} + \frac{2}{5}$

$= \frac{1}{2} \cdot \frac{5}{5} + \frac{2}{5} \cdot \frac{2}{2}$

$= \frac{5}{10} + \frac{4}{10}$

$= \frac{5+4}{10}$

$= \frac{9}{10}$

31. $\dfrac{x^4 - 16y^2}{(x^2 + 4y^2)(x - 2y)}$

$= \dfrac{(x^2 + 4y^2)(x + 2y)(x - 2y)}{(x^2 + 4y^2)(x - 2y)}$

$= \dfrac{(x^2 + 4y^2)(x - 2y)}{(x^2 + 4y^2)(x - 2y)} \cdot \dfrac{x + 2y}{1}$

$= x + 2y$

33. $\dfrac{(t + 2)^3}{(t + 1)^3} \cdot \dfrac{t^2 + 2t + 1}{t^2 + 4t + 4} \cdot \dfrac{t + 1}{t + 2}$

$= \dfrac{(t + 2)(t + 2)(t + 2)(t + 1)(t + 1)(t + 1)}{(t + 1)(t + 1)(t + 1)(t + 2)(t + 2)(t + 2)}$

$= \dfrac{(t + 2)^3}{(t + 2)^3} \cdot \dfrac{(t + 1)^3}{(t + 1)^3}$

$= 1$

35. $\dfrac{x - 7}{x^3 - 9x^2 + 14x} = \dfrac{x - 7}{x(x - 7)(x - 2)}$

The denominator cannot equal 0. If $x(x - 7)(x - 2) = 0$, then $x = 0$, 7, or 2. The replacements which are not sensible are 0, 7, and 2.

Exercise Set 9.2

1. The reciprocal of $\frac{4}{x}$ is $\frac{x}{4}$ because $\frac{4}{x} \cdot \frac{x}{4} = 1$.

3. The reciprocal of $x^2 - y^2$ is $\dfrac{1}{x^2 - y^2}$ because $\dfrac{x^2 - y^2}{1} \cdot \dfrac{1}{x^2 - y^2} = 1$.

5. The reciprocal of $\dfrac{x^2 + 2x - 5}{x^2 - 4x + 7}$ is $\dfrac{x^2 - 4x + 7}{x^2 + 2x - 5}$ because $\dfrac{x^2 + 2x - 5}{x^2 - 4x + 7} \cdot \dfrac{x^2 - 4x + 7}{x^2 + 2x - 5} = 1$.

7. $\frac{2}{5} \div \frac{4}{3} = \frac{2}{5} \cdot \frac{3}{4} = \frac{2 \cdot 3}{5 \cdot 4} = \frac{6}{20} = \frac{3}{10}$

9. $\frac{2}{x} \div \frac{8}{x} = \frac{2}{x} \cdot \frac{x}{8} = \frac{2 \cdot x}{x \cdot 8} = \frac{2 \cdot x}{2 \cdot 4 \cdot x} = \frac{2x}{2x} \cdot \frac{1}{4} = \frac{1}{4}$

11. $\frac{x^2}{y} \div \frac{x^3}{y^3} = \frac{x^2}{y} \cdot \frac{y^3}{x^3} = \frac{x^2y^3}{x^3y} = \frac{x^2y \cdot y^2}{x^2y \cdot x} = \frac{x^2y}{x^2y} \cdot \frac{y^2}{x} = \frac{y^2}{x}$

13. $\dfrac{a + 2}{a - 3} \div \dfrac{a - 1}{a + 3}$

$= \dfrac{a + 2}{a - 3} \cdot \dfrac{a + 3}{a - 1}$

$= \dfrac{(a + 2)(a + 3)}{(a - 3)(a - 1)}$

15. $\dfrac{x^2 - 1}{x} \div \dfrac{x + 1}{x - 1}$

$= \dfrac{x^2 - 1}{x} \cdot \dfrac{x - 1}{x + 1}$

$= \dfrac{(x^2 - 1)(x - 1)}{x(x + 1)}$

$= \dfrac{(x - 1)(x + 1)(x - 1)}{x(x + 1)}$

$= \dfrac{x + 1}{x + 1} \cdot \dfrac{(x - 1)(x - 1)}{x}$

$= \dfrac{(x - 1)^2}{x}$

17. $\dfrac{x + 1}{6} \div \dfrac{x + 1}{3}$

$= \dfrac{x + 1}{6} \cdot \dfrac{3}{x + 1}$

$= \dfrac{(x + 1) \cdot 3}{6 \cdot (x + 1)}$

$= \dfrac{3(x + 1)}{2 \cdot 3(x + 1)}$

$= \dfrac{3(x + 1)}{3(x + 1)} \cdot \dfrac{1}{2}$

$= \dfrac{1}{2}$

19. $\dfrac{x^2 - 9}{4x + 12} \div \dfrac{x - 3}{6}$

$= \dfrac{x^2 - 9}{4x + 12} \cdot \dfrac{6}{x - 3}$

$= \dfrac{(x^2 - 9) \cdot 6}{(4x + 12)(x - 3)}$

$= \dfrac{(x - 3)(x + 3) \cdot 3 \cdot 2}{2 \cdot 2(x + 3)(x - 3)}$

$= \dfrac{2(x - 3)(x + 3)}{2(x - 3)(x + 3)} \cdot \dfrac{3}{2}$

$= \dfrac{3}{2}$

21. $\dfrac{x + y}{x - y} \div \dfrac{x^2 + y}{x^2 - y^2}$

$= \dfrac{x + y}{x - y} \cdot \dfrac{x^2 - y^2}{x^2 + y}$

$= \dfrac{(x + y)(x^2 - y^2)}{(x - y)(x^2 + y)}$

$= \dfrac{(x + y)(x + y)(x - y)}{(x - y)(x^2 + y)}$

$= \dfrac{x - y}{x - y} \cdot \dfrac{(x + y)(x + y)}{x^2 + y}$

$= \dfrac{(x + y)^2}{x^2 + y}$

23. $\frac{x^2 - x - 20}{x^2 + 7x + 12} \div \frac{x^2 - 10x + 25}{x^2 + 6x + 9}$

$= \frac{x^2 - x - 20}{x^2 + 7x + 12} \cdot \frac{x^2 + 6x + 9}{x^2 - 10x + 25}$

$= \frac{(x^2 - x - 20)(x^2 + 6x + 9)}{(x^2 + 7x + 12)(x^2 - 10x + 25)}$

$= \frac{(x - 5)(x + 4)(x + 3)(x + 3)}{(x + 4)(x + 3)(x - 5)(x - 5)}$

$= \frac{(x - 5)(x + 4)(x + 3)}{(x - 5)(x + 4)(x + 3)} \cdot \frac{x + 3}{x - 5}$

$= \frac{x + 3}{x - 5}$

25. $\frac{c^2 + 10c + 21}{c^2 - 2c - 15} \div (c^2 + 2c - 35)$

$= \frac{c^2 + 10c + 21}{c^2 - 2c - 15} \cdot \frac{1}{c^2 + 2c - 35}$

$= \frac{(c^2 + 10c + 21)\cdot 1}{(c^2 - 2c - 15)(c^2 + 2c - 35)}$

$= \frac{(c + 7)(c + 3)}{(c - 5)(c + 3)(c + 7)(c - 5)}$

$= \frac{(c + 7)(c + 3)}{(c + 7)(c + 3)} \cdot \frac{1}{(c - 5)(c - 5)}$

$= \frac{1}{(c - 5)^2}$

27. $\frac{(t + 5)^3}{(t - 5)^3} \div \frac{(t + 5)^2}{(t - 5)^2}$

$= \frac{(t + 5)^3}{(t - 5)^3} \cdot \frac{(t - 5)^2}{(t + 5)^2}$

$= \frac{(t + 5)^3(t - 5)^2}{(t - 5)^3(t + 5)^2}$

$= \frac{(t + 5)^2(t - 5)^2}{(t + 5)^2(t - 5)^2} \cdot \frac{t + 5}{t - 5}$

$= \frac{t + 5}{t - 5}$

29. The reciprocal of $\frac{2}{5}$ is $\frac{5}{2}$ because $\frac{2}{5} \cdot \frac{5}{2} = 1$.

31. $x^2 + 3x + 2 = (x + 1)(x + 2)$

33. $\frac{x^2 - x + xy - y}{x^2 + 6x - 7} \div \frac{x^2 + 2xy + y^2}{4x + 4y}$

$= \frac{x^2 - x + xy - y}{x^2 + 6x - 7} \cdot \frac{4x + 4y}{x^2 + 2xy + y^2}$

$= \frac{x(x - 1) + y(x - 1)}{x^2 + 6x - 7} \cdot \frac{4x + 4y}{x^2 + 2xy + y^2}$

$= \frac{(x + y)(x - 1)\cdot 4(x + y)}{(x + 7)(x - 1)(x + y)(x + y)}$

$= \frac{(x + y)(x + y)(x - 1)}{(x + y)(x + y)(x - 1)} \cdot \frac{4}{x + 7}$

$= \frac{4}{x + 7}$

35. $\left(\frac{y^2 + 5y + 6}{y^2} \cdot \frac{3y^3 + 6y^2}{y^2 - y - 12}\right) \div \frac{y^2 - y}{y^2 - 2y - 8}$

$= \frac{y^2 + 5y + 6}{y^2} \cdot \frac{3y^3 + 6y^2}{y^2 - y - 12} \cdot \frac{y^2 - 2y - 8}{y^2 - y}$

$= \frac{(y + 3)(y + 2)(3y^2)(y + 2)(y - 4)(y + 2)}{y^2(y - 4)(y + 3)(y)(y - 1)}$

$= \frac{y^2(y - 4)(y + 3)}{y^2(y - 4)(y + 3)} \cdot \frac{3(y + 2)(y + 2)(y + 2)}{y(y - 1)}$

$= \frac{3(y + 2)^3}{y(y - 1)}$

Exercise Set 9.3

1. $\frac{5}{12} + \frac{7}{12} = \frac{5 + 7}{12} = \frac{12}{12} = 1$

3. $\frac{1}{3 + x} + \frac{5}{3 + x} = \frac{1 + 5}{3 + x} = \frac{6}{3 + x}$

5. $\frac{x^2 + 7x}{x^2 - 5x} + \frac{x^2 - 4x}{x^2 - 5x}$

$= \frac{(x^2 + 7x) + (x^2 - 4x)}{x^2 - 5x}$

$= \frac{2x^2 + 3x}{x^2 - 5x}$

$= \frac{x(2x + 3)}{x(x - 5)}$

$= \frac{x}{x} \cdot \frac{2x + 3}{x - 5}$

$= \frac{2x + 3}{x - 5}$

7. $\frac{7}{8} + \frac{5}{-8}$

$= \frac{7}{8} + \frac{-1}{-1} \cdot \frac{5}{-8}$

$= \frac{7}{8} + \frac{-5}{8}$

$= \frac{7 + (-5)}{8}$

$= \frac{2}{8}$

$= \frac{1}{4}$

9. $\frac{3}{t} + \frac{4}{-t}$

$= \frac{3}{t} + \frac{-1}{-1} \cdot \frac{4}{-t}$

$= \frac{3}{t} + \frac{-4}{t}$

$= \frac{3 + (-4)}{t}$

$= \frac{-1}{t}$

$= -\frac{1}{t}$

11. $\frac{2x + 7}{x - 6} + \frac{3x}{6 - x}$

$= \frac{2x + 7}{x - 6} + \frac{-1}{-1} \cdot \frac{3x}{6 - x}$

$= \frac{2x + 7}{x - 6} + \frac{-3x}{x - 6}$

$= \frac{(2x + 7) + (-3x)}{x - 6}$

$= \frac{-x + 7}{x - 6}$

13. $\frac{y^2}{y - 3} + \frac{9}{3 - y}$

$= \frac{y^2}{y - 3} + \frac{-1}{-1} \cdot \frac{9}{3 - y}$

$= \frac{y^2}{y - 3} + \frac{-9}{y - 3}$

$= \frac{y^2 + (-9)}{y - 3}$

$= \frac{y^2 - 9}{y - 3}$

$= \frac{(y + 3)(y - 3)}{y - 3}$

$= \frac{y + 3}{1} \cdot \frac{y - 3}{y - 3}$

$= y + 3$

15. $\frac{b - 7}{b^2 - 16} + \frac{7 - b}{16 - b^2}$

$= \frac{b - 7}{b^2 - 16} + \frac{-1}{-1} \cdot \frac{7 - b}{16 - b^2}$

$= \frac{b - 7}{b^2 - 16} + \frac{b - 7}{b^2 - 16}$

$= \frac{(b - 7) + (b - 7)}{b^2 - 16}$

$= \frac{2b - 14}{b^2 - 16}$

17. $\frac{z}{(y + z)(y - z)} + \frac{y}{(z + y)(z - y)}$

$= \frac{z}{(y + z)(y - z)} + \frac{y}{(z + y)(z - y)} \cdot \frac{-1}{-1}$

$= \frac{z}{(y + z)(y - z)} + \frac{-y}{(z + y)(y - z)}$

$= \frac{z - y}{(y + z)(y - z)}$

$= \frac{-(y - z)}{(y + z)(y - z)}$

$= \frac{-1}{y + z} \cdot \frac{y - z}{y - z}$

$= -\frac{1}{y + z}$

19. $\frac{x + 3}{x - 5} + \frac{2x - 1}{5 - x} + \frac{2(3x - 1)}{x - 5}$

$= \frac{x + 3}{x - 5} + \frac{-1}{-1} \cdot \frac{2x - 1}{5 - x} + \frac{2(3x - 1)}{x - 5}$

$= \frac{x + 3}{x - 5} + \frac{1 - 2x}{x - 5} + \frac{2(3x - 1)}{x - 5}$

$= \frac{(x + 3) + (1 - 2x) + (6x - 2)}{x - 5}$

$= \frac{5x + 2}{x - 5}$

21. $\frac{2(4x + 1)}{5x - 7} + \frac{3(x - 2)}{7 - 5x} + \frac{-10x - 1}{5x - 7}$

$= \frac{2(4x + 1)}{5x - 7} + \frac{-1}{-1} \cdot \frac{3(x - 2)}{7 - 5x} + \frac{-10x - 1}{5x - 7}$

$= \frac{2(4x + 1)}{5x - 7} + \frac{-3(x - 2)}{5x - 7} + \frac{-10x - 1}{5x - 7}$

$= \frac{(8x + 2) + (-3x + 6) + (-10x - 1)}{5x - 7}$

$= \frac{-5x + 7}{5x - 7}$

$= \frac{-1(5x - 7)}{5x - 7}$

$= \frac{-1}{1} \cdot \frac{5x - 7}{5x - 7}$

$= -1$

23. $\frac{x + 1}{(x + 3)(x - 3)} + \frac{4(x - 3)}{(x - 3)(x + 3)} + \frac{(x - 1)(x - 3)}{(3 - x)(x + 3)}$

$= \frac{x + 1}{(x + 3)(x - 3)} + \frac{4(x - 3)}{(x - 3)(x + 3)} + \frac{-1}{-1} \cdot \frac{(x-1)(x-3)}{(3-x)(x+3)}$

$= \frac{x + 1}{(x + 3)(x - 3)} + \frac{4(x - 3)}{(x - 3)(x + 3)} + \frac{-1(x^2-4x+3)}{(x-3)(x+3)}$

$= \frac{(x + 1) + (4x - 12) + (-x^2 + 4x - 3)}{(x + 3)(x - 3)}$

$= \frac{-x^2 + 9x - 14}{(x + 3)(x - 3)}$

25. $\frac{7}{8} - \frac{3}{8} = \frac{7 - 3}{8} = \frac{4}{8} = \frac{1}{2}$

27. $\frac{x}{x - 1} - \frac{1}{x - 1} = \frac{x - 1}{x - 1} = 1$

29. $\frac{x + 1}{x^2 - 2x + 1} - \frac{5 - 3x}{x^2 - 2x + 1}$

$= \frac{(x + 1) - (5 - 3x)}{x^2 - 2x + 1}$

$= \frac{x + 1 - 5 + 3x}{x^2 - 2x + 1}$

$= \frac{4x - 4}{x^2 - 2x + 1}$

$= \frac{4(x - 1)}{(x - 1)(x - 1)}$

$= \frac{4}{x - 1} \cdot \frac{x - 1}{x - 1}$

$= \frac{4}{x - 1}$

31. $\frac{11}{6} - \frac{5}{-6}$

$= \frac{11}{6} - \frac{-1}{-1} \cdot \frac{5}{-6}$

$= \frac{11}{6} - \frac{-5}{6}$

$= \frac{11 - (-5)}{6}$

$= \frac{11 + 5}{6}$

$= \frac{16}{6}$

$= \frac{8}{3}$

33. $\frac{5}{a} - \frac{8}{-a}$

$= \frac{5}{a} - \frac{-1}{-1} \cdot \frac{8}{-a}$

$= \frac{5}{a} - \frac{-8}{a}$

$= \frac{5 - (-8)}{a}$

$= \frac{5 + 8}{a}$

$= \frac{13}{a}$

35. $\frac{x}{4} - \frac{3x - 5}{-4}$

$= \frac{x}{4} - \frac{-1}{-1} \cdot \frac{3x - 5}{-4}$

$= \frac{x}{4} - \frac{5 - 3x}{4}$

$= \frac{x - (5 - 3x)}{4}$

$= \frac{x - 5 + 3x}{4}$

$= \frac{4x - 5}{4}$

37. $\frac{3 - x}{x - 7} - \frac{2x - 5}{7 - x}$

$= \frac{3 - x}{x - 7} - \frac{-1}{-1} \cdot \frac{2x - 5}{7 - x}$

$= \frac{3 - x}{x - 7} - \frac{5 - 2x}{x - 7}$

$= \frac{(3 - x) - (5 - 2x)}{x - 7}$

$= \frac{3 - x - 5 + 2x}{x - 7}$

$= \frac{x - 2}{x - 7}$

39. $\frac{x - 8}{x^2 - 16} - \frac{x - 8}{16 - x^2}$

$= \frac{x - 8}{x^2 - 16} - \frac{-1}{-1} \cdot \frac{x - 8}{16 - x^2}$

$= \frac{x - 8}{x^2 - 16} - \frac{8 - x}{x^2 - 16}$

$= \frac{(x - 8) - (8 - x)}{x^2 - 16}$

$= \frac{x - 8 - 8 + x}{x^2 - 16}$

$= \frac{2x - 16}{x^2 - 16}$

41. $\frac{4 - x}{x - 9} - \frac{3x - 8}{9 - x}$

$= \frac{4 - x}{x - 9} - \frac{-1}{-1} \cdot \frac{3x - 8}{9 - x}$

$= \frac{4 - x}{x - 9} - \frac{8 - 3x}{x - 9}$

$= \frac{(4 - x) - (8 - 3x)}{x - 9}$

$= \frac{4 - x - 8 + 3x}{x - 9}$

$= \frac{2x - 4}{x - 9}$

43. $\frac{2(x - 1)}{2x - 3} - \frac{3(x + 2)}{2x - 3} - \frac{x - 1}{3 - 2x}$

$= \frac{2(x - 1)}{2x - 3} - \frac{3(x + 2)}{2x - 3} - \frac{-1}{-1} \cdot \frac{x - 1}{3 - 2x}$

$= \frac{2(x - 1)}{2x - 3} - \frac{3(x + 2)}{2x - 3} - \frac{1 - x}{2x - 3}$

$= \frac{(2x - 2) - (3x + 6) - (1 - x)}{2x - 3}$

$= \frac{2x - 2 - 3x - 6 - 1 + x}{2x - 3}$

$= \frac{-9}{2x - 3}$

45. $\frac{3(2x + 5)}{x - 1} - \frac{3(2x - 3)}{1 - x} + \frac{6x - 1}{x - 1}$

$= \frac{3(2x + 5)}{x - 1} - \frac{-1}{-1} \cdot \frac{3(2x - 3)}{1 - x} + \frac{6x - 1}{x - 1}$

$= \frac{3(2x + 5)}{x - 1} - \frac{-3(2x - 3)}{x - 1} + \frac{6x - 1}{x - 1}$

$= \frac{(6x + 15) - (-6x + 9) + (6x - 1)}{x - 1}$

$= \frac{6x + 15 + 6x - 9 + 6x - 1}{x - 1}$

$= \frac{18x + 5}{x - 1}$

47. $\frac{x - y}{x^2 - y^2} + \frac{x + y}{x^2 - y^2} - \frac{2x}{x^2 - y^2}$

$= \frac{x - y + x + y - 2x}{x^2 - y^2}$

$= \frac{0}{x^2 - y^2}$

$= 0$

49. $\frac{10}{2y - 1} - \frac{6}{1 - 2y} + \frac{y}{2y - 1} + \frac{y - 4}{1 - 2y}$

$= \frac{10}{2y - 1} - \frac{-1}{-1} \cdot \frac{6}{1 - 2y} + \frac{y}{2y - 1} + \frac{-1}{-1} \cdot \frac{y - 4}{1 - 2y}$

$= \frac{10}{2y - 1} - \frac{-6}{2y - 1} + \frac{y}{2y - 1} + \frac{4 - y}{2y - 1}$

$= \frac{10 + 6 + y + 4 - y}{2y - 1}$

$= \frac{20}{2y - 1}$

51. $2x^2 - 3x + 1 = (2x - 1)(x - 1)$

53. $12 = 2 \cdot 2 \cdot 3$

$30 = 2 \cdot 3 \cdot 5$

LCM = $2 \cdot 2 \cdot 3 \cdot 5$, or 60

55. $\frac{x^2}{3x^2 - 5x - 2} - \frac{2x}{3x + 1} \cdot \frac{1}{x - 2}$

$= \frac{x^2}{(3x + 1)(x - 2)} - \frac{2x}{(3x + 1)(x - 2)}$

$= \frac{x^2 - 2x}{(3x + 1)(x - 2)}$

$= \frac{x(x - 2)}{(3x + 1)(x - 2)}$

$= \frac{x}{3x + 1} \cdot \frac{x - 2}{x - 2}$

$= \frac{x}{3x + 1}$

57. $\frac{x}{(x - y)(y - z)} - \frac{x}{(y - x)(z - y)}$

$= \frac{x}{(x - y)(y - z)} - \frac{x}{(-1)(x - y)(-1)(y - z)}$

$= \frac{x}{(x - y)(y - z)} - \frac{x}{(x - y)(y - z)}$

$= \frac{x - x}{(x - y)(y - z)}$

$= \frac{0}{(x - y)(y - z)}$

$= 0$

Exercise Set 9.4

1. $12 = 2 \cdot 2 \cdot 3$

$27 = 3 \cdot 3 \cdot 3$

LCM = $2 \cdot 2 \cdot 3 \cdot 3 \cdot 3$, or 108

3. $8 = 2 \cdot 2 \cdot 2$

$9 = 3 \cdot 3$

LCM = $2 \cdot 2 \cdot 2 \cdot 3 \cdot 3$, or 72

5. $6 = 2 \cdot 3$

$9 = 3 \cdot 3$

$21 = 3 \cdot 7$

LCM = $2 \cdot 3 \cdot 3 \cdot 7$, or 126

7. $24 = 2 \cdot 2 \cdot 2 \cdot 3$

$36 = 2 \cdot 2 \cdot 3 \cdot 3$

$40 = 2 \cdot 2 \cdot 2 \cdot 5$

LCM = $2 \cdot 2 \cdot 2 \cdot 3 \cdot 3 \cdot 5$, or 360

9. $28 = 22 \cdot 7$

$42 = 2 \cdot 3 \cdot 7$

$60 = 2 \cdot 2 \cdot 3 \cdot 5$

LCM = $2 \cdot 2 \cdot 3 \cdot 5 \cdot 7$, or 420

11. $\frac{7}{60} + \frac{6}{75} = \frac{7}{60} + \frac{2}{25}$

$= \frac{7}{2 \cdot 2 \cdot 3 \cdot 5} + \frac{2}{5 \cdot 5}$

LCM = $2 \cdot 2 \cdot 3 \cdot 5 \cdot 5$, or 300

$= \frac{7}{2 \cdot 2 \cdot 3 \cdot 5} \cdot \frac{5}{5} + \frac{2}{5 \cdot 5} \cdot \frac{2 \cdot 2 \cdot 3}{2 \cdot 2 \cdot 3}$

$= \frac{35 + 24}{2 \cdot 2 \cdot 3 \cdot 5 \cdot 5}$

$= \frac{59}{300}$

13. $\frac{5}{24} + \frac{3}{20} + \frac{7}{30} = \frac{5}{2 \cdot 2 \cdot 2 \cdot 3} + \frac{3}{2 \cdot 2 \cdot 5} + \frac{7}{2 \cdot 3 \cdot 5}$

LCM = $2 \cdot 2 \cdot 2 \cdot 3 \cdot 5$, or 120

$= \frac{5}{2 \cdot 2 \cdot 2 \cdot 3} \cdot \frac{5}{5} + \frac{3}{2 \cdot 2 \cdot 5} \cdot \frac{2 \cdot 3}{2 \cdot 3} + \frac{7}{2 \cdot 3 \cdot 5} \cdot \frac{2 \cdot 2}{2 \cdot 2}$

$= \frac{25 + 18 + 28}{2 \cdot 2 \cdot 2 \cdot 3 \cdot 5}$

$= \frac{71}{120}$

15. $\frac{1}{20} + \frac{1}{30} + \frac{2}{45}$

$= \frac{1}{2 \cdot 2 \cdot 5} + \frac{1}{2 \cdot 3 \cdot 5} + \frac{2}{3 \cdot 3 \cdot 5}$

LCM = $2 \cdot 2 \cdot 3 \cdot 3 \cdot 5$, or 180

$= \frac{1}{2 \cdot 2 \cdot 5} \cdot \frac{3 \cdot 3}{3 \cdot 3} + \frac{1}{2 \cdot 3 \cdot 5} \cdot \frac{2 \cdot 3}{2 \cdot 3} + \frac{2}{3 \cdot 3 \cdot 5} \cdot \frac{2 \cdot 2}{2 \cdot 2}$

$= \frac{9 + 6 + 8}{2 \cdot 2 \cdot 3 \cdot 3 \cdot 5}$

$= \frac{23}{180}$

17. $2a^2b = 2 \cdot a \cdot a \cdot b$

$8ab^2 = 2 \cdot 2 \cdot 2 \cdot a \cdot b \cdot b$

LCM = $2 \cdot 2 \cdot 2 \cdot a \cdot a \cdot b \cdot b = 8a^2b^2$

19. $c^2d = c\cdot c\cdot d$

$cd^2 = c\cdot d\cdot d$

$c^3d = c\cdot c\cdot c\cdot d$

LCM = $c\cdot c\cdot c\cdot d\cdot d = c^3d^2$

21. $4(x - 1) = 2\cdot 2(x - 1)$

$8(1 - x) = 2\cdot 2\cdot 2(1 - x)$

LCM = $2\cdot 2\cdot 2(x - 1) = 8(x - 1)$

or

$= 2\cdot 2\cdot 2(1 - x) = 8(1 - x)$

23. $a + 1 = a + 1$

$(a - 1)^2 = (a - 1)(a - 1)$

$(a^2 - 1) = (a + 1)(a - 1)$

LCM = $(a + 1)(a - 1)(a - 1)$, or $(a + 1)(a - 1)^2$

25. $2 + 3k = 2 + 3k$

$9k^2 - 4 = (3k + 2)(3k - 2)$

$2 - 3k = 2 - 3k$

LCM = $(3k + 2)(3k - 2)$, or $(3k + 2)(2 - 3k)$

27. $9x^3 - 9x^2 - 18x = 3\cdot 3\cdot x\cdot(x - 2)(x + 1)$

$6x^5 - 24x^4 + 24x^3 = 2\cdot 3\cdot x\cdot x\cdot x(x - 2)(x - 2)$

LCM = $2\cdot 3\cdot 3\cdot x\cdot x\cdot x\cdot(x + 1)(x - 2)(x - 2)$

$= 18x^3(x + 1)(x - 2)^2$

29. $x^5 + 2x^4 + x^3 = x\cdot x\cdot x(x + 1)(x + 1)$

$2x^3 - 2x = 2x(x + 1)(x - 1)$

$5x - 5 = 5(x - 1)$

LCM = $2\cdot 5\cdot x\cdot x\cdot x(x + 1)(x + 1)(x - 1)$

$= 10x^3(x + 1)^2(x - 1)$

31. $6x^2 + 4x = 2x(3x + 2)$

33. $x^2 - 5x + 6 = 0$

$(x - 3)(x - 2) = 0$

$x - 3 = 0$ or $x - 2 = 0$

$x = 3$ or $x = 2$

The solutions are 3 and 2.

35. One expression is a multiple of the other.

Exercise Set 9.5

1. $\frac{2}{x} + \frac{5}{x^2}$

LCM = $x\cdot x$, or x^2

$= \frac{2}{x}\cdot\frac{x}{x} + \frac{5}{x^2}$

$= \frac{2x}{x^2} + \frac{5}{x^2}$

$= \frac{2x + 5}{x^2}$

3. $\frac{x + y}{xy^2} + \frac{3x + y}{x^2y}$

LCM = x^2y^2

$= \frac{x + y}{xy^2}\cdot\frac{x}{x} + \frac{3x + y}{x^2y}\cdot\frac{y}{y}$

$= \frac{x^2 + xy + 3xy + y^2}{x^2y^2}$

$= \frac{x^2 + 4xy + y^2}{x^2y^2}$

5. $\frac{2}{x - 1} + \frac{2}{x + 1}$

LCM = $(x - 1)(x + 1)$

$= \frac{2}{x - 1}\cdot\frac{x + 1}{x + 1} + \frac{2}{x + 1}\cdot\frac{x - 1}{x - 1}$

$= \frac{2x + 2}{(x - 1)(x + 1)} + \frac{2x - 2}{(x + 1)(x - 1)}$

$= \frac{2x + 2 + 2x - 2}{(x - 1)(x + 1)}$

$= \frac{4x}{(x - 1)(x + 1)}$

7. $\frac{2x}{x^2 - 16} + \frac{x}{x - 4}$

$= \frac{2x}{(x + 4)(x - 4)} + \frac{x}{x - 4}$

LCM = $(x + 4)(x - 4)$

$= \frac{2x}{(x + 4)(x - 4)} + \frac{x}{x - 4}\cdot\frac{x + 4}{x + 4}$

$= \frac{2x}{(x + 4)(x - 4)} + \frac{x^2 + 4x}{(x + 4)(x - 4)}$

$= \frac{2x + x^2 + 4x}{(x + 4)(x - 4)}$

$= \frac{x^2 + 6x}{(x + 4)(x - 4)}$

9. $\frac{3}{x - 1} + \frac{2}{(x - 1)^2}$

LCM = $(x - 1)^2$

$= \frac{3}{x - 1}\cdot\frac{x - 1}{x - 1} + \frac{2}{(x - 1)^2}$

$= \frac{3x - 3}{(x - 1)^2} + \frac{2}{(x - 1)^2}$

$= \frac{3x - 3 + 2}{(x - 1)^2}$

$= \frac{3x - 1}{(x - 1)^2}$

11. $\frac{x}{x^2 + 2x + 1} + \frac{1}{x^2 + 5x + 4}$

$= \frac{x}{(x + 1)(x + 1)} + \frac{1}{(x + 4)(x + 1)}$

LCM = $(x + 1)(x + 1)(x + 4)$

$= \frac{x}{(x + 1)(x + 1)} \cdot \frac{x + 4}{x + 4} + \frac{1}{(x + 4)(x + 1)} \cdot \frac{x + 1}{x + 1}$

$= \frac{x^2 + 4x}{(x + 1)(x + 1)(x + 4)} + \frac{x + 1}{(x + 4)(x + 1)(x + 1)}$

$= \frac{x^2 + 4x + x + 1}{(x + 1)(x + 1)(x + 4)}$

$= \frac{x^2 + 5x + 1}{(x + 1)^2(x + 4)}$

13. $\frac{x + 3}{x - 5} + \frac{x - 5}{x + 3}$

LCM = $(x - 5)(x + 3)$

$= \frac{x + 3}{x - 5} \cdot \frac{x + 3}{x + 3} + \frac{x - 5}{x + 3} \cdot \frac{x - 5}{x - 5}$

$= \frac{x^2 + 6x + 9}{(x - 5)(x + 3)} + \frac{x^2 - 10x + 25}{(x + 3)(x - 5)}$

$= \frac{x^2 + 6x + 9 + x^2 - 10x + 25}{(x - 5)(x + 3)}$

$= \frac{2x^2 - 4x + 34}{(x - 5)(x + 3)}$

15. $\frac{a}{a^2 - 1} + \frac{2a}{a^2 - a}$

$= \frac{a}{(a + 1)(a - 1)} + \frac{2a}{a(a - 1)}$

LCM = $a(a + 1)(a - 1)$

$= \frac{a}{(a + 1)(a - 1)} \cdot \frac{a}{a} + \frac{2a}{a(a - 1)} \cdot \frac{a + 1}{a + 1}$

$= \frac{a^2}{a(a + 1)(a - 1)} + \frac{2a^2 + 2a}{a(a + 1)(a - 1)}$

$= \frac{3a^2 + 2a}{a(a + 1)(a - 1)}$

$= \frac{a(3a + 2)}{a(a + 1)(a - 1)}$

$= \frac{a}{a} \cdot \frac{3a + 2}{(a + 1)(a - 1)}$

$= \frac{3a + 2}{(a + 1)(a - 1)}$

17. $\frac{6}{x - y} + \frac{4x}{y^2 - x^2}$

$= \frac{6}{x - y} + \frac{-1}{-1} \cdot \frac{4x}{y^2 - x^2}$

$= \frac{6}{x - y} + \frac{-4x}{x^2 - y^2}$

$= \frac{6}{x - y} + \frac{-4x}{(x - y)(x + y)}$

LCM = $(x - y)(x + y)$

$= \frac{6}{x - y} \cdot \frac{x + y}{x + y} + \frac{-4x}{(x - y)(x + y)}$

$= \frac{6x + 6y - 4x}{(x - y)(x + y)}$

$= \frac{2x + 6y}{(x - y)(x + y)}$

19. $\frac{10}{x^2 + x - 6} + \frac{3x}{x^2 - 4x + 4}$

$= \frac{10}{(x + 3)(x - 2)} + \frac{3x}{(x - 2)(x - 2)}$

LCM = $(x + 3)(x - 2)(x - 2)$

$= \frac{10}{(x + 3)(x - 2)} \cdot \frac{x - 2}{x - 2} + \frac{3x}{(x - 2)(x - 2)} \cdot \frac{x + 3}{x + 3}$

$= \frac{10x - 20}{(x + 3)(x - 2)(x - 2)} + \frac{3x^2 + 9x}{(x + 3)(x - 2)(x - 2)}$

$= \frac{10x - 20 + 3x^2 + 9x}{(x + 3)(x - 2)(x - 2)}$

$= \frac{3x^2 + 19x - 20}{(x + 3)(x - 2)^2}$

21. [Rectangle with sides labeled ℓ, w, ℓ, w]

We let ℓ = length and w = width.

The perimeter, $2\ell + 2w$, is 642 ft.

$2\ell + 2w = 642$, or $\ell + w = 321$

The length is 15 ft greater than the width.

$\ell = w + 15$

The resulting system is

$\ell + w = 321$

$\ell = w + 15$

Substitute $w + 15$ for ℓ and solve for w.

$\ell + w = 321$

$(w + 15) + w = 321$ (Substituting)

$2w + 15 = 321$

$2w = 306$

$w = 153$

Substitute 153 for w and solve for ℓ.

$\ell = w + 15$

$\ell = 153 + 15$

$\ell = 168$

The perimeter is $2\cdot168 + 2\cdot153$, or 642. The length, 168, is 15 more than the width, $168 = 153 + 15$. The values check.

The area is 168×153, or 25,704 ft^2.

23. $5t - 45 - 8t < -4t + 67$

$-45 - 3t < -4t + 67$

$45 - 45 - 3t < 45 - 4t + 67$

$-3t < 112 - 4t$

$-3t + 4t < 112 - 4t + 4t$

$t < 112$

The solution set is $\{t|t < 112\}$.

25. $P = 2\left(\frac{y + 4}{3}\right) + 2\left(\frac{y - 2}{5}\right)$

$= \frac{2y + 8}{3} + \frac{2y - 4}{5}$

LCM = $3 \cdot 5 = 15$

$= \frac{2y + 8}{3} \cdot \frac{5}{5} + \frac{2y - 4}{5} \cdot \frac{3}{3}$

$= \frac{10y + 40}{15} + \frac{6y - 12}{15}$

$= \frac{10y + 40 + 6y - 12}{15}$

$= \frac{16y + 28}{15}$

$A = \frac{y + 4}{3} \cdot \frac{y - 2}{5}$

$= \frac{(y + 4)(y - 2)}{3 \cdot 5}$

$= \frac{y^2 + 2y - 8}{15}$

27. $\frac{5}{z + 2} + \frac{4z}{z^2 - 4} + 2$

$= \frac{5}{z + 2} + \frac{4z}{(z + 2)(z - 2)} + \frac{2}{1}$

LCM = $(z + 2)(z - 2)$

$= \frac{5}{z + 2} \cdot \frac{z - 2}{z - 2} + \frac{4z}{(z + 2)(z - 2)} + \frac{2}{1} \cdot \frac{(z + 2)(z - 2)}{(z + 2)(z - 2)}$

$= \frac{5z - 10}{(z + 2)(z - 2)} + \frac{4z}{(z + 2)(z - 2)} + \frac{2z^2 - 8}{(z + 2)(z - 2)}$

$= \frac{5z - 10 + 4z + 2z^2 - 8}{(z + 2)(z - 2)}$

$= \frac{2z^2 + 9z - 18}{(z + 2)(z - 2)}$

Exercise Set 9.6

1. $\frac{x - 2}{6} - \frac{x + 1}{3}$

LCM = $2 \cdot 3$, or 6

$= \frac{x - 2}{6} - \frac{x + 1}{3} \cdot \frac{2}{2}$

$= \frac{x - 2}{6} - \frac{2x + 2}{6}$

$= \frac{x - 2 - (2x + 2)}{6}$

$= \frac{x - 2 - 2x - 2}{6}$

$= \frac{-x - 4}{6}$

3. $\frac{4z - 9}{3z} - \frac{3z - 8}{4z}$

LCM = $3 \cdot 4 \cdot z$, or $12z$

$= \frac{4z - 9}{3z} \cdot \frac{4}{4} - \frac{3z - 8}{4z} \cdot \frac{3}{3}$

$= \frac{16z - 36}{12z} - \frac{9z - 24}{12z}$

$= \frac{16z - 36 - (9z - 24)}{12z}$

$= \frac{16z - 36 - 9z + 24}{12z}$

$= \frac{7z - 12}{12z}$

5. $\frac{4x + 2t}{3xt^2} - \frac{5x - 3t}{x^2t}$

LCM = $3x^2t^2$

$= \frac{4x + 2t}{3xt^2} \cdot \frac{x}{x} - \frac{5x - 3t}{x^2t} \cdot \frac{3t}{3t}$

$= \frac{4x^2 + 2tx}{3x^2t^2} - \frac{15xt - 9t^2}{3x^2t^2}$

$= \frac{4x^2 + 2tx - (15xt - 9t^2)}{3x^2t^2}$

$= \frac{4x^2 + 2tx - 15xt + 9t^2}{3x^2t^2}$

$= \frac{4x^2 - 13xt + 9t^2}{3x^2t^2}$

7. $\frac{5}{x + 5} - \frac{3}{x - 5}$

LCM = $(x + 5)(x - 5)$

$= \frac{5}{x + 5} \cdot \frac{x - 5}{x - 5} - \frac{3}{x - 5} \cdot \frac{x + 5}{x + 5}$

$= \frac{5x - 25}{(x + 5)(x - 5)} - \frac{3x + 15}{(x + 5)(x - 5)}$

$= \frac{5x - 25 - (3x + 15)}{(x + 5)(x - 5)}$

$= \frac{5x - 25 - 3x - 15}{(x + 5)(x - 5)}$

$= \frac{2x - 40}{(x + 5)(x - 5)}$

9. $\frac{3}{2t^2 - 2t} - \frac{5}{2t - 2}$

$= \frac{3}{2t(t - 1)} - \frac{5}{2(t - 1)}$

LCM = $2t(t - 1)$

$= \frac{3}{2t(t - 1)} - \frac{5}{2(t - 1)} \cdot \frac{t}{t}$

$= \frac{3}{2t(t - 1)} - \frac{5t}{2t(t - 1)}$

$= \frac{3 - 5t}{2t(t - 1)}$

11. $\dfrac{2s}{t^2 - s^2} - \dfrac{s}{t - s}$

LCM = $(t - s)(t + s)$

$= \dfrac{2s}{(t - s)(t + s)} - \dfrac{s}{t - s} \cdot \dfrac{t + s}{t + s}$

$= \dfrac{2s}{(t - s)(t + s)} - \dfrac{st + s^2}{(t - s)(t + s)}$

$= \dfrac{2s - (st + s^2)}{(t - s)(t + s)}$

$= \dfrac{2s - st - s^2}{(t - s)(t + s)}$

13. $\dfrac{4y}{y^2 - 1} - \dfrac{2}{y} - \dfrac{2}{y + 1}$

$= \dfrac{4y}{(y + 1)(y - 1)} - \dfrac{2}{y} - \dfrac{2}{y + 1}$

LCM = $y(y + 1)(y - 1)$

$= \dfrac{4y}{(y + 1)(y - 1)} \cdot \dfrac{y}{y} - \dfrac{2}{y} \cdot \dfrac{(y+1)(y-1)}{(y+1)(y-1)} - \dfrac{2}{y + 1} \cdot \dfrac{y(y - 1)}{y(y - 1)}$

$= \dfrac{4y^2 - (2y^2 - 2) - (2y^2 - 2y)}{y(y + 1)(y - 1)}$

$= \dfrac{4y^2 - 2y^2 + 2 - 2y^2 + 2y}{y(y + 1)(y - 1)}$

$= \dfrac{2y + 2}{y(y + 1)(y - 1)}$

$= \dfrac{2(y + 1)}{y(y + 1)(y - 1)}$

$= \dfrac{2}{y(y - 1)} \cdot \dfrac{y + 1}{y + 1}$

$= \dfrac{2}{y(y - 1)}$

15. $\dfrac{2z}{1 - 2z} + \dfrac{3z}{2z + 1} - \dfrac{3}{4z^2 - 1}$

$= \dfrac{-1}{-1} \cdot \dfrac{2z}{1 - 2z} + \dfrac{3z}{2z + 1} - \dfrac{3}{4z^2 - 1}$

$= \dfrac{-2z}{2z - 1} + \dfrac{3z}{2z + 1} - \dfrac{3}{(2z - 1)(2z + 1)}$

LCM = $(2z - 1)(2z + 1)$

$= \dfrac{-2z}{2z - 1} \cdot \dfrac{2z + 1}{2z + 1} + \dfrac{3z}{2z + 1} \cdot \dfrac{2z - 1}{2z - 1} - \dfrac{3}{(2z - 1)(2z + 1)}$

$= \dfrac{(-4z^2 - 2z) + (6z^2 - 3z) - 3}{(2z - 1)(2z + 1)}$

$= \dfrac{2z^2 - 5z - 3}{(2z - 1)(2z + 1)}$

$= \dfrac{(z - 3)(2z + 1)}{(2z - 1)(2z + 1)}$

$= \dfrac{z - 3}{2z - 1} \cdot \dfrac{2z + 1}{2z + 1}$

$= \dfrac{z - 3}{2z - 1}$

17. $\dfrac{5}{3 - 2x} + \dfrac{3}{2x - 3} - \dfrac{x - 3}{2x^2 - x - 3}$

$= \dfrac{-1}{-1} \cdot \dfrac{5}{3 - 2x} + \dfrac{3}{2x - 3} - \dfrac{x - 3}{2x^2 - x - 3}$

$= \dfrac{-5}{2x - 3} + \dfrac{3}{2x - 3} - \dfrac{x - 3}{(2x - 3)(x + 1)}$

LCM = $(2x - 3)(x + 1)$

$= \dfrac{-5}{2x - 3} \cdot \dfrac{x + 1}{x + 1} + \dfrac{3}{2x - 3} \cdot \dfrac{x + 1}{x + 1} - \dfrac{x - 3}{(2x - 3)(x + 1)}$

$= \dfrac{(-5x - 5) + (3x + 3) - (x - 3)}{(2x - 3)(x + 1)}$

$= \dfrac{-5x - 5 + 3x + 3 - x + 3}{(2x - 3)(x + 1)}$

$= \dfrac{-3x + 1}{(2x - 3)(x + 1)}$

19. $\dfrac{3}{2c - 1} - \dfrac{1}{c + 2} - \dfrac{5}{2c^2 + 3c - 2}$

$= \dfrac{3}{2c - 1} - \dfrac{1}{c + 2} - \dfrac{5}{(2c - 1)(c + 2)}$

LCM = $(2c - 1)(c + 2)$

$= \dfrac{3}{2c - 1} \cdot \dfrac{c + 2}{c + 2} - \dfrac{1}{c + 2} \cdot \dfrac{2c - 1}{2c - 1} - \dfrac{5}{(2c - 1)(c + 2)}$

$= \dfrac{(3c + 6) - (2c - 1) - 5}{(2c - 1)(c + 2)}$

$= \dfrac{3c + 6 - 2c + 1 - 5}{(2c - 1)(c + 2)}$

$= \dfrac{c + 2}{(2c - 1)(c + 2)}$

$= \dfrac{1}{2c - 1} \cdot \dfrac{c + 2}{c + 2}$

$= \dfrac{1}{2c - 1}$

21. $x^{-7} \div x^{-8} = x^{-7-(-8)} = x^{-7+8} = x^1 = x$

23. $3x^4 \cdot 10x^8 = 3 \cdot 10 \cdot x^{4+8} = 30x^{12}$

25. $\dfrac{1}{2xy - 6x + ay - 3a} - \dfrac{ay + xy}{(a^2 - 4x^2)(y^2 - 6y + 9)}$

$= \dfrac{1}{(2x + a)(y - 3)} - \dfrac{ay + xy}{(a + 2x)(a - 2x)(y - 3)(y - 3)}$

LCM = $(a + 2x)(a - 2x)(y - 3)^2$

$= \dfrac{1}{(2x + a)(y - 3)} \cdot \dfrac{(a - 2x)(y - 3)}{(a - 2x)(y - 3)} - \dfrac{ay + xy}{(a + 2x)(a - 2x)(y - 3)(y - 3)}$

$= \dfrac{ay - 3a - 2xy + 6x - (ay + xy)}{(a + 2x)(a - 2x)(y - 3)^2}$

$= \dfrac{-3a - 3xy + 6x}{(a + 2x)(a - 2x)(y - 3)^2}$

27. $\dfrac{3z^2}{z^4 - 4} - \dfrac{3 - 5z^2}{2z^4 + z^2 - 6}$

$= \dfrac{3z^2}{(z^2 + 2)(z^2 - 2)} - \dfrac{3 - 5z^2}{(2z^2 - 3)(z^2 + 2)}$

LCM = $(z^2 + 2)(z^2 - 2)(2z^2 - 3)$

$= \dfrac{3z^2}{(z^2 + 2)(z^2 - 2)} \cdot \dfrac{2z^2-3}{2z^2-3} - \dfrac{3 - 5z^2}{(2z^2 - 3)(z^2 + 2)} \cdot \dfrac{z^2-2}{z^2-2}$

$= \dfrac{(6z^4 - 9z^2) - (3z^2 - 6 - 5z^4 + 10z^2)}{(z^2 + 2)(z^2 - 2)(2z^2 - 3)}$

$= \dfrac{6z^4 - 9z^2 - 3z^2 + 6 + 5z^4 - 10z^2}{(z^2 + 2)(z^2 - 2)(2z^2 - 3)}$

$= \dfrac{11z^4 - 22z^2 + 6}{(z^2 + 2)(z^2 - 2)(2z^2 - 3)}$

Exercise Set 9.7

1. $\dfrac{1}{4} + \dfrac{1}{6} = \dfrac{1}{t}$, LCM = 12t

$12t(\frac{1}{4} + \frac{1}{6}) = 12t \cdot \frac{1}{t}$

$12t \cdot \frac{1}{4} + 12t \cdot \frac{1}{6} = 12t \cdot \frac{1}{t}$

$3t + 2t = 12$

$5t = 12$

$t = \frac{12}{5}$

Check:

$\frac{1}{4} + \frac{1}{6}$	$= \frac{1}{t}$
$\frac{1}{4} + \frac{1}{6}$	$\dfrac{1}{\frac{12}{5}}$
$\frac{3}{12} + \frac{2}{12}$	$1 \cdot \frac{5}{12}$
$\frac{5}{12}$	$\frac{5}{12}$

This checks, so the solution is $\frac{12}{5}$.

3. $\dfrac{2}{3} - \dfrac{4}{5} = \dfrac{x}{15}$, LCM = 15

$15(\frac{2}{3} - \frac{4}{5}) = 15 \cdot \frac{x}{15}$

$15 \cdot \frac{2}{3} - 15 \cdot \frac{4}{5} = 15 \cdot \frac{x}{15}$

$10 - 12 = x$

$-2 = x$

Check:

$\frac{2}{3} - \frac{4}{5}$	$= \frac{x}{15}$
$\frac{2}{3} - \frac{4}{5}$	$\frac{-2}{15}$
$\frac{10}{15} - \frac{12}{15}$	$-\frac{2}{15}$
$-\frac{2}{15}$	

This checks, so the solution is -2.

5. $\dfrac{5}{x} = \dfrac{6}{x} - \dfrac{1}{3}$, LCM = 3x

$3x \cdot \frac{5}{x} = 3x(\frac{6}{x} - \frac{1}{3})$

$3x \cdot \frac{5}{x} = 3x \cdot \frac{6}{x} - 3x \cdot \frac{1}{3}$

$15 = 18 - x$

$-3 = -x$

$3 = x$

Check:

$\frac{5}{x}$	$= \frac{6}{x} - \frac{1}{3}$
$\frac{5}{3}$	$\frac{6}{3} - \frac{1}{3}$
	$\frac{5}{3}$

This checks, so the solution is 3.

7. $\dfrac{x - 7}{x + 2} = \dfrac{1}{4}$, LCM = $4(x + 2)$

$4(x + 2) \cdot \frac{x - 7}{x + 2} = 4(x + 2) \cdot \frac{1}{4}$

$4(x - 7) = x + 2$

$4x - 28 = x + 2$

$3x = 30$

$x = 10$

Check:

$\frac{x - 7}{x + 2}$	$= \frac{1}{4}$
$\frac{10 - 7}{10 + 2}$	$\frac{1}{4}$
$\frac{3}{12}$	
$\frac{1}{4}$	

This checks, so the solution is 10.

9. $\dfrac{2}{x + 1} = \dfrac{1}{x - 2}$, LCM = $(x + 1)(x - 2)$

$(x + 1)(x - 2) \cdot \frac{2}{x + 1} = (x + 1)(x - 2) \cdot \frac{1}{x - 2}$

$2(x - 2) = x + 1$

$2x - 4 = x + 1$

$x = 5$

This checks, so the solution is 5.

11. $\dfrac{x}{8} - \dfrac{x}{12} = \dfrac{1}{8}$, LCM = 24

$24(\frac{x}{8} - \frac{x}{12}) = 24 \cdot \frac{1}{8}$

$24 \cdot \frac{x}{8} - 24 \cdot \frac{x}{12} = 24 \cdot \frac{1}{8}$

$3x - 2x = 3$

$x = 3$

This checks, so the solution is 3.

13. $$\frac{a - 3}{3a + 2} = \frac{1}{5}, \quad \text{LCM} = 5(3a + 2)$$

$$5(3a + 2) \cdot \frac{a - 3}{3a + 2} = 5(3a + 2) \cdot \frac{1}{5}$$
$$5(a - 3) = 3a + 2$$
$$5a - 15 = 3a + 2$$
$$2a = 17$$
$$a = \frac{17}{2}$$

This checks, so the solution is $\frac{17}{2}$.

15. $$\frac{x - 1}{x - 5} = \frac{4}{x - 5}, \quad \text{LCM} = x - 5$$

$$(x - 5) \cdot \frac{x - 1}{x - 5} = (x - 5) \cdot \frac{4}{x - 5}$$
$$x - 1 = 4$$
$$x = 5$$

The number 5 is not a solution because it makes a denominator zero. Thus, there is no solution.

17. $$x + \frac{4}{x} = -5, \quad \text{LCM} = x$$

$$x(x + \frac{4}{x}) = x(-5)$$
$$x \cdot x + x \cdot \frac{4}{x} = x(-5)$$
$$x^2 + 4 = -5x$$
$$x^2 + 5x + 4 = 0$$
$$(x + 4)(x + 1) = 0$$
$$x + 4 = 0 \text{ or } x + 1 = 0$$
$$x = -4 \text{ or } \quad x = -1$$

Both of these check, so the two solutions are -4 and -1.

19. $$\frac{x + 1}{3} - \frac{x - 1}{2} = 1, \quad \text{LCM} = 6$$

$$6(\frac{x + 1}{3} - \frac{x - 1}{2}) = 6 \cdot 1$$
$$6 \cdot \frac{x + 1}{3} - 6 \cdot \frac{x - 1}{2} = 6 \cdot 1$$
$$2(x + 1) - 3(x - 1) = 6$$
$$2x + 2 - 3x + 3 = 6$$
$$-x + 5 = 6$$
$$-x = 1$$
$$x = -1$$

This checks, so the solution is -1.

21. $$\frac{1}{x} + \frac{2}{x} + \frac{3}{x} = 2, \quad \text{LCM} = x$$

$$x(\frac{1}{x} + \frac{2}{x} + \frac{3}{x}) = x \cdot 2$$
$$x \cdot \frac{1}{x} + x \cdot \frac{2}{x} + x \cdot \frac{3}{x} = x \cdot 2$$
$$1 + 2 + 3 = 2x$$
$$6 = 2x$$
$$3 = x$$

This checks, so the solution is 3.

23. $$\frac{y + 3}{y} = \frac{5}{4}, \quad \text{LCM} = 4y$$

$$4y \cdot \frac{y + 3}{y} = 4y \cdot \frac{5}{4}$$
$$4(y + 3) = y \cdot 5$$
$$4y + 12 = 5y$$
$$12 = y$$

This checks, so the solutions is 12.

25. $$\frac{x - 2}{x - 3} = \frac{x - 1}{x + 1},$$
$$\text{LCM} = (x - 3)(x + 1)$$

$$(x - 3)(x + 1) \cdot \frac{x - 2}{x - 3} = (x - 3)(x + 1) \cdot \frac{x - 1}{x + 1}$$
$$(x + 1)(x - 2) = (x - 3)(x - 1)$$
$$x^2 - x - 2 = x^2 - 4x + 3$$
$$-x - 2 = -4x + 3$$
$$3x = 5$$
$$x = \frac{5}{3}$$

This checks, so the solution is $\frac{5}{3}$.

27. $$\frac{6x - 2}{2x - 1} = \frac{9x}{3x + 1},$$
$$\text{LCM} = (2x - 1)(3x + 1)$$

$$(2x - 1)(3x + 1) \cdot \frac{6x - 2}{2x - 1} = (2x - 1)(3x + 1) \cdot \frac{9x}{3x + 1}$$
$$(3x + 1)(6x - 2) = (2x - 1) \cdot 9x$$
$$18x^2 - 2 = 18x^2 - 9x$$
$$-2 = -9x$$
$$\frac{2}{9} = x$$

This checks, so the solution is $\frac{2}{9}$.

29. $\frac{1}{x + 3} + \frac{1}{x - 3} = \frac{1}{x^2 - 9}$,

LCM = $(x + 3)(x - 3)$

$$(x+3)(x-3)(\frac{1}{x + 3} + \frac{1}{x - 3}) = (x+3)(x-3)\cdot\frac{1}{(x+3)(x-3)}$$
$$(x - 3) + (x + 3) = 1$$
$$2x = 1$$
$$x = \frac{1}{2}$$

This checks, so the solution is $\frac{1}{2}$.

31. $\frac{x}{x + 4} - \frac{4}{x - 4} = \frac{x^2 + 16}{x^2 - 16}$,

LCM = $(x + 4)(x - 4)$

$$(x+4)(x-4)(\frac{x}{x + 4} - \frac{4}{x - 4}) = (x+4)(x-4)\cdot\frac{x^2 + 16}{(x+4)(x-4)}$$
$$x(x - 4) - 4(x + 4) = x^2 + 16$$
$$x^2 - 4x - 4x - 16 = x^2 + 16$$
$$-8x - 16 = 16$$
$$-8x = 32$$
$$x = -4$$

The number -4 is not a solution because it makes a denominator zero. Thus, there is no solution.

33. Graph $2x + y = 6$.

The intercepts are (3,0) and (0,6). Plot these points and draw the line. A third point should be used as a check.

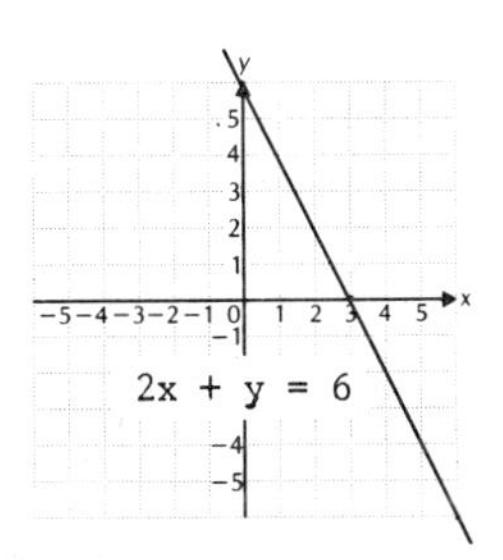

35.
$$42 - 30x < -16x - 14$$
$$42 - 30x - 42 < -16x - 14 \cdot 42$$
$$-30x < -16x - 56$$
$$16x - 30x < 16x - 16x - 56$$
$$-14x < -56$$
$$-\frac{1}{14} \cdot -14x > -\frac{1}{14} \cdot (-56)$$
$$x > 4$$

The solution set is $\{x | x > 4\}$.

37.
$$\frac{x}{x^2 + 3x - 4} + \frac{x + 1}{x^2 + 6x + 8} = \frac{2x}{x^2 + x - 2}$$
$$\frac{x}{(x + 4)(x - 1)} + \frac{x + 1}{(x + 4)(x + 2)} = \frac{2x}{(x + 2)(x - 1)}$$

LCM = $(x + 4)(x - 1)(x + 2)$

$$(x+4)(x-1)(x+2)[\frac{x}{(x + 4)(x - 1)} + \frac{x + 1}{(x + 4)(x + 2)}] =$$
$$(x+4)(x-1)(x+2)\cdot\frac{2x}{(x + 2)(x - 1)}$$
$$(x + 2)x + (x - 1)(x + 1) = (x + 4)2x$$
$$x^2 + 2x + x^2 - 1 = 2x^2 + 8x$$
$$2x^2 + 2x - 1 = 2x^2 + 8x$$
$$2x - 1 = 8x$$
$$-1 = 6x$$
$$-\frac{1}{6} = x$$

This checks, so the solution is $-\frac{1}{6}$.

39.
$$\frac{3a - 5}{a^2 + 4a + 3} + \frac{2a + 2}{a + 3} = \frac{a - 3}{a + 1}$$
$$\frac{3a - 5}{(a + 3)(a + 1)} + \frac{2a + 2}{a + 3} = \frac{a - 3}{a + 1}$$

LCM = $(a + 3)(a + 1)$

$$(a+3)(a+1)[\frac{3a - 5}{(a + 3)(a + 1)} + \frac{2a + 2}{a + 3}] =$$
$$(a+3)(a+1)\cdot\frac{a - 3}{a + 1}$$
$$(3a - 5) + (a + 1)(2a + 2) = (a + 3)(a - 3)$$
$$3a - 5 + 2a^2 + 4a + 2 = a^2 - 9$$
$$2a^2 + 7a - 3 = a^2 - 9$$
$$a^2 + 7a + 6 = 0$$
$$(a + 6)(a + 1) = 0$$

$a + 6 = 0$ or $a + 1 = 0$

$a = -6$ or $a = -1$

The number -1 is not a solution because is makes a denominator zero. The number -6 checks and is the solution.

Exercise Set 9.8

1. Let x represent the number. Then the reciprocal of the number is $\frac{1}{x}$.

The reciprocal of 4	plus	the reciprocal of 5	is	the reciprocal of the number.
↓	↓	↓	↓	↓
$\frac{1}{4}$	+	$\frac{1}{5}$	=	$\frac{1}{x}$

We solve the equation.

$$\frac{1}{4} + \frac{1}{5} = \frac{1}{x}, \quad \text{LCM} = 20x$$

$$20x\left(\frac{1}{4} + \frac{1}{5}\right) = 20x \cdot \frac{1}{x}$$

$$5x + 4x = 20$$

$$9x = 20$$

$$x = \frac{20}{9}$$

The number to be checked is $\frac{20}{9}$. Its reciprocal is $\frac{9}{20}$. The sum of $\frac{1}{4}$ and $\frac{1}{5}$ is $\frac{5}{20} + \frac{4}{20}$, or $\frac{9}{20}$, so the value checks.

The number is $\frac{20}{9}$.

3. Let x represent the smaller number. Then x + 5 represents the larger number.

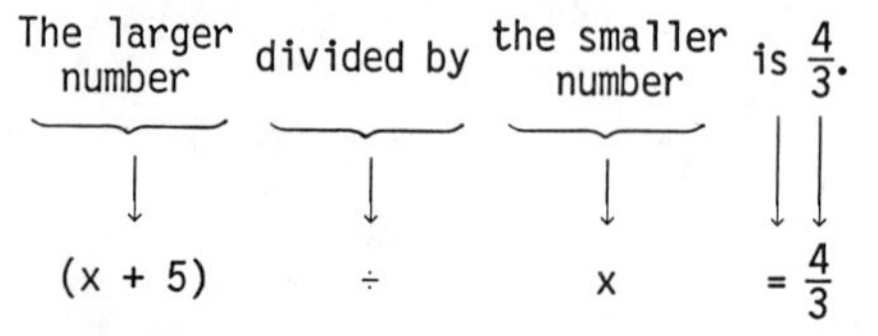

We solve the equation.

$$\frac{x + 5}{x} = \frac{4}{3}, \quad \text{LCM} = 3x$$

$$3x\left(\frac{x + 5}{x}\right) = 3x \cdot \frac{4}{3}$$

$$3(x + 5) = 4x$$

$$3x + 15 = 4x$$

$$15 = x$$

If the smaller number is 15, then the larger is 15 + 5, or 20. The quotient of 20 divided by 15 is $\frac{20}{15}$, or $\frac{4}{3}$. The values check.

The numbers are 15 and 20.

5. We first make a drawing.

Fast car 350 km
r + 40 km/h t hours

Slow car 150 km
r km/h t hours

We let r represent the speed of the slow car. Then r + 40 is the speed of the fast car. The cars travel the same length of time, so we can just use t for time. We organize the information in a chart.

	Distance	Speed	Time
Fast car	350	r + 40	t
Slow car	150	r	t

Solving d = rt for t, we get $t = \frac{d}{r}$. From the rows of the chart we get two different expressions for t. They are

$$t = \frac{350}{r + 40} \text{ and } t = \frac{150}{r}$$

Since the times are the same, we get

$$\frac{350}{r + 40} = \frac{150}{r}$$

We solve the equation.

We multiply by the LCM, r(r + 40).

$$r(r + 40) \cdot \frac{350}{r + 40} = r(r + 40) \cdot \frac{150}{r}$$

$$350r = 150(r + 40)$$

$$350r = 150r + 6000$$

$$200r = 6000$$

$$r = 30$$

If r = 30 km/h, then r + 40 is 70 km/h. The time for the fast car is $\frac{350}{70}$, or 5 hr. The time for the slow car is $\frac{150}{30}$, or 5 hr. The times are the same. The values check.

The speed of the fast car is 70 km/h. The speed of the slow car is 30 km/h.

7. We first make a drawing.

120 mi
r mph t hours
120 mi
2r mph t-3 hours

We let r represent the speed going. Then 2r is the speed returning. We let t represent the time going. Then t - 3 represents the time returning. We organize the information in a chart.

	Distance	Speed	Time
Going	120	r	t
Returning	120	2r	t - 3

Solving d = rt for r, we get $r = \frac{d}{t}$. From the rows of the chart we get two different expressions for r. They are

$r = \frac{120}{t}$ and $2r = \frac{120}{t - 3}$, or $r = \frac{120}{2(t - 3)}$

We use the substitution method to solve the system.

$$\frac{120}{t} = \frac{120}{2(t - 3)}, \quad \text{LCM} = 2t(t - 3)$$

$$2t(t - 3) \cdot \frac{120}{t} = 2t(t - 3) \cdot \frac{120}{2(t - 3)}$$

$$2(t - 3) \cdot 120 = t \cdot 120$$

$$240t - 720 = 120t$$

$$120t = 720$$

$$t = 6$$

Substitute 6 for t in either equation of the system and solve for r.

$r = \frac{120}{t}$

$r = \frac{120}{6}$ (Substituting 6 for t)

$r = 20$

If r = 20 and t = 6, then 2r = 2·20, or 40 mph and t - 3 = 6 - 3, or 3 hr. The distance going is 6·20, or 120 mi. The distance returning is 40·3, or 120 mi. The numbers check.

The speed going is 20 mph.

9. We first make a drawing.

330 km r - 14 km/h t hours
Freight train
400 km r km/h t hours
Passenger train

Let r represent the speed of the passenger train. Then r - 14 is the speed of the freight train. The trains travel the same length of time. We organize the information in a chart.

	Distance	Speed	Time
Freight	330	r - 14	t
Passenger	400	r	t

Solving d = rt for t, we get $t = \frac{d}{r}$. Using $t = \frac{d}{r}$ in each row of the chart we get two different expressions for t.

$t = \frac{330}{r - 14}$ and $t = \frac{400}{r}$

Since the times are the same, we get

$\frac{330}{r - 14} = \frac{400}{r}$

We solve the equation.

We multiply by the LCM, r(r - 14).

$$r(r - 14) \cdot \frac{330}{r - 14} = r(r - 14) \cdot \frac{400}{4}$$

$$330r = 400(r - 14)$$

$$330r = 400r - 5600$$

$$5600 = 70r$$

$$80 = r$$

When r = 80, r - 14 = 66. The time for the freight train is $\frac{330}{66}$, or 5 hours. The time for the passenger train is $\frac{400}{80}$, or 5 hours. The times are the same. The values check.

The speed of the freight train is 66 km/h. The speed of the passenger train is 80 km/h.

11. The job takes painter A 4 hours working alone and painter B 5 hours working alone. Then in 1 hour A does $\frac{1}{4}$ of the job and B does $\frac{1}{5}$ of the job. Working together, they can do $\frac{1}{4} + \frac{1}{5}$, or $\frac{9}{20}$ of the job in 1 hour. In two hours, A does $2(\frac{1}{4})$ of the job and B does $2(\frac{1}{5})$ of the job. Working together they can do $2(\frac{1}{4}) + 2(\frac{1}{5})$, or $\frac{9}{10}$ of the job in 2 hours. In 3 hours they can do $3(\frac{1}{4}) + 3(\frac{1}{5})$, or $\frac{27}{20}$ or $1\frac{7}{20}$ of the job which is more of the job than needs to be done. The answer is somewhere between 2 hr and 3 hr.

If they work together t hours, then painter A does $t(\frac{1}{4})$ of the job and painter B does $t(\frac{1}{5})$ of the job. We want some number t such that

$t(\frac{1}{4}) + t(\frac{1}{5}) = 1.$

We solve the equation.

$$\frac{t}{4} + \frac{t}{5} = 1, \quad \text{LCM} = 20$$

$$20(\frac{t}{4} + \frac{t}{5}) = 20 \cdot 1$$

$$5t + 4t = 20$$

$$9t = 20$$

$$t = \frac{20}{9}, \text{ or } 2\frac{2}{9}$$

The check can be done by repeating the computations. We also have another check. In the familiarization step we learned the time must be between 2 hr and 3 hr. The answer, $2\frac{2}{9}$ hr, is between 2 hr and 3 hr and is less than 4 hours, the time it takes painter A alone.

Working together, it takes then $2\frac{2}{9}$ hr to complete the job.

13. The job takes worker A 12 hours working alone and worker B 9 hours working alone. Then in 1 hour A does $\frac{1}{12}$ of the job and B does $\frac{1}{9}$ of the job. Working together they can do $\frac{1}{12} + \frac{1}{9}$, or $\frac{7}{36}$ of the job in 1 hour. In two hours, A does $2(\frac{1}{12})$ of the job and B does $2(\frac{1}{9})$ of the job. Working together they can do $2(\frac{1}{12}) + 2(\frac{1}{9})$, or $\frac{14}{36}$ of the job in two hours. In 3 hours they can do $3(\frac{1}{12}) + 3(\frac{1}{9})$, or $\frac{21}{36}$ of the job. In 4 hours, they can do $\frac{28}{36}$. In 5 hours, they can do $\frac{35}{36}$. In 6 hours they can do $\frac{42}{36}$, or $1\frac{1}{6}$ which is more of the job than needs to be done. The answer is somewhere between 5 hr and 6 hr.

13. (continued)

If they work together t hours, then worker A does $t(\frac{1}{12})$ of the job and worker B does $t(\frac{1}{9})$ of the job. We want some number t such that

$t(\frac{1}{12}) + t(\frac{1}{9}) = 1.$

We solve the equation.

$$\frac{t}{12} + \frac{t}{9} = 1, \quad \text{LCM} = 36$$

$$36(\frac{t}{12} + \frac{t}{9}) = 36 \cdot 1$$

$$3t + 4t = 36$$

$$7t = 36$$

$$t = \frac{36}{7}, \text{ or } 5\frac{1}{7}$$

The check can be done by repeating the computations. We also have another check. In the familiarization step we learned the time must be between 5 hr and 6 hr. The answer, $5\frac{1}{7}$ hr, is between 5 hr and 6 hr and is less than 9 hours, the time it takes worker B alone.

Working together, it takes them $5\frac{1}{7}$ hr to complete the job.

15. $\frac{54 \text{ days}}{6 \text{ days}} = 9$

17. $\frac{4.6 \text{ km}}{2 \text{ hr}} = 2.3 \frac{\text{km}}{\text{hr}}$, or 2.3 km/h

19. The coffee beans from 14 trees are required to produce 7.7 kilograms of coffee, and we wish to find how many trees are required to produce 320 kilograms of coffee. We can set up ratios:

$\frac{T}{320} \quad \frac{14}{7.7}$

Assuming the two ratios are the same, we can translate to a proportion.

Trees → $\frac{T}{320} = \frac{14}{7.7}$ ← Trees
Kilograms → ← Kilograms

19. (continued)

We solve the equation.

We multiply by 320 to get T alone.

$$320 \cdot \frac{T}{320} = 320 \cdot \frac{14}{7.7}$$

$$T = \frac{4480}{7.7}$$

$$T = 581\frac{9}{11}$$

$$\approx 581.8$$

$\frac{581.8}{320} \approx 1.8$ $\frac{14}{7.7} \approx 1.8$

The ratios are the same.

Thus, $581\frac{9}{11}$ trees are required to produce 320 kilograms of coffee.

21. A student travels 234 kilometers in 14 days, and we wish to find how far the student would travel in 42 days. We can set up ratios:

$\frac{K}{42}$ $\frac{234}{14}$

Assuming the rates are the same, we can translate to a proportion.

Kilometers ⟶ $\frac{K}{42} = \frac{234}{14}$ ⟵ Kilometers
Days ⟶ ⟵ Days

We solve the equation.

We multiply by 42 to get K alone.

$$42 \cdot \frac{K}{42} = 42 \cdot \frac{234}{14}$$

$$K = \frac{9828}{14}$$

$$K = 702$$

$\frac{702}{42} \approx 16.7$ $\frac{234}{14} \approx 16.7$

The ratios are the same.

The student would travel 702 kilometers in 42 days.

23. 10 cm^3 of human blood contains 1.2 grams of hemoglobin, and we wish to find how many grams of hemoglobin are contained in 16 cm^3 of the same blood. We can set up ratios:

$\frac{H}{16}$ $\frac{1.2}{10}$

Assuming the ratios are the same, we can translate to a proportion.

Grams ⟶ $\frac{H}{16} = \frac{1.2}{10}$ ⟵ Grams
cm^3 ⟶ ⟵ cm^3

We solve the equation.

We multiply by 16 to get H alone.

$$16 \cdot \frac{H}{16} = 16 \cdot \frac{1.2}{10}$$

$$H = \frac{19.2}{10}$$

$$H = 1.92$$

$\frac{1.92}{16} = 0.12$ $\frac{1.2}{10} = 0.12$

The ratios are the same.

Thus 16 cm^3 of the same blood would contain 1.92 grams of hemoglobin.

25. The ratio of trout tagged to the total number of trout in the lake, T, is $\frac{112}{T}$. Of the 82 trout caught later, there were 32 trout tagged. The ratio of trout tagged to trout caught is $\frac{32}{82}$.

Assuming the two ratios are the same, we can translate to a proportion.

Trout tagged originally ⟶ $\frac{112}{T} = \frac{32}{82}$ ⟵ Tagged trout caught later
Trout in lake ⟶ ⟵ Trout caught later

We solve the proportion.

We multiply by the LCM, 82T.

$$82T \cdot \frac{112}{T} = 82T \cdot \frac{32}{82}$$

$$82 \cdot 112 = T \cdot 32$$

$$9184 = 32T$$

$$\frac{9184}{32} = T$$

$$287 = T$$

$\frac{112}{287} \approx 0.39$ $\frac{32}{82} \approx 0.39$

The ratios are the same.

There are 287 trout in the lake.

27. The ratio of the weight of an object on the moon to the weight of an object on the earth is 0.16 to 1.

a) We wish to find out how much a 12-ton rocket would weigh on the moon.

b) We wish to find out how much a 90-kilogram astronaut would weigh on the moon.

We can set up ratios.

$\frac{0.16}{1}$ $\frac{T}{12}$ $\frac{K}{90}$

Assuming the ratios are the same, we can translate to proportions.

a) Wgt. on moon ⟶ $\frac{0.16}{1} = \frac{T}{12}$ ⟵ Wgt. on moon
Wgt. on earth ⟶ ⟵ Wgt. on earth

b) Wgt. on moon ⟶ $\frac{0.16}{1} = \frac{K}{90}$ ⟵ Wgt. on moon
Wgt. on earth ⟶ ⟵ Wgt. on earth

We solve each proportion.

a) $\frac{0.16}{1} = \frac{T}{12}$
$12(0.16) = T$
$1.92 = T$

b) $\frac{0.16}{1} = \frac{K}{90}$
$90(0.16) = K$
$14.4 = K$

$\frac{0.16}{1} = 0.16$ $\frac{1.92}{12} = 0.16$ $\frac{14.4}{90} = 0.16$

The ratios are the same.

a) A 12-ton rocket would weigh 1.92 tons on the moon.

b) A 90-kilogram astronaut would weigh 14.4 kilograms on the moon.

29. We first make a drawing.

Upstream 24 mi 10 - r mph ⟶

Downstream 24 mi 10 + r mph ⟵

We let r represent the speed of the current and organize the information in a chart.

	Distance	Speed	Time
Upstream	24	10 - r	$\frac{24}{10 - r}$
Downstream	24	10 + r	$\frac{24}{10 + r}$
		Total	5

29. (continued)

The total time is 5 hours. This translates to

$$\frac{24}{10 - r} + \frac{24}{10 + r} = 5$$

We solve the equation.

We multiply by the LCM, $(10 - r)(10 + r)$.

$$(10-r)(10+r)\left(\frac{24}{10 - r} + \frac{24}{10 + r}\right) = (10-r)(10+r)5$$
$$24(10 + r) + 24(10 - r) = 5(100 - r^2)$$
$$240 + 24r + 240 - 24r = 500 - 5r^2$$
$$480 = 500 - 5r^2$$
$$5r^2 = 20$$
$$r^2 = 4$$
$$r = 2 \text{ or } -2$$

We only check 2, since -2 cannot be the speed of the current. If $r = 2$, then the speed upstream is $10 - 2$, or 8 mph and the time is $\frac{24}{8}$, or 3 hours. If $r = 2$, then the speed downstream is $10 + 2$, or 12 mph and the time is $\frac{24}{12}$, or 2 hours. The sum of 3 hr and 2 hr is 5 hr. This checks.

The speed of the current is 2 mph.

31. Let x represent the numerator and $x + 1$ represent the denominator of the original fraction. The fraction is $\frac{x}{x + 1}$. If 2 is subtracted from both the numerator and the denominator, the resulting fraction is $\frac{x - 2}{x + 1 - 2}$, or $\frac{x - 2}{x - 1}$.

The resulting fraction is $\frac{1}{2}$.

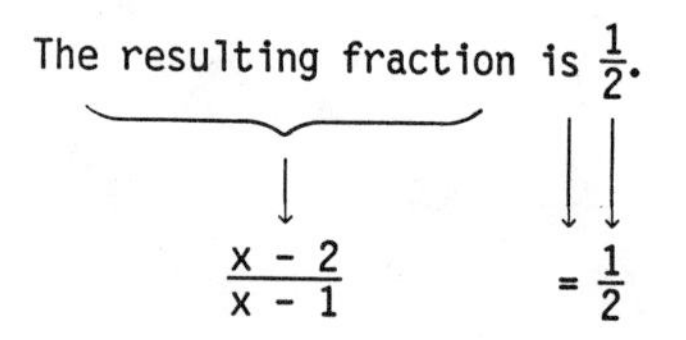

$$\frac{x - 2}{x - 1} = \frac{1}{2}$$

We solve the equation.

$$\frac{x - 2}{x - 1} = \frac{1}{2}, \quad \text{LCM} = 2(x - 1)$$
$$2(x - 1) \cdot \frac{x - 2}{x - 1} = 2(x - 1) \cdot \frac{1}{2}$$
$$2(x - 2) = x - 1$$
$$2x - 4 = x - 1$$
$$x = 3$$

If $x = 3$, then $x + 1 = 4$ and the original fraction is $\frac{3}{4}$. If 2 is subtracted from both numerator and denominator, the resulting fraction is $\frac{3 - 2}{4 - 2}$, or $\frac{1}{2}$. The value checks.

The original fraction was $\frac{3}{4}$.

Exercise Set 9.9

1. $S = 2\pi rh$

$\frac{S}{2\pi h} = r$ (Multiplying by $\frac{1}{2\pi h}$)

3. $A = \frac{1}{2}bh$

$2A = bh$ (Multiplying by 2)

$\frac{2A}{h} = b$ (Multiplying by $\frac{1}{h}$)

5. $S = 180(n - 2)$

$S = 180n - 360$ (Removing parentheses)

$S + 360 = 180n$ (Adding 360)

$\frac{S + 360}{180} = n$ (Multiplying by $\frac{1}{180}$)

7. $V = \frac{1}{3}k(B + b + 4M)$

$3V = k(B + b + 4M)$ (Multiplying by 3)

$3V = kB + kb + 4kM$ (Removing parentheses)

$3V - kB - 4kM = kb$ (Adding $-kb - 4kM$)

$\frac{3V - kB - 4kM}{k} = b$ (Multiplying by $\frac{1}{k}$)

9. $S(r - 1) = r\ell - a$

$Sr - S = r\ell - a$ (Removing parentheses)

$Sr - r\ell = S - a$ (Adding $S - r\ell$)

$r(S - \ell) = S - a$ (Factoring out r)

$r = \frac{S - a}{S - \ell}$ (Multiplying by $\frac{1}{S - \ell}$)

11. $A = \frac{1}{2}h(b_1 + b_2)$

$2A = h(b_1 + b_2)$ (Multiplying by 2)

$\frac{2A}{b_1 + b_2} = h$ (Multiplying by $\frac{1}{b_1 + b_2}$)

13. $r = \frac{v^2pL}{a}$

$ar = v^2pL$ (Multiplying by a)

$a = \frac{v^2pL}{r}$ (Multiplying by $\frac{1}{r}$)

15. $\frac{1}{p} + \frac{1}{q} = \frac{1}{f}$, LCM = pqf

$pqf(\frac{1}{p} + \frac{1}{q}) = pqf \cdot \frac{1}{f}$ (Multiplying by pqf)

$qf + pf = pq$ (Simplifying)

$qf = pq - pf$ (Adding $-pf$)

$qf = p(q - f)$ (Factoring out p)

$\frac{qf}{q - f} = p$ (Multiplying by $\frac{1}{q - f}$)

17. $\frac{A}{P} = 1 + r$

$A = P(1 + r)$ (Multiplying by P)

19. $\frac{1}{R} = \frac{1}{r_1} + \frac{1}{r_2}$, LCM = Rr_1r_2

$Rr_1r_2 \cdot \frac{1}{R} = Rr_1r_2(\frac{1}{r_1} + \frac{1}{r_2})$ (Multiplying by Rr_1r_2)

$r_1r_2 = Rr_2 + Rr_1$ (Simplifying)

$r_1r_2 = R(r_2 + r_1)$ (Factoring out R)

$\frac{r_1r_2}{r_2 + r_1} = R$ (Multiplying by $\frac{1}{r_2 + r_1}$)

21. $\frac{A}{B} = \frac{C}{D}$, LCM = BD

$BD \cdot \frac{A}{B} = BD \cdot \frac{C}{D}$ (Multiplying by BD)

$DA = BC$ (Simplifying)

$D = \frac{BC}{A}$ (Multiplying by $\frac{1}{A}$)

23. $h_1 = q(1 + \frac{h_2}{p})$

$h_1 = q + \frac{qh_2}{p}$ (Removing parentheses)

$h_1 - q = \frac{qh_2}{p}$ (Adding $-q$)

$p(h_1 - q) = qh_2$ (Factoring out p)

$\frac{p(h_1 - q)}{q} = h_2$ (Multiplying by $\frac{1}{q}$)

25. $C = \frac{Ka - b}{a}$

$Ca = Ka - b$ (Multiplying by a)

$Ca + b = Ka$ (Adding b)

$b = Ka - Ca$ (Adding $-Ca$)

$b = a(K - C)$ (Factoring out a)

$\frac{b}{K - C} = a$ (Multiplying by $\frac{1}{K - C}$)

27. $(5x^4 - 6x^3 + 23x^2 - 79x + 24) - (-18x^4 - 56x^3 + 84x - 17)$

$= 5x^4 - 6x^3 + 23x^2 - 79x + 24 + 18x^4 + 56x^3 - 84x + 17$

$= (5 + 18)x^4 + (-6 + 56)x^3 + 23x^2 + (-79 - 84)x + (24 + 17)$

$= 23x^4 + 50x^3 + 23x^2 - 163x + 41$

29. $(6y^2 - 3)(5y^2 + 4)$

$= 6y^2\cdot 5y^2 + 6y^2\cdot 4 - 3\cdot 5y^2 - 3\cdot 4$

$= 30y^4 + 24y^2 - 15y^2 - 12$

$= 30y^4 + 9y^2 - 12$

31. $u = -F(E - \frac{P}{T})$

$u = -FE + \frac{FP}{T}$ (Removing parentheses)

$u + FE = \frac{FP}{T}$ (Adding FE)

$T(u + FE) = FP$ (Multiplying by T)

$T = \frac{FP}{u + FE}$ (Multiplying by $\frac{1}{u + FE}$)

33. $C = \frac{5}{9}(F - 32)$

$C = \frac{5}{9}(C - 32)$ (Substituting C for F)

$9C = 5(C - 32)$

$9C = 5C - 160$

$4C = -160$

$C = -40$

Thus, $-40°C = -40°F$.

Exercise Set 9.10

1. $$\frac{1 + \frac{9}{16}}{1 - \frac{3}{4}} = \frac{1 \cdot \frac{16}{16} + \frac{9}{16}}{1 \cdot \frac{4}{4} - \frac{3}{4}} = \frac{\frac{16 + 9}{16}}{\frac{4 - 3}{4}} = \frac{\frac{25}{16}}{\frac{1}{4}} = \frac{25}{16} \cdot \frac{4}{1} = \frac{25}{4}$$

3. $$\frac{1 - \frac{3}{5}}{1 + \frac{1}{5}} = \frac{1 \cdot \frac{5}{5} - \frac{3}{5}}{1 \cdot \frac{5}{5} + \frac{1}{5}} = \frac{\frac{5 - 3}{5}}{\frac{5 + 1}{5}} = \frac{\frac{2}{5}}{\frac{6}{5}} = \frac{2}{5} \cdot \frac{5}{6} = \frac{2}{6} = \frac{1}{3}$$

5. $$\frac{\frac{1}{x} + 3}{\frac{1}{x} - 5} = \frac{\frac{1}{x} + 3 \cdot \frac{x}{x}}{\frac{1}{x} - 5 \cdot \frac{x}{x}} = \frac{\frac{1 + 3x}{x}}{\frac{1 - 5x}{x}} = \frac{1 + 3x}{x} \cdot \frac{x}{1 - 5x} = \frac{x}{x} \cdot \frac{1 + 3x}{1 - 5x} = \frac{1 + 3x}{1 - 5x}$$

7. $$\frac{\frac{1}{2} + \frac{3}{4}}{\frac{5}{8} - \frac{5}{6}} = \frac{\frac{1}{2} \cdot \frac{2}{2} + \frac{3}{4}}{\frac{5}{8} \cdot \frac{3}{3} - \frac{5}{6} \cdot \frac{4}{4}} = \frac{\frac{2 + 3}{4}}{\frac{15 - 20}{24}} = \frac{\frac{5}{4}}{\frac{-5}{24}} = \frac{5}{4} \cdot \frac{24}{-5} = -6$$

9. $$\frac{\frac{2}{y} + \frac{1}{2y}}{y + \frac{y}{2}} = \frac{\frac{2}{y} \cdot \frac{2}{2} + \frac{1}{2y}}{y \cdot \frac{2}{2} + \frac{y}{2}} = \frac{\frac{4 + 1}{2y}}{\frac{2y + y}{2}} = \frac{\frac{5}{2y}}{\frac{3y}{2}} = \frac{5}{2y} \cdot \frac{2}{3y} = \frac{2}{2} \cdot \frac{5}{3y^2} = \frac{5}{3y^2}$$

11. $$\frac{8 + \frac{8}{d}}{1 + \frac{1}{d}} = \frac{8 \cdot \frac{d}{d} + \frac{8}{d}}{1 \cdot \frac{d}{d} + \frac{1}{d}} = \frac{\frac{8d + 8}{d}}{\frac{d + 1}{d}} = \frac{8d + 8}{d} \cdot \frac{d}{d + 1} = \frac{8(d + 1)}{d} \cdot \frac{d}{d + 1} = \frac{d(d + 1)}{d(d + 1)} \cdot \frac{8}{1} = 8$$

13. $$\frac{\frac{1}{5} - \frac{1}{a}}{\frac{5 - a}{5}} = \frac{\frac{1}{5} \cdot \frac{a}{a} - \frac{1}{a} \cdot \frac{5}{5}}{\frac{5 - a}{5}} = \frac{\frac{a - 5}{5a}}{\frac{5 - a}{5}} = \frac{a - 5}{5a} \cdot \frac{5}{5 - a} = \frac{-(5 - a)}{5a} \cdot \frac{5}{5 - a} = \frac{5(5 - a)}{5(5 - a)} \cdot \frac{-1}{a} = -\frac{1}{a}$$

15. $$\frac{\frac{x}{x - y}}{\frac{x^2}{x^2 - y^2}} = \frac{x}{x - y} \cdot \frac{x^2 - y^2}{x^2} = \frac{x}{x - y} \cdot \frac{(x - y)(x + y)}{x \cdot x} = \frac{x(x - y)}{x(x - y)} \cdot \frac{x + y}{x} = \frac{x + y}{x}$$

17. $\dfrac{x - 3 + \frac{2}{x}}{x - 4 + \frac{3}{x}} = \dfrac{x \cdot \frac{x}{x} - 3 \cdot \frac{x}{x} + \frac{2}{x}}{x \cdot \frac{x}{x} - 4 \cdot \frac{x}{x} + \frac{3}{x}}$

$= \dfrac{\frac{x^2 - 3x + 2}{x}}{\frac{x^2 - 4x + 3}{x}}$

$= \dfrac{x^2 - 3x + 2}{x} \cdot \dfrac{x}{x^2 - 4x + 3}$

$= \dfrac{(x - 2)(x - 1)}{x} \cdot \dfrac{x}{(x - 3)(x - 1)}$

$= \dfrac{x(x - 1)}{x(x - 1)} \cdot \dfrac{x - 2}{x - 3}$

$= \dfrac{x - 2}{x - 3}$

19. $3.47 \times 10^{-5} = 0.0000347$

21. $100m^2 - 81$

$= (10m)^2 - 9^2$

$= (10m + 9)(10m - 9)$

23. $\dfrac{2}{x - 1} - \dfrac{1}{3x - 2}$

LCM $= (x - 1)(3x - 2)$

$= \dfrac{2}{x - 1} \cdot \dfrac{3x - 2}{3x - 2} - \dfrac{1}{3x - 2} \cdot \dfrac{x - 1}{x - 1}$

$= \dfrac{6x - 4}{(x - 1)(3x - 2)} - \dfrac{x - 1}{(x - 1)(3x - 2)}$

$= \dfrac{6x - 4 - (x - 1)}{(x - 1)(3x - 2)}$

$= \dfrac{6x - 4 - x + 1}{(x - 1)(3x - 2)}$

$= \dfrac{5x - 3}{(x - 1)(3x - 2)}$

The reciprocal of $\dfrac{5x - 3}{(x - 1)(3x - 2)}$ is $\dfrac{(x - 1)(3x - 2)}{5x - 3}$.

25. $1 + \dfrac{1}{1 + \dfrac{1}{1 + \dfrac{1}{1 + \frac{1}{x}}}} = 1 + \dfrac{1}{1 + \dfrac{1}{1 + \dfrac{1}{\frac{x + 1}{x}}}}$

$= 1 + \dfrac{1}{1 + \dfrac{1}{1 + \frac{x}{x + 1}}}$

$= 1 + \dfrac{1}{1 + \dfrac{1}{\frac{x + 1 + x}{x + 1}}}$

$= 1 + \dfrac{1}{1 + \frac{x + 1}{2x + 1}}$

$= 1 + \dfrac{1}{\frac{2x + 1 + x + 1}{2x + 1}}$

$= 1 + \dfrac{2x + 1}{3x + 2}$

$= \dfrac{3x + 2 + 2x + 1}{3x + 2}$

$= \dfrac{5x + 3}{3x + 2}$

Exercise Set 9.11

1. $\dfrac{u - 2u^2 - u^5}{u}$

$= \dfrac{u}{u} - \dfrac{2u^2}{u} - \dfrac{u^5}{u}$

$= 1 - 2u - u^4$

3. $(15t^3 + 24t^2 - 6t) \div 3t$

$= \dfrac{15t^3 + 24t^2 - 6t}{3t}$

$= \dfrac{15t^3}{3t} + \dfrac{24t^2}{3t} - \dfrac{6t}{3t}$

$= 5t^2 + 8t - 2$

5. $\dfrac{20x^6 - 20x^4 - 5x^2}{-5x^2}$

$= \dfrac{20x^6}{-5x^2} - \dfrac{20x^4}{-5x^2} - \dfrac{5x^2}{-5x^2}$

$= -4x^4 + 4x^2 + 1$

7. $\dfrac{4x^4y - 8x^6y^2 + 12x^8y^6}{4x^4y}$

$= \dfrac{4x^4y}{4x^4y} - \dfrac{8x^6y^2}{4x^4y} + \dfrac{12x^8y^6}{4x^4y}$

$= 1 - 2x^2y + 3x^4y^5$

9.

$$\begin{array}{r l}
 & x - 5 \\
x - 5 \,\big) & x^2 - 10x - 25 \\
 & \underline{x^2 - 5x} \\
 & \quad -5x - 25 \quad [-10x - (-5x) = -5x] \\
 & \quad \underline{-5x + 25} \\
 & \qquad -50 \quad (-25 - 25 = -50)
\end{array}$$

Check:
Multiply the quotient by the divisor.
$(x - 5)(x - 5) = x^2 - 10x + 25$

Add the remainder.
$x^2 - 10x + 25 + (-50) = x^2 - 10x - 25$

The answer can be expressed as

$x - 5$, R -50 or $x - 5 - \frac{50}{x - 5}$.

11.

$$\begin{array}{r l}
 & x + 2 \\
x + 2 \,\big) & x^2 + 4x + 4 \\
 & \underline{x^2 + 2x} \\
 & \quad 2x + 4 \quad (4x - 2x = 2x) \\
 & \quad \underline{2x + 4} \\
 & \qquad 0
\end{array}$$

Check:
Multiply the quotient by the divisor.
$(x + 2)(x + 2) = x^2 + 4x + 4$

The answer is $x + 2$.

13.

$$\begin{array}{r l}
 & x - 2 \\
x + 6 \,\big) & x^2 + 4x - 14 \\
 & \underline{x^2 + 6x} \\
 & \quad -2x - 14 \quad (4x - 6x = -2x) \\
 & \quad \underline{-2x - 12} \\
 & \qquad -2 \quad [-14 - (-12) = -2]
\end{array}$$

Check:
Multiply the quotient by the divisor.
$(x + 6)(x - 2) = x^2 + 4x - 12$

Add the remainder.
$x^2 + 4x - 12 + (-2) = x^2 + 4x - 14$

The answer can be expressed as

$x - 2$, R -2 or $x - 2 - \frac{2}{x + 6}$.

15.

$$\begin{array}{r l}
 & x^4 + x^3 + x^2 + x + 1 \\
x - 1 \,\big) & x^5 + 0x^4 + 0x^3 + 0x^2 + 0x - 1 \\
 & \underline{x^5 - 1x^4} \\
 & \quad 1x^4 + 0x^3 \quad [0x^4 - (-1x^4) = 1x^4] \\
 & \quad \underline{1x^4 - 1x^3} \\
 & \qquad 1x^3 + 0x^2 \quad [0x^3 - (-1x^3) = 1x^3] \\
 & \qquad \underline{1x^3 - 1x^2} \\
 & \qquad\quad 1x^2 + 0x \quad [0x^2-(-1x^2)=1x^2] \\
 & \qquad\quad \underline{1x^2 - 1x} \\
 & \qquad\qquad 1x - 1 \quad [0x-(-1x)=1x] \\
 & \qquad\qquad \underline{1x - 1} \\
 & \qquad\qquad\quad 0
\end{array}$$

The answer is $x^4 + x^3 + x^2 + x + 1$.

17.

$$\begin{array}{r l}
 & t^2 + 1 \\
t - 1 \,\big) & t^3 - t^2 + t - 1 \\
 & \underline{t^3 - t^2} \\
 & \quad 0 + t - 1 \quad [-t^2 - (-t^2) = 0] \\
 & \quad \underline{t - 1} \\
 & \qquad 0
\end{array}$$

The answer is $t^2 + 1$.

19.

$$\begin{array}{r l}
 & x^3 - 6 \\
x^3 - 7 \,\big) & x^6 - 13x^3 + 42 \\
 & \underline{x^6 - 7x^3} \\
 & \quad -6x^3 + 42 \quad [-13x^3 - (-7x^3) = -6x^3] \\
 & \quad \underline{-6x^3 + 42} \\
 & \qquad 0
\end{array}$$

The answer is $x^3 - 6$.

21. $x = y + 2$
$x + y = 6$

Substitute $y + 2$ for x in the second equation and solve for y.

$$\begin{aligned}
x + y &= 6 \\
(y + 2) + y &= 6 \\
2y + 2 &= 6 \\
2y &= 4 \\
y &= 2
\end{aligned}$$

Substitute 2 for y in either equation and solve for x.

$x = y + 2$
$x = 2 + 2$
$x = 4$

The ordered pair (4,2) checks and is the solution.

23. $-4t + 9 + 8t > -23$

$$4t + 9 > -23$$
$$4t > -32$$
$$t > -8$$

The solution set is $\{t \mid t > -8\}$.

25.

$$\begin{array}{r|l} & a + 3 \\ \hline 5a^2 - 7a - 2 & 5a^3 + 8a^2 - 23a - 1 \\ & \underline{5a^3 - 7a^2 - 2a} \\ & 15a^2 - 21a - 1 \\ & \underline{15a^2 - 21a - 6} \\ & 5 \end{array}$$

The answer can be expressed as
$a + 3$, R 5 or $a + 3 + \frac{5}{5a^2 - 7a - 2}$.

27.

$$\begin{array}{r|l} & y^3 - ay^2 + a^2y - a^3 \\ \hline y + a & y^4 + 0y^3 + 0y^2 + 0y + a^2 \\ & \underline{y^4 + ay^3} \\ & -ay^3 + 0y^2 \\ & \underline{-ay^3 - a^2y^2} \\ & a^2y^2 + 0y \\ & \underline{a^2y^2 + a^3y} \\ & -a^3y + a^2 \\ & \underline{-a^3y - a^4} \\ & (a^2+a^4) \end{array}$$

The answer is $y^3 - ay^2 + a^2y - a^3$, R a^2+a^4.

29.

$$\begin{array}{r|l} & x + 5 \\ \hline x - 1 & x^2 + 4x + c \\ & \underline{x^2 - x} \\ & 5x + c \\ & \underline{5x - 5} \\ & 0 \end{array}$$

$\longleftarrow$ The remainder must be 0.

We solve for c.

$$c - (-5) = 0$$
$$c + 5 = 0$$
$$c = -5$$

Exercise Set 9.12

1. We substitute to find k.

$y = \frac{k}{x}$

$25 = \frac{k}{3}$ (Substituting 25 for y and 3 for x)

$75 = k$ (k is the constant of variation.)

The equation of variation is $y = \frac{75}{x}$.

3. We substitute to find k.

$y = \frac{k}{x}$

$8 = \frac{k}{10}$ (Substituting 8 for y and 10 for x)

$80 = k$

The equation of variation is $y = \frac{80}{x}$.

5. We substitute to find k.

$y = \frac{k}{x}$

$0.125 = \frac{k}{8}$ (Substituting 0.125 for y and 8 for x)

$1 = k$

The equation of variation is $y = \frac{1}{x}$.

7. We substitute to find k.

$y = \frac{k}{x}$

$42 = \frac{k}{25}$ (Substituting 42 for y and 25 for x)

$1050 = k$

The equation of variation is $y = \frac{1050}{x}$.

9. We substitute to find k.

$y = \frac{k}{x}$

$0.2 = \frac{k}{0.3}$ (Substituting 0.2 for y and 0.3 for x)

$0.06 = k$

The equation of variation is $y = \frac{0.06}{x}$.

11. We let T represent the time it takes to do a certain job and N represent the number of people working. The problem states that we have inverse variation between the variables T and N. Thus, an equation $T = \frac{k}{N}$, $k > 0$, applies. As the number of people increases, the time it takes to do the job decreases.

We write an equation of variation.
Time varies inversely as number of people.
This translates to $T = \frac{k}{N}$.

a) First find an equation of variation.

$T = \frac{k}{N}$

$16 = \frac{k}{2}$ (Substituting 16 for T and 2 for N)

$32 = k$

The equation of variation is $T = \frac{32}{N}$.

11. (continued)

b) Use the equation to find the amount of time it takes 6 people to do the job.

$T = \frac{32}{N}$

$T = \frac{32}{6}$ (Substituting 6 for N)

$T = 5\frac{1}{3}$

The check might be done by repeating the computations. We might also analyze the results. The number of people increased from 2 to 6. The time decreased from 16 hours to $5\frac{1}{3}$ hours. This is what we would expect with inverse variation.

It would take 6 people $5\frac{1}{3}$ hours to do the job.

13. The problem states that we have inverse variation between the variables V and P. Thus, and equation $V = \frac{k}{P}$, $k > 0$, applies. As the pressure increases, the volume decreases.

We write an equation of variation.

Volume varies inversely as pressure.

This translates to $V = \frac{k}{P}$.

a) First find an equation of variation.

$V = \frac{k}{P}$

$200 = \frac{k}{32}$ (Substituting 200 for V and 32 for P)

$6400 = k$

The equation of variation is $V = \frac{6400}{P}$.

b) Use the equation to find the volume of a gas under a pressure of 20 kg/cm^2.

$V = \frac{6400}{P}$

$V = \frac{6400}{20}$ (Substituting 20 for P)

$V = 320$

Checking can be done be repeating the computations. We can also analyze the results. The pressure decreased from 32 km/cm^2 to 20 km/cm^2. The volume increased from 200 cm^3 to 320 cm^3. This is what we would expect with inverse variation.

The volume is 320 cm^3 under a pressure of 20 km/cm^2.

15. The problem states that we have inverse variation between the variables, t and r. Thus, an equation $t = \frac{k}{r}$, $k > 0$, applies. As the rate increases, the time decreases.

We write an equation of variation.

Time varies inversely as the rate.

This translates to $t = \frac{k}{r}$.

a) First find an equation of variation.

$t = \frac{k}{r}$

$90 = \frac{k}{1200}$ (Substituting 90 for t and 1200 for r)

$108{,}000 = k$

The equation of variation is $t = \frac{108{,}000}{r}$.

b) Use the equation to find the time it will take the pump to empty the tank at 2000 ℓ/min.

$t = \frac{108{,}000}{r}$

$t = \frac{108{,}000}{2000}$ (Substituting 2000 for r)

$t = 54$

The check might be done by repeating the computations. Let us also analyze the results. The rate increased from 1200 ℓ/min to 2000 ℓ/min. The time decreased from 90 min to 54 min. This is what we would expect with inverse variation.

It will take 54 min for the pump to empty the tank at 2000 ℓ/min.

17. The problem states that we have inverse variation between the variables P and W. Thus, an equation $P = \frac{k}{W}$, $k > 0$, applies. As the wavelength increases the pitch decreases.

Pitch varies inversely as wavelength.

This translates to $P = \frac{k}{W}$.

a) First find an equation of variation.

$P = \frac{k}{W}$

$660 = \frac{k}{1.6}$ (Substituting 660 for P and 1.6 for W)

$1056 = k$

The equation of variation is $P = \frac{1056}{W}$.

17. (continued)

b) Use the equation to find the wavelength of a tone which has a pitch of 440 vibrations per second.

$$P = \frac{1056}{W}$$

$$440 = \frac{1056}{W} \quad \text{(Substituting 440 for P)}$$

$$440\ W = 1056$$

$$W = \frac{1056}{440}$$

$$W = 2.4$$

We can check by repeating the computations. Let us also analyze the results. The pitch decreased from 660 vibrations per second to 440 vibrations per second. The wavelength increased from 1.6 ft to 2.4 ft. We would expect these results with inverse variation.

A tone which has a pitch of 440 vibrations per second has a wavelength of 2.4 ft.

19. The term of highest degree is $-7x^4$. Thus the degree of the polynomial is 4.

21. $3x^3 + 21x^2 + 2x + 14$

$= 3x^2(x + 7) + 2(x + 7)$

$= (3x^2 + 2)(x + 7)$

23. No. As the distance increases, the cost increases.

25. No. As the weight increases, the time increases.

Exercise Set 10.1

1. The square roots of 1 are 1 and -1 because $1^2 = 1$ and $(-1)^2 = 1$.

3. The square roots of 16 are 4 and -4 because $4^2 = 16$ and $(-4)^2 = 16$.

5. The square roots of 100 are 10 and -10 because $10^2 = 100$ and $(-10)^2 = 100$.

7. The square roots of 169 are 13 and -13 because $13^2 = 169$ and $(-13)^2 = 169$.

9. $\sqrt{4} = 2$, taking the principal root.

11. $\sqrt{9} = 3$, so $-\sqrt{9} = -3$.

13. $\sqrt{64} = 8$, so $-\sqrt{64} = -8$.

15. $\sqrt{225} = 15$, taking the principal root.

17. $\sqrt{361} = 19$, taking the principal root.

19. $\sqrt{324} = 18$, taking the principal root.

21. $\sqrt{2}$ is irrational, since 2 is not a perfect square.

23. $-\sqrt{8}$ is irrational, since 8 is not a perfect square.

25. $\sqrt{49}$ is rational, since 49 is a perfect square.

27. $\sqrt{98}$ is irrational, since 98 is not a perfect square.

29. $-\frac{2}{3}$ is rational since it can be expressed as the ratio of two integers $(-\frac{2}{3} = \frac{-2}{3})$.

31. 23 is rational since it can be expressed as the ratio of two integers $(23 = \frac{23}{1})$.

33. 0.424242..., or $0.4\overline{242}$ is rational, since the digits "42" repeat.

35. 4.28228222822228... is irrational, since the decimal notation neither ends nor repeats.

37. -1 is rational since it can be expressed as the ratio of two integers $(-1 = \frac{-1}{1})$.

39. -45.6919119111911119... is irrational, since the decimal notation neither ends nor repeats.

41. $14.6789\overline{89}$ is rational, since the digits "89" repeat.

43. $\sqrt{5} = 2.236$

45. $\sqrt{17} = 4.123$

47. $\sqrt{43} = 6.557$

49. $N = 2.5\sqrt{A}$
$N = 2.5\sqrt{25}$
$= 2.5(5)$
$= 12.5$
≈ 13

$N = 2.5\sqrt{A}$
$N = 2.5\sqrt{89}$
$\approx 2.5(9.434)$
≈ 23.585
≈ 24

51. $5^2 = 5 \cdot 5$, or 25

53. $|-8| = 8$

55. $\sqrt{\sqrt{16}} = \sqrt{4} = 2$

57. $-\sqrt{36} < -\sqrt{33} < -\sqrt{25}$
$-6 < -\sqrt{33} < -5$

$-\sqrt{33}$ is between -6 and -5.

59. $\sqrt{4230} = 65.038$

Exercise Set 10.2

1. In the expression $\sqrt{a - 4}$, $a - 4$ is the radicand.

3. In the expression $5\sqrt{t^2 + 1}$, $t^2 + 1$ is the radicand.

5. In the expression $x^2y\sqrt{\frac{3}{x + 2}}$, $\frac{3}{x + 2}$ is the radicand.

7. The expression $\sqrt{-16}$ is meaningless in the real-number system.

9. The expression $-\sqrt{81}$ is not meaningless in the real-number system.
$-\sqrt{81} = -9$ and -9 is a real number.

11. If we replace y by 4, we get
$\sqrt{y} = \sqrt{4}$
which is not meaningless because the radicand is nonnegative. Thus 4 is a sensible replacement.

13. If we replace x by -11, we get
$\sqrt{1 + x} = \sqrt{1 + (-11)} = \sqrt{-10}$
which is meaningless because the radicand is negative. Thus, -11 is not a sensible replacement.

15. The radicand must be greater than or equal to 0.
We solve: $5x \geqslant 0$
$x \geqslant 0$

Any number greater than or equal to 0 is sensible.

17. The radicand must be greater than or equal to 0.
We solve: $t - 5 \geqslant 0$
$t \geqslant 5$

Any number greater than or equal to 5 is sensible.

19. The radicand must be greater than or equal to 0.
We solve: $y + 8 \geqslant 0$
$y \geqslant -8$

Any number greater than or equal to -8 is sensible.

21. The radicand must be greater than or equal to 0.
We solve: $x + 20 \geqslant 0$
$x \geqslant -20$

Any number greater than or equal to -20 is sensible.

23. The radicand must be greater than or equal to 0.
We solve: $2y - 7 \geqslant 0$
$2y \geqslant 7$
$y \geqslant \frac{7}{2}$

Any number greater than or equal to $\frac{7}{2}$ is sensible.

25. Since t^2 is never negative, then $t^2 + 5$ is never negative. All real-numbers are sensible replacements.

27. $\sqrt{t^2} = t$ (Since t is assumed to be nonnegative.)

29. $\sqrt{9x^2} = \sqrt{(3x)^2} = 3x$ (Since 3x is assumed to be nonnegative.)

31. $\sqrt{(ab)^2} = ab$

33. $\sqrt{(34d)^2} = 34d$

35. $\sqrt{(x + 3)^2} = x + 3$

37. $\sqrt{a^2 - 10a + 25} = \sqrt{(a - 5)^2} = a - 5$

39. $\sqrt{4x^2 - 20x + 25} = \sqrt{(2x - 5)^2} = 2x - 5$

41. $\frac{81}{27} = \frac{3 \cdot 27}{27} = 3$

43. $\frac{1}{x} - \frac{1}{x^2} + \frac{2}{x + 1}$, LCM $= x^2(x + 1)$

$= \frac{1}{x} \cdot \frac{x(x + 1)}{x(x + 1)} - \frac{1}{x^2} \cdot \frac{x + 1}{x + 1} + \frac{2}{x + 1} \cdot \frac{x^2}{x^2}$

$= \frac{x^2 + x - x - 1 + 2x^2}{x^2(x + 1)}$

$= \frac{3x^2 - 1}{x^2(x + 1)}$

45. If $\sqrt{x^2} = 6$, then $x^2 = (6)^2$, or 36.
Thus $x = 6$ or $x = -6$.

47. If $m^2 = 49$, then $m = 7$ or $m = -7$.

49. We solve: $t^2 - 4 \geqslant 0$
$t^2 \geqslant 4$

If $t^2 = 4$, then $t = -2$ or 2.
If $t^2 > 4$, then $t < -2$ or $t > 2$.

Thus, the sensible replacements are $\{t \mid t \leqslant -2 \text{ or } t \geqslant 2\}$.

Exercise Set 10.3

1. $\sqrt{2}\,\sqrt{3} = \sqrt{2 \cdot 3} = \sqrt{6}$

3. $\sqrt{3}\,\sqrt{3} = \sqrt{3 \cdot 3} = \sqrt{9}$, or 3

5. $\sqrt{7}\,\sqrt{2} = \sqrt{7 \cdot 2} = \sqrt{14}$

7. $\sqrt{\frac{2}{5}}\,\sqrt{\frac{3}{4}} = \sqrt{\frac{2}{5} \cdot \frac{3}{4}} = \sqrt{\frac{6}{20}} = \sqrt{\frac{3}{10}}$

9. $\sqrt{2}\,\sqrt{x} = \sqrt{2 \cdot x} = \sqrt{2x}$

11. $\sqrt{x}\,\sqrt{x - 3} = \sqrt{x(x - 3)} = \sqrt{x^2 - 3x}$

13. $\sqrt{x + 2}\,\sqrt{x + 1} = \sqrt{(x + 2)(x + 1)} = \sqrt{x^2 + 3x + 2}$

15. $\sqrt{x + y}\,\sqrt{x - y} = \sqrt{(x + y)(x - y)} = \sqrt{x^2 - y^2}$

17. $\sqrt{43}\,\sqrt{2x} = \sqrt{43 \cdot 2x} = \sqrt{86x}$

19. $\sqrt{12} = \sqrt{4 \cdot 3} = \sqrt{4}\,\sqrt{3} = 2\sqrt{3}$

21. $\sqrt{75} = \sqrt{25 \cdot 3} = \sqrt{25}\,\sqrt{3} = 5\sqrt{3}$

23. $\sqrt{200x} = \sqrt{100 \cdot 2x} = \sqrt{100}\,\sqrt{2x} = 10\sqrt{2x}$

25. $\sqrt{16a^2} = \sqrt{16}\,\sqrt{a^2} = 4a$

27. $\sqrt{49t^2} = \sqrt{49}\,\sqrt{t^2} = 7t$

29. $\sqrt{x^3 - 2x^2} = \sqrt{x^2(x - 2)} = \sqrt{x^2}\,\sqrt{x - 2} = x\sqrt{x - 2}$

31. $\sqrt{4x^2 - 4x + 1} = \sqrt{(2x - 1)^2} = 2x - 1$

33. $\sqrt{9a^2 - 18ab + 9b^2} = \sqrt{9(a^2 - 2ab + b^2)}$
$= \sqrt{9}\,\sqrt{(a - b)^2}$
$= 3(a - b)$

35. $\sqrt{125} = \sqrt{25}\,\sqrt{5} = 5\sqrt{5} \approx 5(2.236) \approx 11.180$

37. $\sqrt{360} = \sqrt{36}\,\sqrt{10} = 6\sqrt{10} \approx 6(3.162) \approx 18.972$

39. $\sqrt{300} = \sqrt{100}\,\sqrt{3} = 10\sqrt{3} \approx 10(1.732) \approx 17.320$

41. $\sqrt{122} = \sqrt{2}\,\sqrt{61} \approx (1.414)(7.810) \approx 11.043$

43. $\frac{1}{5} + \frac{1}{7} = \frac{1}{t}$, LCM = 35t

$$35t(\tfrac{1}{5} + \tfrac{1}{7}) = 35t\cdot\tfrac{1}{t}$$
$$7t + 5t = 35$$
$$12t = 35$$
$$t = \frac{35}{12}$$

45. $50 = 2\cdot25 = 2\cdot5\cdot5$

47. $\sqrt{0.01} = \sqrt{(0.1)^2} = x^2$

49. $\sqrt{x^4} = \sqrt{(x^2)^2} = x^2$

51. $\sqrt{49} = 7$

$\sqrt{490} = \sqrt{49\cdot10} = 7\sqrt{10}$

$\sqrt{4900} = 70$

$\sqrt{49{,}000} = \sqrt{4900\cdot10} = 70\sqrt{10}$

$\sqrt{490{,}000} = 700$

Each is $\sqrt{10}$ times the preceding.

Exercise Set 10.4

1. $\sqrt{24} = \sqrt{4\cdot6} = \sqrt{4}\,\sqrt{6} = 2\sqrt{6}$

3. $\sqrt{40} = \sqrt{4\cdot10} = \sqrt{4}\,\sqrt{10} = 2\sqrt{10}$

5. $\sqrt{175} = \sqrt{25\cdot7} = \sqrt{25}\,\sqrt{7} = 5\sqrt{7}$

7. $\sqrt{48x} = \sqrt{16\cdot3x} = \sqrt{16}\,\sqrt{3x} = 4\sqrt{3x}$

9. $\sqrt{28x^2} = \sqrt{4\cdot7\cdot x^2} = \sqrt{4}\,\sqrt{7}\,\sqrt{x^2} = 2x\sqrt{7}$

11. $\sqrt{8x^2 + 8x + 2} = \sqrt{2(4x^2 + 4x + 1)}$

$= \sqrt{2(2x + 1)^2}$

$= \sqrt{2}\,\sqrt{(2x + 1)^2}$

$= \sqrt{2}\,(2x + 1)$

13. $\sqrt{t^6} = t^3$

15. $\sqrt{x^5} = \sqrt{x^4\cdot x} = \sqrt{x^4}\cdot\sqrt{x} = x^2\sqrt{x}$

17. $\sqrt{(y - 2)^8} = (y - 2)^4$

19. $\sqrt{36m^3} = \sqrt{36\cdot m^2\cdot m} = \sqrt{36}\,\sqrt{m^2}\,\sqrt{m} = 6m\sqrt{m}$

21. $\sqrt{448x^6y^3}$

$= \sqrt{64\cdot x^6\cdot y^2\cdot 7y}$

$= \sqrt{64}\,\sqrt{x^6}\,\sqrt{y^2}\,\sqrt{7y}$

$= 8x^3y\sqrt{7y}$

23. $\sqrt{3}\,\sqrt{18}$

$= \sqrt{3\cdot18}$

$= \sqrt{3\cdot3\cdot6}$

$= \sqrt{3\cdot3}\,\sqrt{6}$

$= 3\sqrt{6}$

25. $\sqrt{18}\,\sqrt{14}$

$= \sqrt{18\cdot14}$

$= \sqrt{3\cdot3\cdot2\cdot2\cdot7}$

$= \sqrt{3\cdot3}\,\sqrt{2\cdot2}\,\sqrt{7}$

$= 3\cdot2\,\sqrt{7}$

$= 6\sqrt{7}$

27. $\sqrt{10}\,\sqrt{10}$

$= \sqrt{10\cdot10}$

$= 10$

29. $\sqrt{5b}\,\sqrt{15b}$

$= \sqrt{5b\cdot15b}$

$= \sqrt{5\cdot5\cdot3\cdot b\cdot b}$

$= \sqrt{5\cdot5}\,\sqrt{b\cdot b}\,\sqrt{3}$

$= 5b\sqrt{3}$

31. $\sqrt{ab}\,\sqrt{ac}$

$= \sqrt{ab\cdot ac}$

$= \sqrt{a\cdot a}\,\sqrt{b\cdot c}$

$= a\sqrt{bc}$

33. $\sqrt{18x^2y^3}\,\sqrt{6xy^4}$

$= \sqrt{18x^2y^3\cdot6xy^4}$

$= \sqrt{3\cdot6\cdot6\cdot x^2\cdot x\cdot y^6\cdot y}$

$= \sqrt{6\cdot6}\,\sqrt{x^2}\,\sqrt{y^6}\,\sqrt{3xy}$

$= 6xy^3\,\sqrt{3xy}$

35. $\sqrt{50ab}\,\sqrt{10a^2b^4}$

$= \sqrt{500\cdot a\cdot a^2\cdot b\cdot b^4}$

$= \sqrt{100\ a^2\ b^4\ 5ab}$

$= \sqrt{100}\,\sqrt{a^2}\,\sqrt{b^4}\,\sqrt{5ab}$

$= 10ab^2\,\sqrt{5ab}$

37. $\frac{3}{x-5} + \frac{1}{x+5} = \frac{2}{x^2-25}$,

LCM = $(x - 5)(x + 5)$

$(x-5)(x+5)(\frac{3}{x-5} + \frac{1}{x+5}) = (x-5)(x+5)\cdot\frac{2}{x^2-25}$

$3(x + 5) + (x - 5) = 2$

$3x + 15 + x - 5 = 2$

$4x + 10 = 2$

$4x = -8$

$x = -2$

39. $y = kx$

Find k.

$7 = k\cdot 32$ (Substituting 7 for y and 32 for x)

$\frac{7}{32} = k$

The equation of variation is $y = \frac{7}{32}x$.

41. $\sqrt{x}\,\sqrt{2x}\,\sqrt{10x^5}$

$= \sqrt{2\cdot 2\cdot 5\cdot x^6\cdot x}$

$= \sqrt{2\cdot 2}\,\sqrt{x^6}\,\sqrt{5x}$

$= 2x^3\,\sqrt{5x}$

43. $\sqrt{0.04x^{4n}}$

$= \sqrt{(0.2)^2}\,\sqrt{(x^{2n})^2}$

$= 0.2x^{2n}$

Exercise Set 10.5

1. $\sqrt{\frac{9}{49}} = \frac{\sqrt{9}}{\sqrt{49}} = \frac{3}{7}$

3. $\sqrt{\frac{1}{36}} = \frac{\sqrt{1}}{\sqrt{36}} = \frac{1}{6}$

5. $-\sqrt{\frac{16}{81}} = -\frac{\sqrt{16}}{\sqrt{81}} = -\frac{4}{9}$

7. $\sqrt{\frac{64}{289}} = \frac{\sqrt{64}}{\sqrt{289}} = \frac{8}{17}$

9. $\sqrt{\frac{1690}{1960}} = \sqrt{\frac{169\cdot 10}{196\cdot 10}} = \sqrt{\frac{169}{196}} = \frac{\sqrt{169}}{\sqrt{196}} = \frac{13}{14}$

11. $\sqrt{\frac{36}{a^2}} = \frac{\sqrt{36}}{\sqrt{a^2}} = \frac{6}{a}$

13. $\sqrt{\frac{9a^2}{625}} = \frac{\sqrt{9a^2}}{\sqrt{625}} = \frac{3a}{25}$

15. $\sqrt{\frac{2}{5}} = \sqrt{\frac{2}{5}\cdot\frac{5}{5}} = \sqrt{\frac{10}{25}} = \frac{\sqrt{10}}{\sqrt{25}} = \frac{\sqrt{10}}{5}$

17. $\sqrt{\frac{3}{8}} = \sqrt{\frac{3}{8}\cdot\frac{2}{2}} = \sqrt{\frac{6}{16}} = \frac{\sqrt{6}}{\sqrt{16}} = \frac{\sqrt{6}}{4}$

19. $\sqrt{\frac{1}{2}} = \sqrt{\frac{1}{2}\cdot\frac{2}{2}} = \sqrt{\frac{2}{4}} = \frac{\sqrt{2}}{\sqrt{4}} = \frac{\sqrt{2}}{2}$

21. $\sqrt{\frac{3}{x}} = \sqrt{\frac{3}{x}\cdot\frac{x}{x}} = \sqrt{\frac{3x}{x^2}} = \frac{\sqrt{3x}}{\sqrt{x^2}} = \frac{\sqrt{3x}}{x}$

23. Method 1. Using a calculator.

$\sqrt{\frac{3}{7}} \approx \sqrt{0.428571} \approx 0.655$

Method 2. Using Table 2.

$\sqrt{\frac{3}{7}} = \sqrt{\frac{3}{7}\cdot\frac{7}{7}} = \sqrt{\frac{21}{49}} = \frac{\sqrt{21}}{\sqrt{49}} = \frac{\sqrt{21}}{7} \approx \frac{4.583}{7} \approx 0.655$

25. Method 1. Using a calculator.

$\sqrt{\frac{1}{3}} = \sqrt{0.333333} \approx 0.577$

Method 2. Using Table 2.

$\sqrt{\frac{1}{3}} = \sqrt{\frac{1}{3}\cdot\frac{3}{3}} = \sqrt{\frac{3}{9}} = \frac{\sqrt{3}}{\sqrt{9}} = \frac{\sqrt{3}}{3} \approx \frac{1.732}{3} \approx 0.577$

27. Method 1. Using a calculator.

$\sqrt{\frac{7}{20}} = \sqrt{0.35} \approx 0.592$

Method 2. Using Table 2.

$\sqrt{\frac{7}{20}} = \sqrt{\frac{7}{20}\cdot\frac{5}{5}} = \sqrt{\frac{35}{100}} = \frac{\sqrt{35}}{\sqrt{100}} = \frac{\sqrt{35}}{10} \approx \frac{5.916}{10} \approx 0.592$

29. Method 1. Using a calculator.

$\sqrt{\frac{12}{5}} = \sqrt{2.4} \approx 1.549$

Method 2. Using Table 2.

$\sqrt{\frac{12}{5}} = \sqrt{\frac{12}{5}\cdot\frac{5}{5}} = \sqrt{\frac{60}{25}} = \frac{\sqrt{60}}{\sqrt{25}} = \frac{\sqrt{60}}{5} \approx \frac{7.746}{5} \approx 1.549$

31. $2x^3 - 8x^2 - 10x^3 + 5x^2 + 7$

$= (2 - 10)x^3 + (-8 + 5)x^2 + 7$

$= -8x^3 - 3x^2 + 7$

33. $x + 5 = -\frac{6}{x}$, LCM = x

$x(x + 5) = x(-\frac{6}{x})$

$x^2 + 5x = -6$

$x^2 + 5x + 6 = 0$

$(x + 3)(x + 2) = 0$

$x + 3 = 0$ or $x + 2 = 0$

$x = -3$ or $x = -2$

35. $\sqrt{\frac{5}{1600}} = \frac{\sqrt{5}}{\sqrt{1600}} = \frac{\sqrt{5}}{40}$

37. $\sqrt{\frac{1}{5x^3}} = \sqrt{\frac{1}{5x^3}\cdot\frac{5x}{5x}} = \sqrt{\frac{5x}{25x^4}} = \frac{\sqrt{5x}}{\sqrt{25x^4}} = \frac{\sqrt{5x}}{5x^2}$

39. $T = 2\pi\sqrt{\frac{L}{32}}$ We use 3.14 for π.

Substitute 2 for L.

$T \approx 2(3.14)\sqrt{\frac{2}{32}}$

$\approx 6.28\sqrt{\frac{1}{16}}$

$\approx 6.28(\frac{1}{4})$

≈ 1.57

The period of a 2 ft pendulum is 1.57 sec.

Substitute 8 for L.

$T \approx 2(3.14)\sqrt{\frac{8}{32}}$

$\approx 6.28\sqrt{\frac{1}{4}}$

$\approx 6.28(\frac{1}{2})$

≈ 3.14

The period of an 8 ft pendulum is 3.14 sec.

Substitute 64 for L.

$T \approx 2(3.14)\sqrt{\frac{64}{32}}$

$\approx 6.28\sqrt{2}$

$\approx 6.28(1.414)$

≈ 8.880

The period of a 64 ft pendulum is 8.880 sec.

Substitute 100 for L.

$T \approx 2(3.14)\sqrt{\frac{100}{32}}$

$\approx 6.28\sqrt{3.125}$

$\approx 6.28(1.768)$

≈ 11.103

The period of a 100 ft pendulum is 11.103 sec.

Exercise Set 10.6

1. $\frac{\sqrt{18}}{\sqrt{2}} = \sqrt{\frac{18}{2}} = \sqrt{9} = 3$

3. $\frac{\sqrt{60}}{\sqrt{15}} = \sqrt{\frac{60}{15}} = \sqrt{4} = 2$

5. $\frac{\sqrt{75}}{\sqrt{15}} = \sqrt{\frac{75}{15}} = \sqrt{5}$

7. $\frac{\sqrt{12}}{\sqrt{75}} = \sqrt{\frac{12}{75}} = \sqrt{\frac{4}{25}} = \frac{\sqrt{4}}{\sqrt{25}} = \frac{2}{5}$

9. $\frac{\sqrt{8x}}{\sqrt{2x}} = \sqrt{\frac{8x}{2x}} = \sqrt{4} = 2$

11. $\frac{\sqrt{63y^3}}{\sqrt{7y}} = \sqrt{\frac{63y^3}{7y}} = \sqrt{9y^2} = 3y$

13. $\frac{\sqrt{15x^5}}{\sqrt{3x}} = \sqrt{\frac{15x^5}{3x}} = \sqrt{5x^4} = \sqrt{5}\sqrt{x^4} = x^2\sqrt{5}$

15. $\frac{\sqrt{3x}}{\sqrt{\frac{3x}{4}}} = \sqrt{\frac{3x}{\frac{3x}{4}}} = \sqrt{3x\cdot\frac{4}{3x}} = \sqrt{4} = 2$

17. $\frac{\sqrt{2}}{\sqrt{5}} = \frac{\sqrt{2}}{\sqrt{5}}\cdot\frac{\sqrt{5}}{\sqrt{5}} = \frac{\sqrt{10}}{5}$

19. $\frac{2}{\sqrt{2}} = \frac{2}{\sqrt{2}}\cdot\frac{\sqrt{2}}{\sqrt{2}} = \frac{2\sqrt{2}}{2} = \sqrt{2}$

21. $\frac{\sqrt{48}}{\sqrt{32}} = \sqrt{\frac{48}{32}} = \sqrt{\frac{3}{2}} = \frac{\sqrt{3}}{\sqrt{2}}\cdot\frac{\sqrt{2}}{\sqrt{2}} = \frac{\sqrt{6}}{2}$

23. $\frac{\sqrt{450}}{\sqrt{18}} = \sqrt{\frac{450}{18}} = \sqrt{25} = 5$

25. $\frac{\sqrt{3}}{\sqrt{x}} = \frac{\sqrt{3}}{\sqrt{x}}\cdot\frac{\sqrt{x}}{\sqrt{x}} = \frac{\sqrt{3x}}{x}$

27. $\frac{4y}{\sqrt{3}} = \frac{4y}{\sqrt{3}}\cdot\frac{\sqrt{3}}{\sqrt{3}} = \frac{4y\sqrt{3}}{3}$

29. $\frac{\sqrt{a^3}}{\sqrt{8}} = \frac{\sqrt{a^3}}{\sqrt{8}}\cdot\frac{\sqrt{2}}{\sqrt{2}} = \frac{\sqrt{a^2\cdot a\cdot 2}}{\sqrt{16}} = \frac{a\sqrt{2a}}{4}$

31. $\frac{\sqrt{16a^4b^6}}{\sqrt{128a^6b^6}} = \frac{\sqrt{16a^4b^6}}{\sqrt{128a^6b^6}}\cdot\frac{\sqrt{2}}{\sqrt{2}} = \frac{\sqrt{16a^4b^6\cdot 2}}{\sqrt{256a^6b^6}} = \frac{4a^2b^3\sqrt{2}}{16a^3b^3} = \frac{\sqrt{2}}{4a}$

33. $\frac{x^2 + 2x}{x^2 + 5x + 6} \div \frac{x}{3x + 9}$

$= \frac{x^2 + 2x}{x^2 + 5x + 6}\cdot\frac{3x + 9}{x}$

$= \frac{x(x + 2)}{(x + 3)(x + 2)}\cdot\frac{3(x + 3)}{x}$

$= \frac{x(x + 2)(x + 3)}{x(x + 2)(x + 3)}\cdot 3$

$= 3$

35. $\frac{x^2}{x + 4} = \frac{16}{x + 4}$, LCM = x + 4

$(x + 4)\cdot\frac{x^2}{x + 4} = (x + 4)\cdot\frac{16}{x + 4}$

$x^2 = 16$

$x = \pm 4$

Since -4 makes the denominator 0, -4 cannot be a solution. The number 4 checks and is the solution.

37. $\frac{3\sqrt{15}}{5\sqrt{32}} = \frac{3\sqrt{15}}{5\sqrt{32}}\cdot\frac{\sqrt{2}}{\sqrt{2}} = \frac{3\sqrt{30}}{5\sqrt{64}} = \frac{3\sqrt{30}}{5\cdot 8} = \frac{3\sqrt{30}}{40}$

39. $\frac{\sqrt{\frac{2}{3}}}{\sqrt{\frac{3}{2}}} = \sqrt{\frac{2}{3}\div\frac{3}{2}} = \sqrt{\frac{2}{3}\cdot\frac{2}{3}} = \frac{2}{3}$

Exercise Set 10.7

1. $3\sqrt{2} + 4\sqrt{2}$
$= (3 + 4)\sqrt{2}$
$= 7\sqrt{2}$

3. $6\sqrt{a} - 14\sqrt{a}$
$= (6 - 14)\sqrt{a}$
$= -8\sqrt{a}$

5. $3\sqrt{12} + 2\sqrt{3}$
$= 3\sqrt{4\cdot 3} + 2\sqrt{3}$
$= 3\cdot 2\sqrt{3} + 2\sqrt{3}$
$= 6\sqrt{3} + 2\sqrt{3}$
$= (6 + 2)\sqrt{3}$
$= 8\sqrt{3}$

7. $\sqrt{27} - 2\sqrt{3}$
$= \sqrt{9\cdot 3} - 2\sqrt{3}$
$= 3\sqrt{3} - 2\sqrt{3}$
$= (3 - 2)\sqrt{3}$
$= \sqrt{3}$

9. $\sqrt{72} + \sqrt{98}$
$= \sqrt{36\cdot 2} + \sqrt{49\cdot 2}$
$= 6\sqrt{2} + 7\sqrt{2}$
$= (6 + 7)\sqrt{2}$
$= 13\sqrt{2}$

11. $3\sqrt{18} - 2\sqrt{32} - 5\sqrt{50}$
$= 3\sqrt{9\cdot 2} - 2\sqrt{16\cdot 2} - 5\sqrt{25\cdot 2}$
$= 3\cdot 3\sqrt{2} - 2\cdot 4\sqrt{2} - 5\cdot 5\sqrt{2}$
$= (9 - 8 - 25)\sqrt{2}$
$= -24\sqrt{2}$

13. $\sqrt{4x} + \sqrt{81x^3}$
$= \sqrt{4\cdot x} + \sqrt{81\cdot x^2\cdot x}$
$= 2\sqrt{x} + 9x\sqrt{x}$
$= (2 + 9x)\sqrt{x}$

15. $\sqrt{8x + 8} + \sqrt{2x + 2}$
$= \sqrt{4(2x + 2)} + \sqrt{2x + 2}$
$= 2\sqrt{2x + 2} + 1\sqrt{2x + 2}$
$= (2 + 1)\sqrt{2x + 2}$
$= 3\sqrt{2x + 2}$

17. $3x\sqrt{y^3x} - x\sqrt{yx^3} + y\sqrt{y^3x}$
$= 3x\sqrt{y^2\cdot y\cdot x} - x\sqrt{y\cdot x^2\cdot x} + y\sqrt{y^2\cdot y\cdot x}$
$= 3x\cdot y\sqrt{yx} - x\cdot x\sqrt{yx} + y\cdot y\sqrt{yx}$
$= (3xy - x^2 + y^2)\sqrt{yx}$

19. $\sqrt{3} - \sqrt{\frac{1}{3}}$
$= \sqrt{3} - \sqrt{\frac{1}{3}\cdot\frac{3}{3}}$
$= \sqrt{3} - \frac{\sqrt{3}}{3}$
$= (1 - \frac{1}{3})\sqrt{3}$
$= \frac{2}{3}\sqrt{3}$, or $\frac{2\sqrt{3}}{3}$

21. $5\sqrt{2} + 3\sqrt{\frac{1}{2}}$
$= 5\sqrt{2} + 3\sqrt{\frac{1}{2}\cdot\frac{2}{2}}$
$= 5\sqrt{2} + \frac{3}{2}\sqrt{2}$
$= (5 + \frac{3}{2})\sqrt{2}$
$= \frac{13}{2}\sqrt{2}$, or $\frac{13\sqrt{2}}{2}$

23. $\sqrt{\frac{1}{12}} - \sqrt{\frac{1}{27}}$
$= \sqrt{\frac{1}{12}\cdot\frac{3}{3}} - \sqrt{\frac{1}{27}\cdot\frac{3}{3}}$
$= \frac{\sqrt{3}}{\sqrt{36}} - \frac{\sqrt{3}}{\sqrt{81}}$
$= \frac{\sqrt{3}}{6} - \frac{\sqrt{3}}{9}$
$= (\frac{1}{6} - \frac{1}{9})\sqrt{3}$
$= \frac{1}{18}\sqrt{3}$, or $\frac{\sqrt{3}}{18}$

25. $\frac{x^2 - 25}{3x}\cdot\frac{9x}{x + 5}$
$= \frac{(x + 5)(x - 5)\cdot 3\cdot 3\cdot x}{3x(x + 5)}$
$= \frac{3x(x + 5)}{3x(x + 5)}\cdot\frac{3(x - 5)}{1}$
$= 3(x - 5)$

27. $-3x + 6y = -6$

$-2x + 7y = 2$

We multiply the first equation by -2 and the second equation by 3 and then add.

$$\begin{aligned} 6x - 12y &= 12 \\ -6x + 21y &= 6 \\ \hline 9y &= 18 \quad \text{(Adding)} \\ y &= 2 \end{aligned}$$

Substitute 2 for y in either equation of the original system and solve for x.

$$\begin{aligned} -3x + 6y &= -6 \\ -3x + 6\cdot 2 &= -6 \quad \text{(Substituting)} \\ -3x + 12 &= -6 \\ -3x &= -18 \\ x &= 6 \end{aligned}$$

The ordered pair (6,2) checks. It is the solution.

29. $\sqrt{1 + x^2} + \dfrac{1}{\sqrt{1 + x^2}}$

$= \sqrt{1 + x^2} \cdot \dfrac{\sqrt{1 + x^2}}{\sqrt{1 + x^2}} + \dfrac{1}{\sqrt{1 + x^2}}$

$= \dfrac{\sqrt{1 + x^2} \cdot \sqrt{1 + x^2} + 1}{\sqrt{1 + x^2}}$

$= \dfrac{1 + x^2 + 1}{\sqrt{1 + x^2}}$

$= \dfrac{2 + x^2}{\sqrt{1 + x^2}}$

$= \dfrac{2 + x^2}{\sqrt{1 + x^2}} \cdot \dfrac{\sqrt{1 + x^2}}{\sqrt{1 + x^2}}$

$= \dfrac{(2 + x^2)\sqrt{1 + x^2}}{1 + x^2}$

31. $\dfrac{2}{\sqrt{3} - \sqrt{5}}$

$= \dfrac{2}{\sqrt{3} - \sqrt{5}} \cdot \dfrac{\sqrt{3} + \sqrt{5}}{\sqrt{3} + \sqrt{5}}$

$= \dfrac{2(\sqrt{3} + \sqrt{5}}{3 - 5}$

$= \dfrac{2(\sqrt{3} + \sqrt{5})}{-2}$

$= -\sqrt{3} - \sqrt{5}$

33. Any pairs of numbers a, b such that $a = 0, b \geqslant 0$ or $a \geqslant 0, b = 0$.

Exercise Set 10.8

1. $$\begin{aligned} a^2 + b^2 &= c^2 \\ 8^2 + 15^2 &= c^2 \\ 64 + 225 &= c^2 \\ 289 &= c^2 \\ \sqrt{289} &= c \\ 17 &= c \end{aligned}$$

3. $$\begin{aligned} a^2 + b^2 &= c^2 \\ 4^2 + 4^2 &= c^2 \\ 16 + 16 &= c^2 \\ 32 &= c^2 \\ \sqrt{32} &= c \\ 5.657 &\approx c \end{aligned}$$

5. $$\begin{aligned} a^2 + b^2 &= c^2 \\ 5^2 + b^2 &= 13^2 \\ 25 + b^2 &= 169 \\ b^2 &= 144 \\ b &= 12 \end{aligned}$$

7. $$\begin{aligned} a^2 + b^2 &= c^2 \\ (4\sqrt{3})^2 + b^2 &= 8^2 \\ 16\cdot 3 + b^2 &= 64 \\ 48 + b^2 &= 64 \\ b^2 &= 16 \\ b &= 4 \end{aligned}$$

9. $$\begin{aligned} a^2 + b^2 &= c^2 \\ 10^2 + 24^2 &= c^2 \\ 100 + 576 &= c^2 \\ 676 &= c^2 \\ 26 &= c \end{aligned}$$

11. $$\begin{aligned} a^2 + b^2 &= c^2 \\ 9^2 + b^2 &= 15^2 \\ 81 + b^2 &= 225 \\ b^2 &= 144 \\ b &= 12 \end{aligned}$$

13. $$\begin{aligned} a^2 + b^2 &= c^2 \\ a^2 + 1^2 &= (\sqrt{5})^2 \\ a^2 + 1 &= 5 \\ a^2 &= 4 \\ a &= 2 \end{aligned}$$

15. $a^2 + b^2 = c^2$

$1^2 + b^2 = (\sqrt{3})^2$

$1 + b^2 = 3$

$b^2 = 2$

$b = \sqrt{2}$

$b \approx 1.414$

17. $a^2 + b^2 = c^2$

$a^2 + (5\sqrt{3})^2 = 10^2$

$a^2 + 25\cdot3 = 100$

$a^2 + 75 = 100$

$a^2 = 25$

$a = 5$

19. We first make a drawing.

We label the unknown height h.

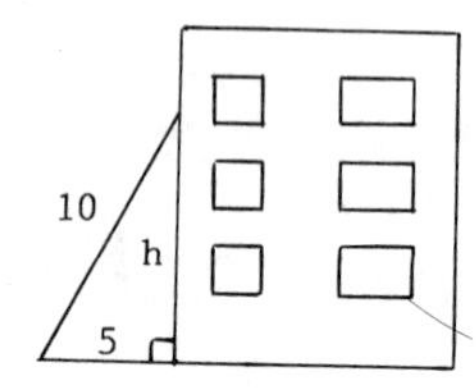

We use the Pythagorean Property.

$a^2 + b^2 = c^2$

$5^2 + h^2 = 10^2$ (Substituting 5 for a, h for b, and 10 for c)

We solve the equation.

$5^2 + h^2 = 10^2$

$25 + h^2 = 100$

$h^2 = 75$

$h = \sqrt{75}$ (Exact)

$h \approx 8.660$ (Approximation)

We check by substituting 5, $\sqrt{75}$, and 10 into the Pythagorean Equation:

$a^2 + b^2$	$= c^2$
$5^2 + (\sqrt{75})^2$	10^2
$25 + 75$	100
100	

The top of the ladder is $\sqrt{75}$ or about 8.660 m from the ground.

21. We first make a drawing.

We label the unknown length w.

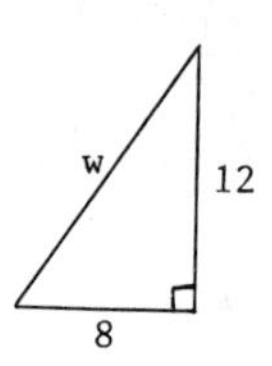

We use the Pythagorean Property.

$a^2 + b^2 = c^2$

$8^2 + 12^2 = w^2$ (Substituting 8 for a, 12 for b, and w for c)

We solve the equation.

$8^2 + 12^2 = w^2$

$64 + 144 = w^2$

$208 = w^2$

$\sqrt{208} = w$

$4\sqrt{13} = w$ (Exact)

$14.422 \approx w$ (Approximation)

We check by substituting 8, 12, and $\sqrt{208}$ into the Pythagorean Equation:

$a^2 + b^2$	$= c^2$
$8^2 + 12^2$	$(\sqrt{208})^2$
$64 + 144$	208
208	

The guy wire is $4\sqrt{13}$ or about 14.422 feet long.

23. We first make a drawing.

We label the length from home to second base d.

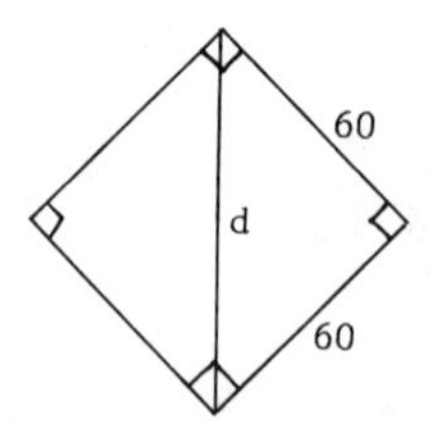

23. (continued)

We use the Pythagorean Property.

$a^2 + b^2 = c^2$

$60^2 + 60^2 = d^2$ (Substituting 60 for a, 60 for b, and d for c)

We solve the equation.

$60^2 + 60^2 = d^2$

$3600 + 3600 = d^2$

$7200 = d^2$

$\sqrt{7200} = d$

$60\sqrt{2} = d$ (Exact)

$84.853 \approx d$ (Approximation)

We check by substituting 60, 60, and $\sqrt{7200}$ into the Pythagorean Equation:

$a^2 + b^2$	$= c^2$
$60^2 + 60^2$	$(\sqrt{7200})^2$
$3600 + 3600$	7200
7200	

It is $60\sqrt{2}$ or about 84.853 feet from home to second base.

25. $\frac{12}{x} = \frac{48}{x + 9}$, LCM = $x(x + 9)$

$x(x + 9)\cdot\frac{12}{x} = x(x + 9)\cdot\frac{48}{x + 9}$

$12(x + 9) = x\cdot48$

$12x + 108 = 48x$

$108 = 36x$

$3 = x$

This checks, so the solution is 3.

27. We first make a drawing.

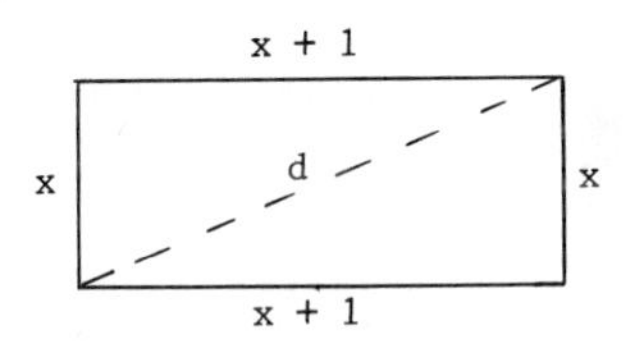

We let x represent the width; then x + 1 represents the length. Using an area formula we can determine the value of x. Then we can use the Pythagorean Property to determine the length of the diagonal labeled d.

27. (continued)

The area of the rectangle is 90 cm^2.
This translates to $(x + 1)x = 90$.

We solve the equation.

$x^2 + x = 90$

$x^2 + x - 90 = 0$

$(x + 10)(x - 9) = 0$

$x + 10 = 0$ or $x - 9 = 0$

$x = -10$ or $x = 9$

Since the width of the rectangle cannot be negative, we only check x = 9. If the width is 9 cm, then the length is 10 cm and the area is $9\cdot10$, or 90 cm^2.

We repeat the previous three steps to determine the length of the diagonal.

We use the Pythagorean Property.

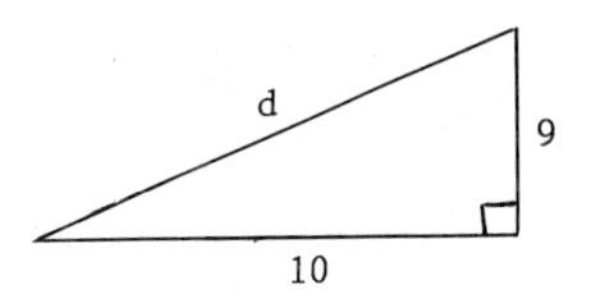

$a^2 + b^2 = c^2$

$10^2 + 9^2 = d^2$ (Substituting 10 for a, 9 for b, and d for c)

$100 + 81 = d^2$

$181 = d^2$

$\sqrt{181} = d$

$13.454 \approx d$ (Using a calculator)

We check by substituting 10, 9, and $\sqrt{181}$ into the Pythagorean Equation:

$a^2 + b^2$	$= c^2$
$10^2 + 9^2$	$(\sqrt{181})^2$
$100 + 81$	181
181	

The length of the diagonal of the rectangle is $\sqrt{181}$, or about 13.454 cm.

29. We first find y in the smaller right triangle. We use the Pythagorean Property.

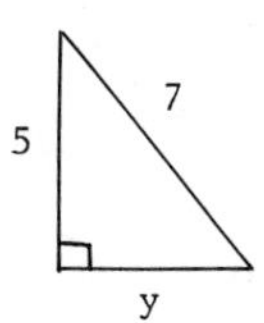

$5^2 + y^2 = 7^2$

$25 + y^2 = 49$

$y^2 = 24$

$y = \sqrt{24}$

29. (continued)

Next we find x in the larger right triangle.

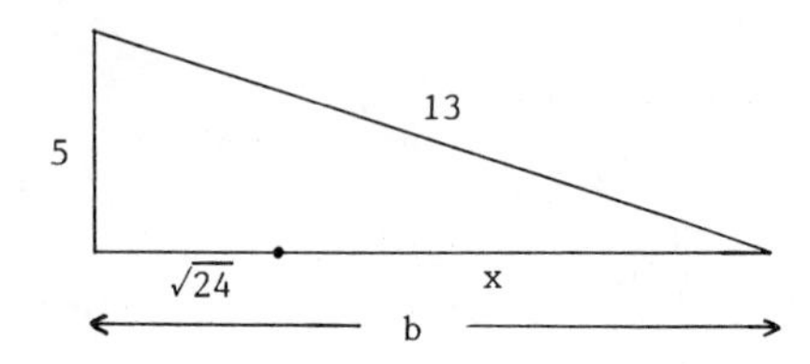

We can express the length of leg b in two ways. Using the Pythagorean Property, we get

$5^2 + b^2 = 13^2$

$25 + b^2 = 169$

$b^2 = 144$

$b = 12$

Leg b can also be expressed as the sum of $\sqrt{24}$ and x.

$x + \sqrt{24} = 12$

$x = 12 - \sqrt{24}$

$x = 12 - 2\sqrt{6}$

$x \approx 7.101$

31. Using the Pythagorean Property we can label the figure with additional information.

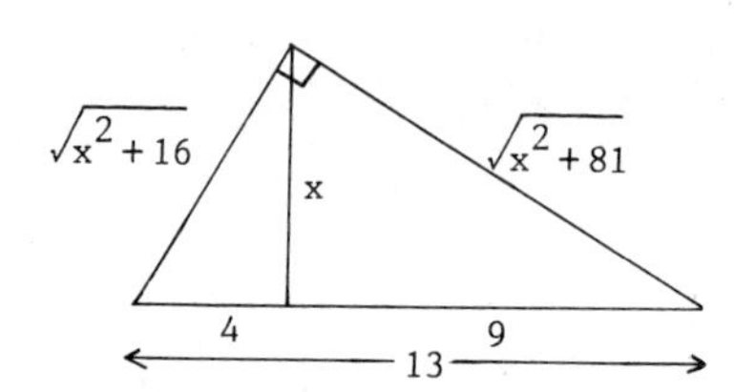

Next we use the Pythagorean Property with the largest right triangle and solve for x.

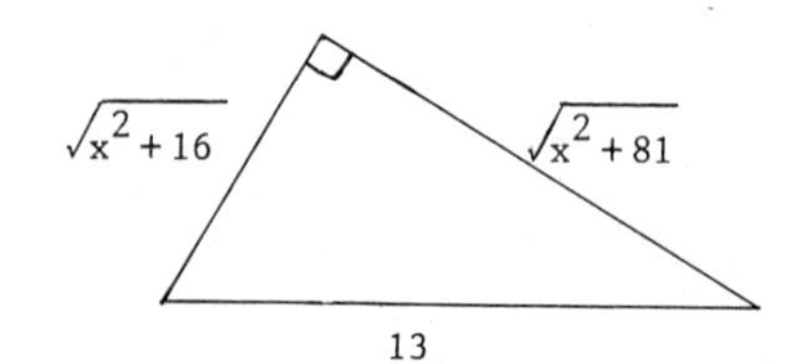

$(\sqrt{x^2 + 16})^2 + (\sqrt{x^2 + 81})^2 = 13^2$

$x^2 + 16 + x^2 + 81 = 169$

$2x^2 + 97 = 169$

$2x^2 = 72$

$x^2 = 36$

$x = 6$

33. $A = \frac{1}{2}\cdot b\cdot h$

$A = \frac{1}{2}\cdot a\cdot\frac{a\sqrt{3}}{2}$ (Substituting $\frac{a\sqrt{3}}{2}$ for h) (See Exercise 32.)

$= \frac{a^2\sqrt{3}}{4}$

Exercise Set 10.9

1. $\sqrt{x} = 5$

$(\sqrt{x})^2 = 5^2$

$x = 25$

Check: $\sqrt{x} = 5$; $\sqrt{25}$ | 5 ; 5

3. $\sqrt{x} = 6.2$

$(\sqrt{x})^2 = (6.2)^2$

$x = 38.44$

Check: $\sqrt{x} = 6.2$; $\sqrt{38.44}$ | 6.2 ; 6.2

5. $\sqrt{x + 3} = 20$

$(\sqrt{x + 3})^2 = 20^2$

$x + 3 = 400$

$x = 397$

Check: $\sqrt{x + 3} = 20$; $\sqrt{397 + 3}$ | 20 ; $\sqrt{400}$; 20

7. $\sqrt{2x + 4} = 25$

$(\sqrt{2x + 4})^2 = 25^2$

$2x + 4 = 625$

$2x = 621$

$x = \frac{621}{2}$

Check: $\sqrt{2x + 4} = 25$; $\sqrt{2\cdot\frac{621}{2} + 4}$ | 25 ; $\sqrt{621 + 4}$; $\sqrt{625}$; 25

9. $3 + \sqrt{x - 1} = 5$

$\sqrt{x - 1} = 2$

$(\sqrt{x - 1})^2 = 2^2$

$x - 1 = 4$

$x = 5$

Check: $3 + \sqrt{x - 1} = 5$; $3 + \sqrt{5 - 1}$ | 5 ; $3 + \sqrt{4}$; $3 + 2$; 5

11. $6 - 2\sqrt{3n} = 0$

$6 = 2\sqrt{3n}$

$6^2 = (2\sqrt{3n})^2$

$36 = 4\cdot 3n$

$36 = 12n$

$3 = n$

Check: $6 - 2\sqrt{3n} = 0$; $6 - 2\sqrt{3\cdot 3}$ | 0 ; $6 - 2\cdot 3$; $6 - 6$; 0

13. $\sqrt{5x - 7} = \sqrt{x + 10}$

$(\sqrt{5x - 7})^2 = (\sqrt{x + 10})^2$

$5x - 7 = x + 10$

$4x = 17$

$x = \frac{17}{4}$

Check: $\sqrt{5x - 7} = \sqrt{x + 10}$; $\sqrt{5\cdot\frac{17}{4} - 7}$ | $\sqrt{\frac{17}{4} + 10}$; $\sqrt{\frac{85}{4} - \frac{28}{4}}$ | $\sqrt{\frac{57}{4}}$; $\sqrt{\frac{57}{4}}$

15. $\sqrt{x} = -7$

There is no solution. The principal square root of x cannot be negative.

17.
$$\sqrt{2y+6} = \sqrt{2y-5}$$
$$(\sqrt{2y+6})^2 = (\sqrt{2y-5})^2$$
$$2y + 6 = 2y - 5$$
$$6 = -5$$

The equation 6 = -5 is false; there is no solution.

19.
$$V = 3.5\sqrt{h}$$
$$V = 3.5\sqrt{9800} \quad \text{(Substituting 9800 for h)}$$
$$= 3.5\sqrt{4900\cdot 2}$$
$$= 3.5(70)\sqrt{2}$$
$$= 245\sqrt{2}$$
$$\approx 346.48$$

You can see 346.48 km to the horizon.

21.
$$V = 3.5\sqrt{h}$$
$$371 = 3.5\sqrt{h} \quad \text{(Substituting 371 for V)}$$
$$\frac{371}{3.5} = \sqrt{h}$$
$$106 = \sqrt{h}$$
$$106^2 = (\sqrt{h})^2$$
$$11{,}236 = h$$

The airplane is 11,236 m high.

23.
$$r = 2\sqrt{5L}$$
$$50 = 2\sqrt{5L} \quad \text{(Substituting 50 for r)}$$
$$25 = \sqrt{5L}$$
$$25^2 = (\sqrt{5L})^2$$
$$625 = 5L$$
$$125 = L$$

The car will skid 125 ft.

$$r = 2\sqrt{5L}$$
$$70 = 2\sqrt{5L} \quad \text{(Substituting 70 for r)}$$
$$35 = \sqrt{5L}$$
$$35^2 = (\sqrt{5L})^2$$
$$1225 = 5L$$
$$245 = L$$

The car will skid 245 ft.

25.
$$A = \frac{\pi r^2 E}{180}$$
$$180A = \pi r^2 E$$
$$\frac{180A}{\pi r^2} = E$$

27. $E = kI$ (Direct variation)

We first find k.

$$6000 = k\cdot 25{,}000 \quad \text{(Substituting \$25,000 for I and \$6000 for E)}$$
$$\frac{6000}{25{,}000} = k$$
$$0.24 = k$$

The equation of variation is

$E = 0.24I$

We substitute $30,000 for I and solve for E.

$E = 0.24(30{,}000)$

$E = 7200$

A family making $30,000 a year will spend $7200 for entertainment.

29.
$$\sqrt{x} = -x$$
$$(\sqrt{x})^2 = (-x)^2$$
$$x = x^2$$
$$0 = x^2 - x$$
$$0 = x(x - 1)$$

$x = 0$ or $x - 1 = 0$

$x = 0$ or $x = 1$

Check: For 0

$\sqrt{x}$	$= -x$
$\sqrt{0}$	-0
0	0

For 1

$\sqrt{x}$	$= -x$
$\sqrt{1}$	-1
1	

Only 0 checks. It is the solution.

31.
$$\sqrt{x-5} + \sqrt{x} = 5$$
$$\sqrt{x} = 5 - \sqrt{x-5}$$
$$(\sqrt{x})^2 = (5 - \sqrt{x-5})^2$$
$$x = 25 - 10\sqrt{x-5} + x - 5$$
$$x = 20 + x - 10\sqrt{x-5}$$
$$10\sqrt{x-5} = 20$$
$$\sqrt{x-5} = 2$$
$$(\sqrt{x-5})^2 = 2^2$$
$$x - 5 = 4$$
$$x = 9$$

Since 9 checks, it is the solution.

Exercise Set 11.1

1. $x^2 = 3x + 2$
$x^2 - 3x - 2 = 0$
$a = 1,\ b = -3,\ c = -2$

3. $7x^2 = 4x - 3$
$7x^2 - 4x + 3 = 0$
$a = 7,\ b = -4,\ c = 3$

5. $2x - 1 = 3x^2 + 7$
$0 = 3x^2 - 2x + 8$
$a = 3,\ b = -2,\ c = 8$

7. $x^2 = 4$
$x = \sqrt{4}$ or $x = -\sqrt{4}$
$x = 2$ or $x = -2$

Check:

For 2: $x^2 = 4$; $2^2 = 4$ ✓ ($4 \mid 4$)

For -2: $x^2 = 4$; $(-2)^2 = 4$ ✓ ($4 \mid 4$)

The solutions are 2 and -2.

9. $x^2 = 49$
$x = \sqrt{49}$ or $x = -\sqrt{49}$
$x = 7$ or $x = -7$
Both check. The solutions are 7 and -7.

11. $x^2 = 7$
$x = \sqrt{7}$ or $x = -\sqrt{7}$

Check:

For $\sqrt{7}$: $x^2 = 7$; $(\sqrt{7})^2 = 7$ ($7 \mid 7$)

For $-\sqrt{7}$: $x^2 = 7$; $(-\sqrt{7})^2 = 7$ ($7 \mid 7$)

The solutions are $\sqrt{7}$ and $-\sqrt{7}$.

13. $3x^2 = 30$
$x^2 = 10$
$x = \sqrt{10}$ or $x = -\sqrt{10}$
Both check. The solutions are $\sqrt{10}$ and $-\sqrt{10}$.

15. $3x^2 = 24$
$x^2 = 8$
$x = \sqrt{8}$ or $x = -\sqrt{8}$
$x = 2\sqrt{2}$ or $x = -2\sqrt{2}$

Check:

For $2\sqrt{2}$: $3x^2 = 24$; $3(2\sqrt{2})^2$, $3\cdot 8$, $24 \mid 24$

For $-2\sqrt{2}$: $3x^2 = 24$; $3(-2\sqrt{2})^2$, $3\cdot 8$, $24 \mid 24$

The solutions are $2\sqrt{2}$ and $-2\sqrt{2}$.

17. $4x^2 - 25 = 0$
$4x^2 = 25$
$x^2 = \frac{25}{4}$
$x = \sqrt{\frac{25}{4}}$ or $x = -\sqrt{\frac{25}{4}}$
$x = \frac{5}{2}$ or $x = -\frac{5}{2}$

Check:

For $\frac{5}{2}$: $4x^2 - 25 = 0$; $4(\frac{5}{2})^2 - 25$, $4\cdot\frac{25}{4} - 25$, $25 - 25$, $0 \mid 0$

For $-\frac{5}{2}$: $4x^2 - 25 = 0$; $4(-\frac{5}{2})^2 - 25$, $4\cdot\frac{25}{4} - 25$, $25 - 25$, $0 \mid 0$

The solutions are $\frac{5}{2}$ and $-\frac{5}{2}$.

19. $3x^2 - 49 = 0$
$3x^2 = 49$
$x^2 = \frac{49}{3}$
$x = \sqrt{\frac{49}{3}}$ or $x = -\sqrt{\frac{49}{3}}$
$x = \sqrt{\frac{49}{3}\cdot\frac{3}{3}}$ or $x = -\sqrt{\frac{49}{3}\cdot\frac{3}{3}}$
$x = \frac{7\sqrt{3}}{3}$ or $x = -\frac{7\sqrt{3}}{3}$
Both check. The solutions are $\frac{7\sqrt{3}}{3}$ and $-\frac{7\sqrt{3}}{3}$.

21. $4y^2 - 3 = 9$
$4y^2 = 12$
$y^2 = 3$
$y = \sqrt{3}$ or $y = -\sqrt{3}$
Both check. The solutions are $\sqrt{3}$ and $-\sqrt{3}$.

23. $s = 16t^2$

$1000 = 16t^2$ (Substituting 1000 for s)

$\frac{1000}{16} = t^2$

$62.5 = t^2$

$\sqrt{62.5} = t$ (Taking the principal square root)

$7.9 \approx t$

25. $s = 16t^2$

$175 = 16t^2$ (Substituting 175 for s)

$\frac{175}{16} = t^2$

$10.9375 = t^2$

$\sqrt{10.9375} = t$ (Taking the principal square root)

$3.3 \approx t$

27. $x^2 - 6x = 0$

$x(x - 6) = 0$

$x = 0$ or $x - 6 = 0$

$x = 0$ or $x = 6$

The solutions are 0 and 6.

29. $\sqrt{20} = \sqrt{4 \cdot 5} = 2\sqrt{5}$

31. $4.82x^2 = 12,000$

$x^2 = \frac{12,000}{4.82}$

$x^2 \approx 2489.627$

$x \approx \sqrt{2489.627}$ or $x \approx -\sqrt{2489.627}$

$x \approx 49.896$ or $x \approx -49.896$

Both check. The solutions are approximately 49.896 and -49.896.

33. $\frac{4}{x^2 - 7} = \frac{6}{x^2}$, LCM $= x^2(x^2 - 7)$

$x^2(x^2 - 7) \cdot \frac{4}{x^2 - 7} = x^2(x^2 - 7) \cdot \frac{6}{x^2}$

$4x^2 = (x^2 - 7) \cdot 6$

$4x^2 = 6x^2 - 42$

$42 = 2x^2$

$21 = x^2$

$x = \sqrt{21}$ or $x = -\sqrt{21}$

Both check. The solutions are $\sqrt{21}$ and $-\sqrt{21}$.

Exercise Set 11.2

1. $x^2 + 7x = 0$

$x(x + 7) = 0$

$x = 0$ or $x + 7 = 0$

$x = 0$ or $x = -7$

The solutions are 0 and -7.

3. $3x^2 + 2x = 0$

$x(3x + 2) = 0$

$x = 0$ or $3x + 2 = 0$

$x = 0$ or $3x = -2$

$x = 0$ or $x = -\frac{2}{3}$

The solutions are 0 and $-\frac{2}{3}$.

5. $4x^2 + 4x = 0$

$4x(x + 1) = 0$

$x = 0$ or $x + 1 = 0$

$x = 0$ or $x = -1$

The solutions are 0 and -1.

7. $55x^2 - 11x = 0$

$11x(5x - 1) = 0$

$x = 0$ or $5x - 1 = 0$

$x = 0$ or $5x = 1$

$x = 0$ or $x = \frac{1}{5}$

The solutions are 0 and $\frac{1}{5}$.

9. $x^2 - 16x + 48 = 0$

$(x - 12)(x - 4) = 0$

$x - 12 = 0$ or $x - 4 = 0$

$x = 12$ or $x = 4$

The solutions are 12 and 4.

11. $x^2 + 4x - 21 = 0$

$(x + 7)(x - 3) = 0$

$x + 7 = 0$ or $x - 3 = 0$

$x = -7$ or $x = 3$

The solutions are -7 and 3.

13. $x^2 + 10x + 25 = 0$

$(x + 5)(x + 5) = 0$

$x + 5 = 0$ or $x + 5 = 0$

$x = -5$ or $x = -5$

The solution is -5.

15. $2x^2 - 13x + 15 = 0$

$(2x - 3)(x - 5) = 0$

$2x - 3 = 0$ or $x - 5 = 0$

$2x = 3$ or $x = 5$

$x = \frac{3}{2}$ or $x = 5$

The solutions are $\frac{3}{2}$ and 5.

17. $3x^2 - 7x = 20$

$3x^2 - 7x - 20 = 0$

$(3x + 5)(x - 4) = 0$

$3x + 5 = 0$ or $x - 4 = 0$

$3x = -5$ or $x = 4$

$x = -\frac{5}{3}$ or $x = 4$

The solutions are $-\frac{5}{3}$ and 4.

19. $t(t - 5) = 14$

$t^2 - 5t = 14$

$t^2 - 5t - 14 = 0$

$(t - 7)(t + 2) = 0$

$t - 7 = 0$ or $t + 2 = 0$

$t = 7$ or $t = -2$

The solutions are 7 and -2.

21. $t(9 + t) = 4(2t + 5)$

$9t + t^2 = 8t + 20$

$t^2 + t - 20 = 0$

$(t + 5)(t - 4) = 0$

$t + 5 = 0$ or $t - 4 = 0$

$t = -5$ or $t = 4$

The solutions are -5 and 4.

23. $d = \frac{n^2 - 3n}{2}$

$d = \frac{6^2 - 3 \cdot 6}{2}$ (Substituting 6 for n)

$= \frac{36 - 18}{2}$

$= \frac{18}{2}$

$= 9$

A hexagon has 9 diagonals.

25. $d = \frac{n^2 - 3n}{2}$

$14 = \frac{n^2 - 3n}{2}$ (Substituting 14 for d)

$28 = n^2 - 3n$

$0 = n^2 - 3n - 28$

$0 = (n - 7)(n + 4)$

$n - 7 = 0$ or $n + 4 = 0$

$n = 7$ or $n = -4$

Since the number of sides cannot be negative, -4 cannot be a solution. We substitute 7 for n in the original equation: $d = \frac{7^2 - 3 \cdot 7}{2} = \frac{49 - 21}{2} = 14$ This checks. This polygon has 7 sides.

27. $(3x + 1)^2$

$= (3x)^2 + 2(3x)(1) + 1^2$

$= 9x^2 + 6x + 1$

29. $\sqrt{17} \approx 4.123$

31. $4m^2 - (m + 1)^2 = 0$

$4m^2 - (m^2 + 2m + 1) = 0$

$4m^2 - m^2 - 2m - 1 = 0$

$3m^2 - 2m - 1 = 0$

$(3m + 1)(m - 1) = 0$

$3m + 1 = 0$ or $m - 1 = 0$

$3m = -1$ or $m = 1$

$m = -\frac{1}{3}$ or $m = 1$

The solutions are $-\frac{1}{3}$ and 1.

33. $0.0025x^2 + 70,400x = 0$

$0.0025x(x + 28,160,000) = 0$

$x = 0$ or $x + 28,160,000 = 0$

$x = 0$ or $x = -28,160,000$

The solutions are 0 and -28,160,000.

35. $y^4 - 4y^2 + 4 = 0$

Let $y^2 = x$. Then $y^4 = x^2$. We substitute.

$x^2 - 4x + 4 = 0$

$(x - 2)(x - 2) = 0$

$x - 2 = 0$ or $x - 2 = 0$

$x = 2$ or $x = 2$

If $x = 2$, $y^2 = 2$ and $y = \sqrt{2}$ or $-\sqrt{2}$.

Exercise Set 11.3

1. $(x + 2)^2 = 25$

$x + 2 = 5$ or $x + 2 = -5$

$x = 3$ or $x = -7$

The solutions are 3 and -7.

3. $(x + 1)^2 = 6$

$x + 1 = \sqrt{6}$ or $x + 1 = -\sqrt{6}$

$x = -1 + \sqrt{6}$ or $x = -1 - \sqrt{6}$

The solutions are $-1 + \sqrt{6}$ and $-1 - \sqrt{6}$, or simply $-1 \pm \sqrt{6}$.

5. $(x - 3)^2 = 6$

$x - 3 = \sqrt{6}$ or $x - 3 = -\sqrt{6}$

$x = 3 + \sqrt{6}$ or $x = 3 - \sqrt{6}$

The solutions are $3 + \sqrt{6}$ and $3 - \sqrt{6}$, or simply $3 \pm \sqrt{6}$.

7. $x^2 - 2x$ $\quad (\frac{-2}{2})^2 = (-1)^2 = 1$

The trinomial $x^2 - 2x + 1$ is the square of $x - 1$.

9. $x^2 + 18x$ $\quad (\frac{18}{2})^2 = 9^2 = 81$

The trinomial $x^2 + 18x + 81$ is the square of $x + 9$

11. $x^2 - x$ $\quad (\frac{-1}{2})^2 = (-\frac{1}{2})^2 = \frac{1}{4}$

The trinomial $x^2 - x + \frac{1}{4}$ is the square of $x - \frac{1}{2}$.

13. $x^2 + 5x$ $\quad (\frac{5}{2})^2 = \frac{25}{4}$

The trinomial $x^2 + 5x + \frac{25}{4}$ is the square of $x + \frac{5}{2}$.

15. $A = P(1 + r)^t$

$1210 = 1000(1 + r)^2$ (Substituting 1210 for A, 1000 for P and 2 for t)

$\frac{1210}{1000} = (1 + r)^2$

$1.21 = (1 + r)^2$

$\sqrt{1.21} = 1 + r$ (Taking the principal square root. Since r must be positive, so would 1 + r be positive.)

$1.1 = 1 + r$

$0.1 = r$

$10\% = r$

The interest rate must be 10% for \$1000 to grow to \$1210 in 2 years.

17. $A = P(1 + r)^t$

$3610 = 2560(1 + r)^2$ (Substituting 3610 for A, 2560 for P and 2 for t)

$\frac{3610}{2560} = (1 + r)^2$

$\frac{361}{256} = (1 + r)^2$

$\sqrt{\frac{361}{256}} = 1 + r$ (Taking the principal square root. Since r must be positive, so would 1 + r be positive.)

$\frac{19}{16} = 1 + r$

$1.1875 = 1 + r$

$0.1875 = r$

$18.75\% = r$

The interest rate must be 18.75% for \$2560 to grow to \$3610 in 2 years.

19. $A = P(1 + r)^t$

$7290 = 6250(1 + r)^2$ (Substituting 7290 for A, 6250 for P and 2 for t)

$\frac{7290}{6250} = (1 + r)^2$

$\frac{729}{625} = (1 + r)^2$

$\sqrt{\frac{729}{625}} = 1 + r$ (Taking the principal square root)

$\frac{27}{25} = 1 + r$

$1.08 = 1 + r$

$0.08 = r$

$8\% = r$

The interest rate must be 8% for \$6250 to grow to \$7290 in 2 years.

21. $A = P(1 + r)^t$

$3600 = 2500(1 + r)^2$ (Substituting 3600 for A, 2500 for P, and 2 for t)

$\frac{3600}{2500} = (1 + r)^2$

$\frac{36}{25} = (1 + r)^2$

$\sqrt{\frac{36}{25}} = 1 + r$ (Taking the principal square root)

$\frac{6}{5} = 1 + r$

$1.2 = 1 + r$

$0.2 = r$

$20\% = r$

The interest rate must be 20% for \$2500 to grow to \$3600 in 2 years.

23. $2x + 5y = 3$
$3x + 2y = 10$

We use the multiplication principle with each equation and then add.

$$\begin{aligned} -4x - 10y &= -6 && \text{(Multiplying by -2)} \\ 15x + 10y &= 50 && \text{(Multiplying by 5)} \\ 11x &= 44 \\ x &= 4 \end{aligned}$$

Substitute 4 for x in either equation of the original system and solve for y.

$$\begin{aligned} 2x + 5y &= 3 \\ 2\cdot4 + 5y &= 3 \\ 8 + 5y &= 3 \\ 5y &= -5 \\ y &= -1 \end{aligned}$$

The ordered pair (4,-1) checks and is the solution.

25. The job takes Jack 5 hours working alone and Jill 10 hours working alone. Then in 1 hour Jack does $\frac{1}{5}$ of the job and Jill does $\frac{1}{10}$ of the job. Working together they can do $\frac{1}{5} + \frac{1}{10}$ of the job in 1 hour. If they work together t hours, then Jack does $t(\frac{1}{5})$ of the job and Jill does $t(\frac{1}{10})$ of the job. We want some number t such that $t(\frac{1}{5}) + t(\frac{1}{10}) = 1$.

We solve the equation. The LCM is 10.

$$\begin{aligned} 10\cdot(\frac{t}{5} + \frac{t}{10}) &= 10\cdot1 \\ 2t + t &= 10 \\ 3t &= 10 \\ t &= \frac{10}{3}, \text{ or } 3\frac{1}{3} \end{aligned}$$

Working together it takes them $3\frac{1}{3}$ hr to paint the shed.

27. $x^2 + 2x + 1 = 81$
$(x + 1)^2 = 81$

$x + 1 = 9$ or $x + 1 = -9$
$x = 8$ or $x = -10$

The solutions are 8 and -10.

29.
$$\begin{aligned} A &= P(1 + r)^t \\ 1267.88 &= 1000(1 + r)^2 && \text{(Substituting)} \\ \frac{1267.88}{1000} &= (1 + r)^2 \\ 1.26788 &= (1 + r)^2 \\ \sqrt{1.26788} &= 1 + r && \text{(Taking the principal square root)} \\ 1.126 &\approx 1 + r \\ 0.126 &\approx r \\ 12.6\% &\approx r \end{aligned}$$

The interest rate is 12.6%.

Exercise Set 11.4

1. $x^2 - 6x - 16 = 0$
$x^2 - 6x = 16$
$x^2 - 6x + 9 = 16 + 9$ $\quad \left[\frac{-6}{2} = -3,\ (-3)^2 = 9\right]$
$(x - 3)^2 = 25$

$x - 3 = 5$ or $x - 3 = -5$
$x = 8$ or $x = -2$

The solutions are 8 and -2.

3. $x^2 + 22x + 21 = 0$
$x^2 + 22x = -21$
$x^2 + 22x + 121 = -21 + 121$ $\quad \left[\frac{22}{2} = 11,\ 11^2 = 121\right]$
$(x + 11)^2 = 100$

$x + 11 = 10$ or $x + 11 = -10$
$x = -1$ or $x = -21$

The solutions are -1 and -21.

5. $x^2 - 2x - 5 = 0$
$x^2 - 2x = 5$
$x^2 - 2x + 1 = 5 + 1$ $\quad \left[\frac{-2}{1} = -1,\ (-1)^2 = 1\right]$
$(x - 1)^2 = 6$

$x - 1 = \sqrt{6}$ or $x - 1 = -\sqrt{6}$
$x = 1 + \sqrt{6}$ or $x = 1 - \sqrt{6}$

The solutions are $1 \pm \sqrt{6}$.

7. $x^2 - 18x + 74 = 0$
$x^2 - 18x = -74$
$x^2 - 18x + 81 = -74 + 81$ $\quad \left[\frac{-18}{2} = -9,\ (-9)^2 = 81\right]$
$(x - 9)^2 = 7$

$x - 9 = \sqrt{7}$ or $x - 9 = -\sqrt{7}$
$x = 9 + \sqrt{7}$ or $x = 9 - \sqrt{7}$

The solutions are $9 \pm \sqrt{7}$.

9. $x^2 + 7x - 18 = 0$

$x^2 + 7x = 18$

$x^2 + 7x + \frac{49}{4} = 18 + \frac{49}{4}$ $\left[(\frac{7}{2})^2 = \frac{49}{4}\right]$

$(x + \frac{7}{2})^2 = \frac{121}{4}$

$x + \frac{7}{2} = \frac{11}{2}$ or $x + \frac{7}{2} = -\frac{11}{2}$

$x = \frac{4}{2}$ or $x = -\frac{18}{2}$

$x = 2$ or $x = -9$

The solutions are 2 and -9.

11. $x^2 + x - 6 = 0$

$x^2 + x = 6$

$x^2 + x + \frac{1}{4} = 6 + \frac{1}{4}$ $\left[(\frac{1}{2})^2 = \frac{1}{4}\right]$

$(x + \frac{1}{2})^2 = \frac{25}{4}$

$x + \frac{1}{2} = \frac{5}{2}$ or $x + \frac{1}{2} = -\frac{5}{2}$

$x = \frac{4}{2}$ or $x = -\frac{6}{2}$

$x = 2$ or $x = -3$

The solutions are 2 and -3.

13. $x^2 - 7x - 2 = 0$

$x^2 - 7x = 2$

$x^2 - 7x + \frac{49}{4} = 2 + \frac{49}{4}$

$(x - \frac{7}{2})^2 = \frac{57}{4}$

$x - \frac{7}{2} = \sqrt{\frac{57}{4}}$ or $x - \frac{7}{2} = -\sqrt{\frac{57}{4}}$

$x - \frac{7}{2} = \frac{\sqrt{57}}{2}$ or $x - \frac{7}{2} = -\frac{\sqrt{57}}{2}$

$x = \frac{7}{2} + \frac{\sqrt{57}}{2}$ or $x = \frac{7}{2} - \frac{\sqrt{57}}{2}$

$x = \frac{7 + \sqrt{57}}{2}$ or $x = \frac{7 - \sqrt{57}}{2}$

The solutions are $\frac{7 \pm \sqrt{57}}{2}$.

15. $x^2 + \frac{3}{2}x - \frac{1}{2} = 0$

$x^2 + \frac{3}{2}x = \frac{1}{2}$

$x^2 + \frac{3}{2}x + \frac{9}{16} = \frac{1}{2} + \frac{9}{16}$

$(x + \frac{3}{4})^2 = \frac{17}{16}$

$x + \frac{3}{4} = \sqrt{\frac{17}{16}}$ or $x + \frac{3}{4} = -\sqrt{\frac{17}{16}}$

$x + \frac{3}{4} = \frac{\sqrt{17}}{4}$ or $x + \frac{3}{4} = -\frac{\sqrt{17}}{4}$

$x = -\frac{3}{4} + \frac{\sqrt{17}}{4}$ or $x = -\frac{3}{4} - \frac{\sqrt{17}}{4}$

$x = \frac{-3 + \sqrt{17}}{4}$ or $x = \frac{-3 - \sqrt{17}}{4}$

The solutions are $\frac{-3 \pm \sqrt{17}}{4}$.

17. $3x^2 + 4x - 1 = 0$

$x^2 + \frac{4}{3}x - \frac{1}{3} = 0$

$x^2 + \frac{4}{3}x = \frac{1}{3}$

$x^2 + \frac{4}{3}x + \frac{4}{9} = \frac{1}{3} + \frac{4}{9}$

$(x + \frac{2}{3})^2 = \frac{7}{9}$

$x + \frac{2}{3} = \sqrt{\frac{7}{9}}$ or $x + \frac{2}{3} = -\sqrt{\frac{7}{9}}$

$x + \frac{2}{3} = \frac{\sqrt{7}}{3}$ or $x + \frac{2}{3} = -\frac{\sqrt{7}}{3}$

$x = \frac{-2 + \sqrt{7}}{3}$ or $x = \frac{-2 - \sqrt{7}}{3}$

The solutions are $\frac{-2 \pm \sqrt{7}}{3}$.

19. $4x^2 + 12x - 7 = 0$

$x^2 + 3x - \frac{7}{4} = 0$

$x^2 + 3x = \frac{7}{4}$

$x^2 + 3x + \frac{9}{4} = \frac{7}{4} + \frac{9}{4}$

$(x + \frac{3}{2})^2 = \frac{16}{4}$

$(x + \frac{3}{2})^2 = 4$

$x + \frac{3}{2} = 2$ or $x + \frac{3}{2} = -2$

$x = \frac{1}{2}$ or $x = -\frac{7}{2}$

The solutions are $\frac{1}{2}$ and $-\frac{7}{2}$.

21. $6x^2 + 11x - 10 = 0$

$$x^2 + \frac{11}{6}x - \frac{10}{6} = 0$$

$$x^2 + \frac{11}{6}x = \frac{10}{6}$$

$$x^2 + \frac{11}{6}x + \frac{121}{144} = \frac{10}{6} + \frac{121}{144}$$

$$(x + \frac{11}{12})^2 = \frac{361}{144}$$

$$x + \frac{11}{12} = \frac{19}{12} \text{ or } x + \frac{11}{12} = -\frac{19}{12}$$

$$x = \frac{8}{12} \text{ or } x = -\frac{30}{12}$$

$$x = \frac{2}{3} \text{ or } x = -\frac{5}{2}$$

The solutions are $\frac{2}{3}$ and $-\frac{5}{2}$.

23. $\sqrt{54} - \sqrt{24}$

$= \sqrt{9\cdot6} - \sqrt{4\cdot6}$

$= 3\sqrt{6} - 2\sqrt{6}$

$= (3 - 2)\sqrt{6}$

$= \sqrt{6}$

25. $7x - 2y = -31$

$4x - 3y = -27$

We multiply the first equation by 3 and the second equation by -2. Then add.

$21x - 6y = -93$

$-8x + 6y = 54$

$13x = -39$ (Adding)

$x = -3$

Substitute -3 for x in either of the original equations and solve for y.

$7x - 2y = -31$

$7(-3) - 2y = -31$

$-21 - 2y = -31$

$-2y = -10$

$y = 5$

The ordered pair (-3,5) checks. It is the solution.

27. $x^2 + qx + 55$

$$(\frac{q}{2})^2 = 55$$

$$\frac{q^2}{4} = 55$$

$$q^2 = 220$$

$q = \sqrt{220}$ or $q = -\sqrt{220}$

$q = 2\sqrt{55}$ or $q = -2\sqrt{55}$

The trinomial $x^2 + 2\sqrt{55}x + 55$ is the square of $x + \sqrt{55}$.

The trinomial $x^2 - 2\sqrt{55}x + 55$ is the square of $x - \sqrt{55}$.

Thus, q can be $2\sqrt{55}$ or $-2\sqrt{55}$.

29. $4x^2 + 4x + c = 0$

$$x^2 + x + \frac{c}{4} = 0$$

$$x^2 + x = -\frac{c}{4}$$

$$x^2 + x + \frac{1}{4} = -\frac{c}{4} + \frac{1}{4}$$

$$(x + \frac{1}{2})^2 = \frac{1 - c}{4}$$

$$x + \frac{1}{2} = \sqrt{\frac{1 - c}{4}} \text{ or } x + \frac{1}{2} = -\sqrt{\frac{1 - c}{4}}$$

$$x = -\frac{1}{2} + \frac{\sqrt{1 - c}}{2} \text{ or } x = -\frac{1}{2} - \frac{\sqrt{1 - c}}{2}$$

The solutions are $-\frac{1}{2} \pm \frac{\sqrt{1 - c}}{2}$, or $\frac{-1 \pm \sqrt{1 - c}}{2}$.

Exercise Set 11.5

1. $x^2 - 4x = 21$

$x^2 - 4x - 21 = 0$

$(x - 7)(x + 3) = 0$

$x - 7 = 0$ or $x + 3 = 0$

$x = 7$ or $x = -3$

The solutions are 7 and -3.

3. $x^2 = 6x - 9$

$x^2 - 6x + 9 = 0$

$(x - 3)(x - 3) = 0$

$x - 3 = 0$ or $x - 3 = 0$

$x = 3$ or $x = 3$

The solution is 3.

5. $3y^2 - 2y - 8 = 0$

$(3y + 4)(y - 2) = 0$

$3y + 4 = 0$ or $y - 2 = 0$

$3y = -4$ or $y = 2$

$y = -\frac{4}{3}$ or $y = 2$

The solutions are $-\frac{4}{3}$ and 2.

7. $x^2 - 9 = 0$

$(x - 3)(x + 3) = 0$

$x - 3 = 0$ or $x + 3 = 0$

$x = 3$ or $x = -3$

The solutions are 3 and -3.

9. $y^2 - 10y + 26 = 4$

$y^2 - 10y + 22 = 0$

$a = 1 \quad b = -10 \quad c = 22$

$$y = \frac{-b \pm \sqrt{b^2 - 4ac}}{2a}$$

$$y = \frac{-(-10) \pm \sqrt{(-10)^2 - 4\cdot1\cdot22}}{2\cdot1}$$

$$y = \frac{10 \pm \sqrt{100 - 88}}{2}$$

$$y = \frac{10 \pm \sqrt{12}}{2}$$

$$y = \frac{10 \pm 2\sqrt{3}}{2}$$

$$y = \frac{2(5 \pm \sqrt{3})}{2}$$

$$y = 5 \pm \sqrt{3}$$

11. $x^2 - 2x = 2$

$x^2 - 2x - 2 = 0$

$a = 1 \quad b = -2 \quad c = -2$

$$x = \frac{-b \pm \sqrt{b^2 - 4ac}}{2a}$$

$$x = \frac{-(-2) \pm \sqrt{(-2)^2 - 4\cdot1\cdot(-2)}}{2\cdot1}$$

$$x = \frac{2 \pm \sqrt{4 + 8}}{2}$$

$$x = \frac{2 \pm \sqrt{12}}{2}$$

$$x = \frac{2 \pm 2\sqrt{3}}{2}$$

$$x = \frac{2(1 \pm \sqrt{3})}{2}$$

$$x = 1 \pm \sqrt{3}$$

13. $4y^2 + 3y + 2 = 0$

$a = 4 \quad b = 3 \quad c = 2$

$$x = \frac{-b \pm \sqrt{b^2 - 4ac}}{2a}$$

$$x = \frac{-3 \pm \sqrt{3^2 - 4\cdot4\cdot2}}{2\cdot4}$$

$$x = \frac{-3 \pm \sqrt{9 - 32}}{8}$$

$$x = \frac{-3 \pm \sqrt{-23}}{8}$$

The discriminate, -23, is negative. There are no real-number solutions.

15. $3p^2 + 2p = 3$

$3p^2 + 2p - 3 = 0$

$a = 3 \quad b = 2 \quad c = -3$

$$p = \frac{-2 \pm \sqrt{2^2 - 4\cdot3\cdot(-3)}}{2\cdot3}$$

$$p = \frac{-2 \pm \sqrt{4 + 36}}{6}$$

$$p = \frac{-2 \pm \sqrt{40}}{6}$$

$$p = \frac{-2 \pm 2\sqrt{10}}{6}$$

$$p = \frac{2(-1 \pm \sqrt{10})}{2\cdot3}$$

$$p = \frac{-1 \pm \sqrt{10}}{3}$$

17. $(y + 4)(y + 3) = 15$

$y^2 + 7y + 12 = 15$

$y^2 + 7y - 3 = 0$

$a = 1 \quad b = 7 \quad c = -3$

$$y = \frac{-7 \pm \sqrt{7^2 - 4\cdot1\cdot(-3)}}{2\cdot1}$$

$$y = \frac{-7 \pm \sqrt{49 + 12}}{2}$$

$$y = \frac{-7 \pm \sqrt{61}}{2}$$

19. $5x + x(x - 7) = 0$

$5x + x^2 - 7x = 0$

$x^2 - 2x = 0$

$x(x - 2) = 0$

$x = 0$ or $x - 2 = 0$

$x = 0$ or $x = 2$

The solutions are 0 and 2.

21. $x^2 - 4x - 7 = 0$

$a = 1 \quad b = -4 \quad c = -7$

$$x = \frac{-(-4) \pm \sqrt{(-4)^2 - 4(1)(-7)}}{2\cdot1}$$

$$= \frac{4 \pm \sqrt{16 + 28}}{2}$$

$$= \frac{4 \pm \sqrt{44}}{2}$$

$$= \frac{4 \pm 2\sqrt{11}}{2}$$

$$= \frac{2(2 \pm \sqrt{11})}{2}$$

$$= 2 \pm \sqrt{11}$$

$$\approx 2 \pm 3.317$$

21. (continued)

$x \approx 2 + 3.317$ or $x \approx 2 - 3.317$

$x \approx 5.3$ or $x \approx -1.3$

The solutions are approximately 5.3 and -1.3.

23. $y^2 - 6y - 1 = 0$

$a = 1 \quad b = -6 \quad c = -1$

$$x = \frac{-(-6) \pm \sqrt{(-6)^2 - 4(1)(-1)}}{2\cdot 1}$$
$$= \frac{6 \pm \sqrt{36 + 4}}{2}$$
$$= \frac{6 \pm \sqrt{40}}{2}$$
$$= \frac{6 \pm 2\sqrt{10}}{2}$$
$$= \frac{2(3 \pm \sqrt{10})}{2}$$
$$= 3 \pm \sqrt{10}$$
$$\approx 3 \pm 3.162$$

$x \approx 3 + 3.162$ or $x \approx 3 - 3.162$

$x \approx 6.2$ or $x \approx -0.2$

The solutions are approximately 6.2 and -0.2.

25. $3x^2 + 4x - 2 = 0$

$a = 3 \quad b = 4 \quad c = -2$

$$x = \frac{-4 \pm \sqrt{4^2 - 4(3)(-2)}}{2\cdot 3}$$
$$= \frac{-4 \pm \sqrt{16 + 24}}{2\cdot 3}$$
$$= \frac{-4 \pm \sqrt{40}}{2\cdot 3}$$
$$= \frac{-4 \pm 2\sqrt{10}}{2\cdot 3}$$
$$= \frac{2(-2 \pm \sqrt{10})}{2\cdot 3}$$
$$= \frac{-2 \pm \sqrt{10}}{3}$$
$$\approx \frac{-2 \pm 3.162}{3}$$

$x \approx \frac{-2 + 3.162}{3}$ or $x \approx \frac{-2 - 3.162}{3}$

$x \approx \frac{1.162}{3}$ or $x \approx \frac{-5.162}{3}$

$x \approx 0.4$ or $x \approx -1.7$

The solutions are approximately 0.4 and -1.7.

27. $\sqrt{3x^2}\sqrt{9x^3}$

$$= \sqrt{27x^5}$$
$$= \sqrt{9\cdot x^4\cdot 3\cdot x}$$
$$= 3x^2\sqrt{3x}$$

29. $\sqrt{3} + \sqrt{\frac{1}{3}}$

$$= \sqrt{3} + \sqrt{\frac{1}{3}\cdot\frac{3}{3}}$$
$$= \sqrt{3} + \frac{\sqrt{3}}{3}$$
$$= \frac{3}{3}\sqrt{3} + \frac{1}{3}\sqrt{3}$$
$$= \frac{4}{3}\sqrt{3}$$

31. $0.8x^2 + 0.16x - 0.09 = 0$

$$80x^2 + 16x - 9 = 0$$
$$(20x + 9)(4x - 1) = 0$$

$20x + 9 = 0$ or $4x - 1 = 0$

$20x = -9$ or $4x = 1$

$x = -\frac{9}{20}$ or $x = \frac{1}{4}$

The solutions are $-\frac{9}{20}$ and $\frac{1}{4}$, or -0.45 and 0.25.

33. $ax^2 + 2x = 3$

$ax^2 + 2x - 3 = 0$

The discriminant is $\sqrt{2^2 - 4\cdot a\cdot(-3)}$, or $\sqrt{4 + 12a}$.

To have real-number solutions $4 + 12a \geqslant 0$.

We solve for a.

$$4 + 12a \geqslant 0$$
$$12a \geqslant -4$$
$$a \geqslant -\frac{4}{12}$$
$$a \geqslant -\frac{1}{3}$$

Exercise Set 11.6

1. $\frac{8}{x + 2} + \frac{8}{x - 2} = 3$, LCM = $(x + 2)(x - 2)$

$$(x + 2)(x - 2)\left(\frac{8}{x + 2} + \frac{8}{x - 2}\right) = (x + 2)(x - 2)\cdot 3$$
$$8(x - 2) + 8(x + 2) = 3(x^2 - 4)$$
$$8x - 16 + 8x + 16 = 3x^2 - 12$$
$$16x = 3x^2 - 12$$
$$0 = 3x^2 - 16x - 12$$
$$0 = (3x + 2)(x - 6)$$

$3x + 2 = 0$ or $x - 6 = 0$

$3x = -2$ or $x = 6$

$x = -\frac{2}{3}$ or $x = 6$

Both numbers check. The solutions are $-\frac{2}{3}$ and 6.

3. $\frac{1}{x} + \frac{1}{x+6} = \frac{1}{4}$, LCM = $4x(x + 6)$

$4x(x + 6)(\frac{1}{x} + \frac{1}{x+6}) = 4x(x + 6)\cdot\frac{1}{4}$

$4(x + 6) + 4x = x(x + 6)$

$4x + 24 + 4x = x^2 + 6x$

$8x + 24 = x^2 + 6x$

$0 = x^2 - 2x - 24$

$0 = (x - 6)(x + 4)$

$x - 6 = 0$ or $x + 4 = 0$

$x = 6$ or $x = -4$

Both numbers check. The solutions are 6 and -4.

5. $1 + \frac{12}{x^2 - 4} = \frac{3}{x - 2}$

LCM = $(x + 2)(x - 2)$

$(x + 2)(x - 2)(1 + \frac{12}{(x+2)(x-2)}) = (x + 2)(x - 2)\cdot\frac{3}{x-2}$

$(x + 2)(x - 2) + 12 = 3(x + 2)$

$x^2 - 4 + 12 = 3x + 6$

$x^2 + 8 = 3x + 6$

$x^2 - 3x + 2 = 0$

$(x - 2)(x - 1) = 0$

$x - 2 = 0$ or $x - 1 = 0$

$x = 2$ or $x = 1$

The number 1 checks, but 2 does not.
Thus the solution is 1.

7. $\frac{r}{r - 1} + \frac{2}{r^2 - 1} = \frac{8}{r + 1}$

LCM = $(r + 1)(r - 1)$

$(r + 1)(r - 1)(\frac{r}{r - 1} + \frac{2}{r^2 - 1}) = (r + 1)(r - 1)\frac{8}{r+1}$

$r(r + 1) + 2 = 8(r - 1)$

$r^2 + r + 2 = 8r - 8$

$r^2 - 7r + 10 = 0$

$(r - 5)(r - 2) = 0$

$r - 5 = 0$ or $r - 2 = 0$

$r = 5$ or $r = 2$

Both numbers check. The solutions are 5 and 2.

9. $\frac{4 - x}{x - 4} + \frac{x + 3}{x - 3} = 0$ LCM = $(x - 4)(x - 3)$

$(x - 4)(x - 3)(\frac{4 - x}{x - 4} + \frac{x + 3}{x - 3}) = (x - 4)(x - 3)\cdot 0$

$(x - 3)(4 - x) + (x - 4)(x + 3) = 0$

$(-x^2 + 7x - 12) + (x^2 - x - 12) = 0$

$6x - 24 = 0$

$6x = 24$

$x = 4$

The number 4 does not check. There is no solution.

11. $\frac{x^2}{x - 4} - \frac{7}{x - 4} = 0$

$\frac{x^2 - 7}{x - 4} = 0$, LCM = $x - 4$

$(x - 4)(\frac{x^2 - 7}{x - 4}) = (x - 4)\cdot 0$

$x^2 - 7 = 0$

$x^2 = 7$

$|x| = \sqrt{7}$

$x = \sqrt{7}$ or $x = -\sqrt{7}$

Both numbers check. The solutions are $\sqrt{7}$ and $-\sqrt{7}$.

13. $x + 2 = \frac{3}{x + 2}$, LCM = $x + 2$

$(x + 2)(x + 2) = (x + 2)(\frac{3}{x + 2})$

$x^2 + 4x + 4 = 3$

$x^2 + 4x + 1 = 0$

$a = 1 \quad b = 4 \quad c = 1$

$x = \frac{-4 \pm \sqrt{4^2 - 4\cdot1\cdot1}}{2\cdot1}$

$x = \frac{-4 \pm \sqrt{16 - 4}}{2}$

$x = \frac{-4 \pm \sqrt{12}}{2}$

$x = \frac{-4 \pm 2\sqrt{3}}{2}$

$x = \frac{2(-2 \pm \sqrt{3})}{2}$

$x = -2 \pm \sqrt{3}$

15. $\frac{1}{x} + \frac{1}{x + 6} = \frac{1}{5}$, LCM = $5x(x + 6)$

$5x(x + 6)(\frac{1}{x} + \frac{1}{x + 6}) = 5x(x + 6)\cdot\frac{1}{5}$

$5(x + 6) + 5x = x(x + 6)$

$5x + 30 + 5x = x^2 + 6x$

$10x + 30 = x^2 + 6x$

$0 = x^2 - 4x - 30$

$a = 1 \quad b = -4 \quad c = -30$

$x = \frac{-(-4) \pm \sqrt{(-4)^2 - 4\cdot1\cdot(-30)}}{2\cdot1}$

$x = \frac{4 \pm \sqrt{16 + 120}}{2}$

$x = \frac{4 \pm \sqrt{136}}{2}$

$x = \frac{4 \pm 2\sqrt{34}}{2}$

$x = \frac{2(2 \pm \sqrt{34})}{2}$

$x = 2 \pm \sqrt{34}$

17.
$$x - 7 = \sqrt{x - 5}$$
$$(x - 7)^2 = (\sqrt{x - 5})^2$$
$$x^2 - 14x + 49 = x - 5$$
$$x^2 - 15x + 54 = 0$$
$$(x - 6)(x - 9) = 0$$

$x - 6 = 0$ or $x - 9 = 0$

$x = 6$ or $x = 9$

Check:

For 6

$x - 7 = \sqrt{x - 5}$	
$6 - 7$	$\sqrt{6 - 5}$
-1	$\sqrt{1}$
	1

For 9

$x - 7 = \sqrt{x - 5}$	
$9 - 7$	$\sqrt{9 - 5}$
2	$\sqrt{4}$
	2

The number 9 checks, but 6 does not. Thus the solution is 9.

19.
$$\sqrt{x + 18} = x - 2$$
$$(\sqrt{x + 18})^2 = (x - 2)^2$$
$$x + 18 = x^2 - 4x + 4$$
$$0 = x^2 - 5x - 14$$
$$0 = (x - 7)(x + 2)$$

$x - 7 = 0$ or $x + 2 = 0$

$x = 7$ or $x = -2$

Check:

For 7

$\sqrt{x + 18} = x - 2$	
$\sqrt{7 + 18}$	$7 - 2$
$\sqrt{25}$	5
5	

For -2

$\sqrt{x + 18} = x - 2$	
$\sqrt{-2 + 18}$	$-2 - 2$
$\sqrt{16}$	-4
4	

The number 7 checks, but -2 does not. Thus the solution is 7.

21.
$$2\sqrt{x - 1} = x - 1$$
$$(2\sqrt{x - 1})^2 = (x - 1)^2$$
$$4(x - 1) = x^2 - 2x + 1$$
$$4x - 4 = x^2 - 2x + 1$$
$$0 = x^2 - 6x + 5$$
$$0 = (x - 5)(x - 1)$$

$x - 5 = 0$ or $x - 1 = 0$

$x = 5$ or $x = 1$

Check:

For 5

$2\sqrt{x - 1} = x - 1$	
$2\sqrt{5 - 1}$	$5 - 1$
$2\sqrt{4}$	4
$2\cdot 2$	
4	

For 1

$2\sqrt{x - 1} = x - 1$	
$2\sqrt{1 - 1}$	$1 - 1$
$2\sqrt{0}$	0
$2\cdot 0$	
0	

Both numbers check. The solutions are 5 and 1.

23.
$$\sqrt{5x + 21} = x + 3$$
$$(\sqrt{5x + 21})^2 = (x + 3)^2$$
$$5x + 21 = x^2 + 6x + 9$$
$$0 = x^2 + x - 12$$
$$0 = (x + 4)(x - 3)$$

$x + 4 = 0$ or $x - 3 = 0$

$x = -4$ or $x = 3$

Check:

For -4

$\sqrt{5x + 21} = x + 3$	
$\sqrt{5(-4) + 21}$	$-4 + 3$
$\sqrt{-20 + 21}$	-1
$\sqrt{1}$	
1	

For 3

$\sqrt{5x + 21} = x + 3$	
$\sqrt{5\cdot 3 + 21}$	$3 + 3$
$\sqrt{15 + 21}$	6
$\sqrt{36}$	
6	

The number 3 checks, but -4 does not. Thus the solution is 3.

25.
$$x = 1 + 6\sqrt{x - 9}$$
$$x - 1 = 6\sqrt{x - 9}$$
$$(x - 1)^2 = (6\sqrt{x - 9})^2$$
$$x^2 - 2x + 1 = 36(x - 9)$$
$$x^2 - 2x + 1 = 36x - 324$$
$$x^2 - 38x + 325 = 0$$
$$(x - 25)(x - 13) = 0$$

$x - 25 = 0$ or $x - 13 = 0$

$x = 25$ or $x = 13$

Check:

For 25

$x = 1 + 6\sqrt{x - 9}$	
25	$1 + 6\sqrt{25 - 9}$
	$1 + 6\sqrt{16}$
	$1 + 6\cdot 4$
	$1 + 24$
	25

For 13

$x = 1 + 6\sqrt{x - 9}$	
13	$1 + 6\sqrt{13 - 9}$
	$1 + 6\sqrt{4}$
	$1 + 6\cdot 2$
	$1 + 12$
	13

Both numbers check. The solutions are 25 and 13.

27.
$$\sqrt{x^2 + 6} - x + 3 = 0$$
$$\sqrt{x^2 + 6} = x - 3$$
$$(\sqrt{x^2 + 6})^2 = (x - 3)^2$$
$$x^2 + 6 = x^2 - 6x + 9$$
$$6 = -6x + 9$$
$$6x = 3$$
$$x = \frac{1}{2}$$

27. (continued)

Check:

$$\begin{array}{l|l} \sqrt{x^2 + 6} - x + 3 & 0 \\ \hline \sqrt{(\frac{1}{2})^2 + 6} - \frac{1}{2} + 3 & 0 \\ \sqrt{\frac{25}{4}} - \frac{1}{2} + 3 & \\ \frac{5}{2} - \frac{1}{2} + 3 & \\ 2 + 3 & \\ 5 & \end{array}$$

The number $\frac{1}{2}$ does not check.
There is no solution.

29. $\sqrt{(p + 6)(p + 1)} - 2 = p + 1$

$$\sqrt{p^2 + 7p + 6} = p + 3$$
$$(\sqrt{p^2 + 7p + 6})^2 = (p + 3)^2$$
$$p^2 + 7p + 6 = p^2 + 6p + 9$$
$$7p + 6 = 6p + 9$$
$$p = 3$$

Check:

$$\begin{array}{l|l} \sqrt{(p + 6)(p + 1)} - 2 & p + 1 \\ \hline \sqrt{(3 + 6)(3 + 1)} - 2 & 3 + 1 \\ \sqrt{9\cdot 4} - 2 & 4 \\ \sqrt{36} - 2 & \\ 6 - 2 & \\ 4 & \end{array}$$

The number 3 checks.
It is the solution.

31. $\frac{7}{1 + x} - 1 = \frac{5x}{x^2 + 3x + 2}$

$$\frac{7}{1 + x} - 1 = \frac{5x}{(x + 2)(x + 1)}$$

LCM = $(x + 1)(x + 2)$

$$(x + 1)(x + 2)(\frac{7}{1 + x} - 1) = (x+1)(x+2)\cdot\frac{5x}{(x+2)(x+1)}$$
$$7(x + 2) - (x + 1)(x + 2) = 5x$$
$$7x + 14 - x^2 - 3x - 2 = 5x$$
$$-x^2 + 4x + 12 = 5x$$
$$0 = x^2 + x - 12$$
$$0 = (x + 4)(x - 3)$$

$x + 4 = 0$ or $x - 3 = 0$

$x = -4$ or $x = 3$

Both numbers check. The solutions are -4 and 3.

33. $x + 1 + 3\sqrt{x + 1} = 4$

$$3\sqrt{x + 1} = 3 - x$$
$$(3\sqrt{x + 1})^2 = (3 - x)^2$$
$$9(x + 1) = 9 - 6x + x^2$$
$$9x + 9 = x^2 - 6x + 9$$
$$0 = x^2 - 15x$$
$$0 = x(x - 15)$$

$x = 0$ or $x - 15 = 0$

$x = 0$ or $x = 15$

The number 0 checks, but 15 does not.
Thus the solution is 0.

Exercise Set 11.7

1. $N = 2.5\sqrt{A}$

$$N^2 = (2.5\sqrt{A})^2$$
$$N^2 = 6.25A$$
$$\frac{N^2}{6.25} = A$$

3. $Q = \sqrt{\frac{aT}{c}}$

$$Q^2 = (\sqrt{\frac{aT}{c}})^2$$
$$Q^2 = \frac{aT}{c}$$
$$Q^2c = aT$$
$$\frac{Q^2c}{a} = T$$

5. $E = mc^2$

$$\frac{E}{m} = c^2$$
$$\sqrt{\frac{E}{m}} = c$$

7. $Q = ad^2 - cd$

$$Q = ad^2 - cd - Q$$

$a = a$ $b = -c$ $c = -Q$

$$d = \frac{-(-c) \pm \sqrt{(-c)^2 - 4\cdot a\cdot(-Q)}}{2a}$$
$$d = \frac{c \pm \sqrt{c^2 + 4aQ}}{2a}$$

9. $c^2 = a^2 + b^2$

$$c^2 - b^2 = a^2$$
$$\sqrt{c^2 - b^2} = a$$

11. $S = \frac{1}{2} gt^2$

$$2S = gt^2$$
$$\frac{2S}{g} = t^2$$
$$\sqrt{\frac{2S}{g}} = t$$

13. $A = \pi r^2 + 2\pi rh$

$0 = \pi r^2 + 2\pi hr - A$

$a = \pi \quad b = 2\pi h \quad c = -A$

$r = \dfrac{-2\pi h \pm \sqrt{(2\pi h)^2 - 4\cdot\pi\cdot(-A)}}{2\cdot\pi}$

$r = \dfrac{-2\pi h \pm \sqrt{4\pi^2h^2 + 4\pi A}}{2\pi}$

$r = \dfrac{-2\pi h \pm 2\sqrt{\pi^2h^2 + \pi A}}{2\pi}$

$r = \dfrac{-\pi h \pm \sqrt{\pi^2h^2 + \pi A}}{\pi}$

15. $A = \dfrac{\pi r^2 S}{360}$

$360A = \pi r^2 S$

$\dfrac{360A}{\pi S} = r^2$

$\sqrt{\dfrac{360A}{\pi S}} = r$

$\sqrt{\dfrac{36\cdot 10A}{\pi S}} = r$

$6\sqrt{\dfrac{10A}{\pi S}} = r$

17. $c = \sqrt{a^2 + b^2}$

$c^2 = a^2 + b^2$

$c^2 - b^2 = a^2$

$\sqrt{c^2 - b^2} = a$

19. $h = \dfrac{a}{2}\sqrt{3}$

$2h = a\sqrt{3}$

$\dfrac{2h}{\sqrt{3}} = a$

or $a = \dfrac{2h\sqrt{3}}{3}$

21. $\dfrac{x - 7}{x^2 - 9} - \dfrac{x - 7}{9 - x^2}$

$= \dfrac{x - 7}{x^2 - 9} - \dfrac{x - 7}{9 - x^2}\cdot\dfrac{-1}{-1}$

$= \dfrac{x - 7}{x^2 - 9} - \dfrac{7 - x}{x^2 - 9}$

$= \dfrac{x - 7 - 7 + x}{x^2 - 9}$

$= \dfrac{2x - 14}{x^2 - 9}$

23. $\sqrt{8x^3}\,\sqrt{2x^3y^4} = \sqrt{16x^6y^4} = 4x^3y^2$

25. $n = aT^2 - 4T + m$

$0 = aT^2 - 4T + m - n$

$a = a \quad b = -4 \quad c = m - n$

$T = \dfrac{-(-4) \pm \sqrt{(-4)^2 - 4\cdot a\cdot(m - n)}}{2\cdot a}$

$T = \dfrac{4 \pm \sqrt{16 - 4am + 4an}}{2a}$

$T = \dfrac{4 \pm 2\sqrt{4 - am + an}}{2a}$

$T = \dfrac{2 \pm \sqrt{4 - am + an}}{a}$

27. $y = ax^2 + bx + c$

$0 = ax^2 + bx + (c - y)$

$a = a \quad b = b \quad c = c - y$

We use the quadratic formula.

$x = \dfrac{-b \pm \sqrt{b^2 - 4a(c - y)}}{2a}$

Exercise Set 11.8

1. We first make a drawing and label it with both known and unknown information.

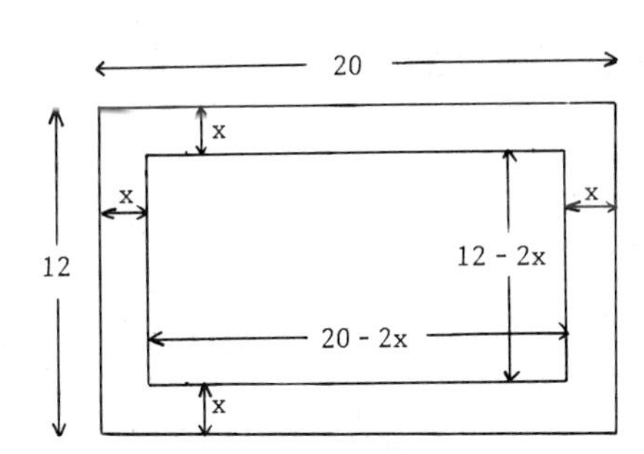

We let x represent the width of the frame. The length of the frame is 20 cm and the width is 12 cm. The length of the picture is 20 - 2x, and the width of the picture is 12 - 2x.

Recall that area is length × width. Thus, we have two expressions for the area of the picture: (20 - 2x)(12 - 2x) and 84. This gives us a translation.

$(20 - 2x)(12 - 2x) = 84$

We solve the equation.

$240 - 64x + 4x^2 = 84$

$4x^2 - 64x + 156 = 0$

$x^2 - 16x + 39 = 0$

$(x - 13)(x - 3) = 0$

$x - 13 = 0$ or $x - 3 = 0$

$x = 13$ or $x = 3$

1. (continued)

The number 13 is not a solution because when $x = 13$, $20 - 2x = -6$, and the length of the picture cannot be negative. When $x = 3$, $20 - 2x = 14$. This is the length. When $x = 3$, $12 - 2x = 6$. This is the width. The area is 14×6, or 84. This checks.

The width of the frame is 3 cm.

3. We first make a drawing and label it.

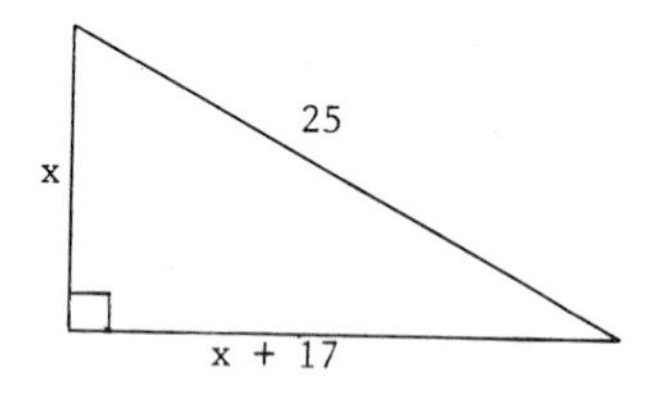

We let x represent the length of one leg. Then x + 1 represents the length of the other leg.

We use the Pythagorean equation.

$$x^2 + (x + 17)^2 = 25^2$$

We solve the equation.

$$x^2 + x^2 + 34x + 289 = 625$$
$$2x^2 + 34x - 336 = 0$$
$$x^2 + 17x - 168 = 0$$
$$(x - 7)(x + 24) = 0$$

$$x - 7 = 0 \text{ or } x + 24 = 0$$
$$x = 7 \text{ or } \quad x = -24$$

Since the length of a leg cannot be negative, -24 does not check. But 7 does check. If the smaller leg is 7, the other leg is 7 + 17, or 24. Then, $7^2 + 24^2 = 49 + 576 = 625$, and $\sqrt{625} = 25$, the length of the hypotenuse.

The legs measure 7 ft and 24 ft.

5. We first make a drawing.

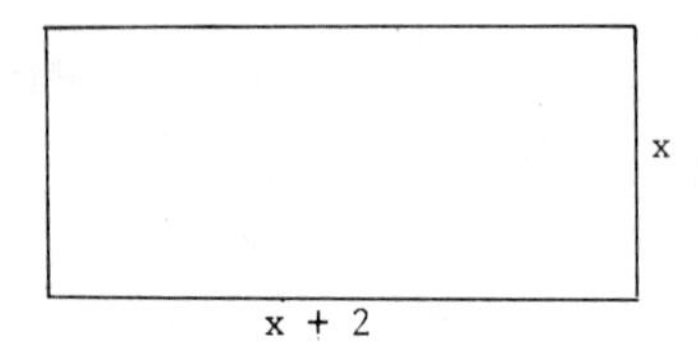

We let x represent the width. Then x + 2 represents the length. The area is length × width. Thus, we have two expressions for the area of the rectangle: $(x + 2)x$ and 80. This gives us a translation.

$$(x + 2)x = 80$$

5. (continued)

We solve the equation.

$$x^2 + 2x = 80$$
$$x^2 + 2x - 80 = 0$$
$$(x + 10)(x - 8) = 0$$

$$x + 10 = 0 \quad \text{or } x - 8 = 0$$
$$x = -10 \text{ or } \quad x = 8$$

Since the length of a side cannot be negative, -10 does not check. But 8 does check. If the width is 8, then the length is 8 + 2, or 10. The area is 10×8, or 80. This checks.

The length is 10 cm, and the width is 8 cm.

7. We first make a drawing.

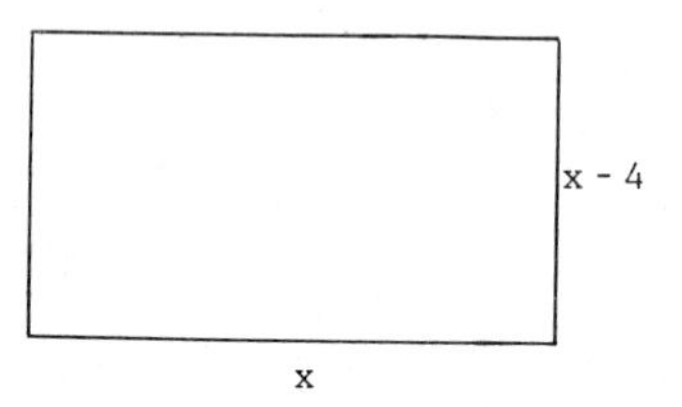

We let x represent the length. Then x - 4 represents the width. The area is length × width. Thus, we have two expressions for the area of the rectangle: $x(x - 4)$ and 320. This gives us a translation.

$$x(x - 4) = 320$$

We solve the equation.

$$x^2 - 4x = 320$$
$$x^2 - 4x - 320 = 0$$
$$(x - 20)(x + 16) = 0$$

$$x - 20 = 0 \quad \text{or } x + 16 = 0$$
$$x = 20 \text{ or } \quad x = -16$$

Since the length of a side cannot be negative, -16 does not check. But 20 does check. If the length is 20, then the width is 20 - 4, or 16. The area is 20×16, or 320. This checks.

The length is 20 cm, and the width is 16 cm.

9. We first make a drawing.

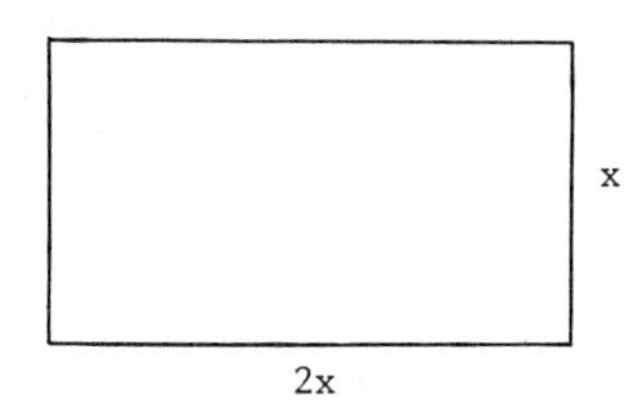

We let x represent the width. Then 2x represents the length. The area is length × width. Thus, we have two expressions for the area of the rectangle: $2x \cdot x$ and 50. This gives us a translation.

$2x \cdot x = 50$

We solve the equation.

$2x^2 = 50$

$x^2 = 25$

$x = 5$ or $x = -5$

Since the length of a side cannot be negative, -5 does not check. But 5 does check. If the width is 5, then the length is $2 \cdot 5$, or 10. The area is 10×5, or 50. This checks.

The length is 10 m, and the width is 5 m.

11. We first make a drawing.

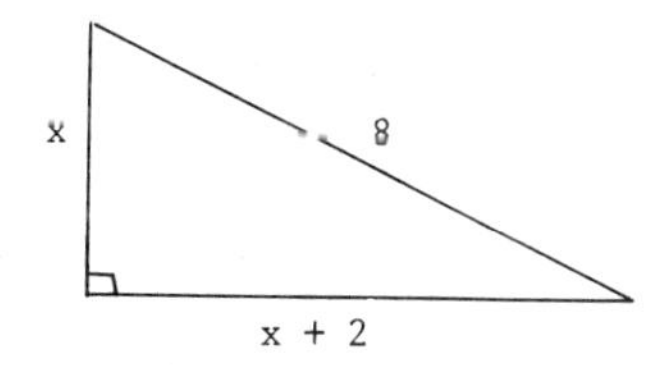

We let x represent the length of one leg. Then x + 2 represents the length of the other leg.

We use the Pythagorean equation.

$x^2 + (x + 2)^2 = 8^2$

We solve the equation.

$x^2 + x^2 + 4x + 4 = 64$

$2x^2 + 4x - 60 = 0$

$x^2 + 2x - 30 = 0$

$a = 1 \qquad b = 2 \qquad c = -30$

$$x = \frac{-2 \pm \sqrt{2^2 - 4 \cdot 1 \cdot (-30)}}{2 \cdot 1}$$

$$x = \frac{-2 \pm \sqrt{4 + 120}}{2}$$

$$x = \frac{-2 \pm \sqrt{124}}{2}$$

$$x = \frac{-2 \pm 2\sqrt{31}}{2}$$

$x = -1 \pm \sqrt{31}$

11. (continued)

$x = -1 + \sqrt{31}$ or $x = -1 - \sqrt{31}$

$x \approx -1 + 5.568$ or $x \approx -1 - 5.568$

$x \approx 4.568$ or $x \approx -6.568$

$x \approx 4.6$ or $x \approx -6.6$

Since the length of a leg cannot be negative, -6.6 does not check. But 4.6 does check. If the smaller leg is 4.6, the other leg is 4.6 + 2, or 6.6. Then $4.6^2 + 6.6^2 = 21.16 + 43.56 = 64.72$ and using a calculator, $\sqrt{64.72} \approx 8.04 \approx 8$. Note that our check does not come out exact since we are using an approximation.

One leg is about 4.6 m, and the other is about 6.6 m long.

13. We first make a drawing.

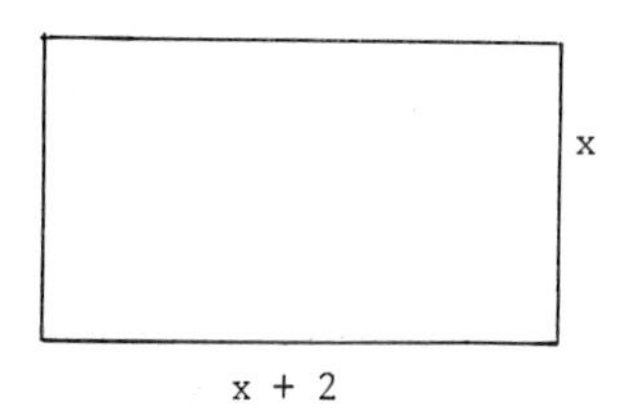

We let x represent the width and x + 2 the length. The area is length × width. We have two expressions for the area of the rectangle: $(x + 2)x$ and 20. This gives us a translation.

$(x + 2)x = 20$

We solve the equation.

$x^2 + 2x = 20$

$x^2 + 2x - 20 = 0$

$a = 1 \qquad b = 2 \qquad c = -20$

$$x = \frac{-2 \pm \sqrt{2^2 - 4 \cdot 1 \cdot (-20)}}{2 \cdot 1}$$

$$x = \frac{-2 \pm \sqrt{4 + 80}}{2}$$

$$x = \frac{-2 \pm \sqrt{84}}{2}$$

$$x = \frac{-2 \pm 2\sqrt{21}}{2}$$

$x = -1 \pm \sqrt{21}$

$x = -1 + \sqrt{21}$ or $x = -1 - \sqrt{21}$

$x \approx -1 + 4.583$ or $x \approx -1 - 4.583$

$x \approx 3.583$ or $x \approx -5.583$

$x \approx 3.6$ or $x \approx -5.6$

Since the length of a side cannot be negative, -5.6 does not check. But 3.6 does check. If the width is 3.6, then the length is 3.6 + 2, or 5.6. The area is 5.6(3.6), or $20.16 \approx 20$.

The length is about 5.6 in., and the width is about 3.6 in.

15. We first make a drawing.

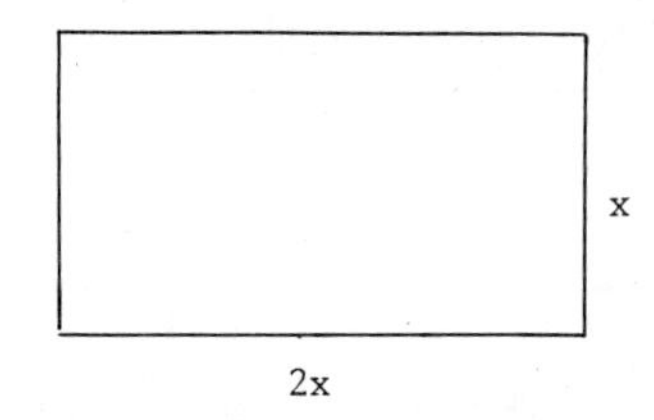

We let x represent the width and 2x the length. The area is length × width. We have two expressions for the area of the rectangle: $2x \cdot x$ and 10. This gives us a translation.

$2x \cdot x = 10$

We solve the equation.

$2x^2 = 10$

$x^2 = 5$

$x = \sqrt{5}$ or $x = -\sqrt{5}$

$x \approx 2.236$ or $x \approx -2.236$

$x \approx 2.2$ or $x \approx -2.2$

Since the length cannot be negative, -2.2 does not check. But 2.2 does check. If the width is 2.2 m, then the length is 2(2.2) or 4.4 m. The area is 4.4(2.2), or $9.68 \approx 10$.

The length is about 4.4 m, and the width is about 2.2 m.

17. We first make a drawing.

Upstream
r - 3 km/h
40 km

Downstream
r + 3 km/h
40 km

We let r represent the speed of the boat in still water. Then r - 3 is the speed of the boat traveling upstream and r + 3 is the speed of the boat traveling downstream. Using $t = \frac{d}{t}$, we can express the time for each part of the trip. We organize the information in a chart.

	Distance	Speed	Time
Upstream	40	r - 3	$\frac{40}{r-3}$
Downstream	40	r + 3	$\frac{40}{r+3}$
		Total	14

17. (continued)

Since the total time is 14 hours, we have

$$\frac{40}{r-3} + \frac{40}{r+3} = 14$$

We solve the equation.
We multiply by (r - 3)(r + 3), the LCM.

$$(r-3)(r+3)\left(\frac{40}{r-3} + \frac{40}{r+3}\right) = (r-3)(r+3)\cdot 14$$

$$40(r+3) + 40(r-3) = 14(r^2 - 9)$$

$$40r + 120 + 40r - 120 = 14r^2 - 126$$

$$80r = 14r^2 - 126$$

$$0 = 14r^2 - 80r - 126$$

$$0 = 7r^2 - 40r - 63$$

$$0 = (7r + 9)(r - 7)$$

$7r + 9 = 0$ or $r - 7 = 0$

$7r = -9$ or $r = 7$

$r = -\frac{9}{7}$ or $r = 7$

Since speed cannot be negative, $-\frac{9}{7}$ cannot be a solution. If the speed of the boat is 7 km/h, the speed upstream is 7 - 3, or 4 km/h, and the speed downstream is 7 + 3, or 10 km/h. The time upstream is $\frac{40}{4}$, or 10 hr. The time downstream is $\frac{40}{10}$, or 4 hr. The total time is 14 hr. This checks.

The speed of the boat in still water is 7 km/h.

19. We first make a drawing.

Upstream
r - 4 mph
4 mi

Downstream
r + 4 mph
12 mi

We let r represent the speed of the boat in still water. Then r - 4 is the speed of the boat traveling upstream and r + 4 is the speed of the boat traveling downstream. Using $t = \frac{d}{r}$, we can express the time for each part of the trip. We organize the information in a chart.

19. (continued)

	Distance	Speed	Time
Upstream	4	r - 4	$\frac{4}{r - 4}$
Downstream	12	r + 4	$\frac{12}{r + 4}$
		Total	2

Since the total time is 2 hours, we have

$$\frac{4}{r - 4} + \frac{12}{r + 4} = 2$$

We solve the equation.

We multiply by (r - 4)(r + 4), the LCM.

$$(r - 4)(r + 4)\left(\frac{4}{r - 4} + \frac{12}{r + 4}\right) = (r - 4)(r + 4)\cdot 2$$

$$4(r + 4) + 12(r - 4) = 2(r^2 - 16)$$

$$4r + 16 + 12r - 48 = 2r^2 - 32$$

$$16r - 32 = 2r^2 - 32$$

$$0 = 2r^2 - 16r$$

$$0 = 2r(r - 8)$$

$2r = 0$ or $r - 8 = 0$

$r = 0$ or $r = 8$

If r = 0, then the speed upstream, 0 - 4, would be negative. Since speed cannot be negative, 0 cannot be a solution. If the speed of the boat is 8 mph, the speed upstream is 8 - 4, or 4 mph and the speed downstream is 8 + 4, or 12 mph. The time upstream is $\frac{4}{4}$, or 1 hr. The time downstream is $\frac{12}{12}$, or 1 hr. The total time is 2 hr. This checks.

The speed of the boat in still water is 8 mph.

21. We first make a drawing.

Upstream
10 - r km/h
12 km

Downstream
10 + r km/h
28 km

21. (continued)

We let r represent the speed of the current. Then 10 - r is the speed of the boat traveling upstream and 10 + r is the speed of the boat traveling downstream. Using $t = \frac{d}{r}$, we can express the time for each part of the trip. We organize the information in a chart.

	Distance	Speed	Time
Upstream	12	10 - r	$\frac{12}{10 - r}$
Downstream	28	10 + r	$\frac{28}{10 + r}$
		Total	4

Since the total time is 4 hours, we have

$$\frac{12}{10 - r} + \frac{28}{10 + r} = 4$$

We solve the equation.

We multiply by (10 - r)(10 + r), the LCM.

$$(10 - r)(10 + r)\left(\frac{12}{10 - r} + \frac{28}{10 + r}\right) = (10-r)(10+r)\cdot 4$$

$$12(10 + r) + 28(10 - r) = 4(100 - r^2)$$

$$120 + 12r + 280 - 28r = 400 - 4r^2$$

$$400 - 16r = 400 - 4r^2$$

$$4r^2 - 16r = 0$$

$$r^2 - 4r = 0$$

$$r(r - 4) = 0$$

$r = 0$ or $r - 4 = 0$

$r = 0$ or $r = 4$

Since a stream is defined to be a flow of running water, its rate must be greater than 0. Thus, 0 cannot be a solution. If the speed of the current is 4 km/h, the speed upstream is 10 - 4, or 6 km/h and the speed downstream is 10 + 4, or 14 km/h. The time upstream is $\frac{12}{6}$, or 2 hours. The time downstream is $\frac{28}{14}$, or 2 hours. The total time is 4 hours.

The speed of the stream is 4 km/h.

23. First we make a drawing.

Against the wind
200 - r mph
738 miles

With the wind
200 + r mph
1062 miles

We let r represent the speed of the wind. Then the speed of the plane flying against the wind is 200 - r and the speed of the plane flying with the wind is 200 + r. Using $t = \frac{d}{r}$, we can express the time for each part of the trip. We organize the information in a chart.

	Distance	Speed	Time
Against the wind	738	200 - r	$\frac{738}{200 - r}$
With the wind	1062	200 + r	$\frac{1062}{200 + r}$
		Total	9

Since the total time is 9 hours, we have

$$\frac{738}{200 - r} + \frac{1062}{200 + r} = 9$$

We solve the equation.
We multiply by (200 - r)(200 + r), the LCM.

$$(200-r)(200+r)\left(\frac{738}{200-r} + \frac{1062}{200+r}\right) = (200-r)(200+r)\cdot 9$$

$$738(200 + r) + 1062(200 - r) = 9(40{,}000 - r^2)$$

$$147{,}600 + 738r + 212{,}400 - 1062r = 360{,}000 - 9r^2$$

$$360{,}000 - 324r = 360{,}000 - 9r^2$$

$$9r^2 - 324r = 0$$

$$9r(r - 36) = 0$$

$9r = 0$ or $r - 36 = 0$
$r = 0$ or $r = 36$

In this problem we assume there is a wind. Thus, the speed of the wind must be greater than 0 and the number 0 cannot be a solution. If the speed of the wind is 36 mph, the speed of the airplane against the wind is 200 - 36, or 164 mph and the speed with the wind is 200 + 36, or 236 mph. The time against the wind is $\frac{738}{164}$, or $4\frac{1}{2}$ hr. The time with the wind is $\frac{1062}{236}$, or $4\frac{1}{2}$ hr. The total time is 9 hours. The value checks.

The speed of the wind is 36 mph.

25.

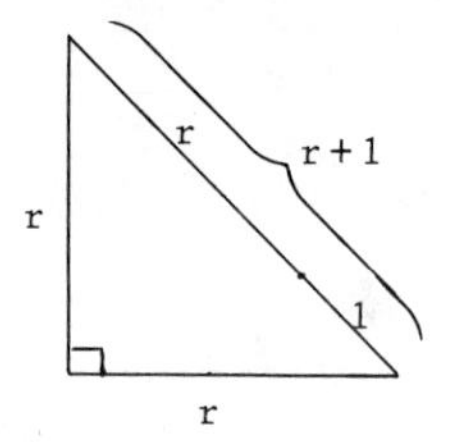

$r^2 + r^2 = (r + 1)^2$ (Using the Pythagorean equation)

$$2r^2 = r^2 + 2r + 1$$

$$r^2 - 2r - 1 = 0$$

$a = 1$ $b = -2$ $c = -1$

$$r = \frac{-(-2) \pm \sqrt{(-2)^2 - 4\cdot 1\cdot(-1)}}{2\cdot 1}$$

$$r = \frac{2 \pm \sqrt{4 + 4}}{2}$$

$$r = \frac{2 \pm \sqrt{8}}{2}$$

$$r = \frac{2 \pm 2\sqrt{2}}{2}$$

$$r = 1 \pm \sqrt{2}$$

$r = 1 + \sqrt{2}$ or $x = 1 - \sqrt{2}$

Since $1 - \sqrt{2}$ is negative, it cannot be the length of a side. Thus it cannot be a solution. If the length of a side is $1 + \sqrt{2}$, then the length of a hypotenuse is $2 + \sqrt{2}$. These values check using the Pythagorean equation. Thus, $r = 1 + \sqrt{2}$, or ≈ 2.41.

27. The radius of a 10" pizza is 5". The radius of a d" pizza is $(\frac{d}{2})$". The area of a circle is πr^2.

Area of d" pizza = Area of 10" pizza + Area of 10" pizza

$$\pi\left(\frac{d}{2}\right)^2 = \pi\cdot 5^2 + \pi\cdot 5^2$$

We solve the equation.

$$\frac{d^2}{4}\pi = 25\pi + 25\pi = 50\pi$$

$$\frac{d^2}{4} = 50$$

$$d^2 = 200$$

$d = \sqrt{200}$ or $d = -\sqrt{200}$
$d = 10\sqrt{2}$ or $d = -10\sqrt{2}$
$d \approx 14.14$ or $d \approx -14.14$

27. (continued)

Since the diameter cannot be negative, -14.14 is not a solution. If $d = 10\sqrt{2}$, or 14.14, then $r = 5\sqrt{2}$ and the area is $\pi(5\sqrt{2})^2$, or 50π. The area of the two 10" pizzas is $2\cdot\pi\cdot 5^2$, or 50π. The value checks.

The diameter of the pizza should be $10\sqrt{2}$, or $\approx$ 14.14 in.

The area of two 10" pizzas is approximately the same as a 14" pizza. Thus, you get more to eat with two 10" pizzas.

Exercise Set 11.9

1. The graph of $y = x^2$ opens upward since the coefficient of x^2, which is 1, is positive.

To graph $y = x^2$ we choose some numbers for x and then compute the corresponding values of y.

When $x = -2$, $y = (-2)^2 = 4$.

When $x = -1$, $y = (-1)^2 = 1$.

When $x = 0$, $y = 0^2 = 0$.

When $x = 1$, $y = 1^2 = 1$.

When $x = 2$, $y = 2^2 = 4$.

We plot the ordered pairs resulting from the computations and connect the points with a smooth curve.

x	y
-2	4
-1	1
0	0
1	1
2	4

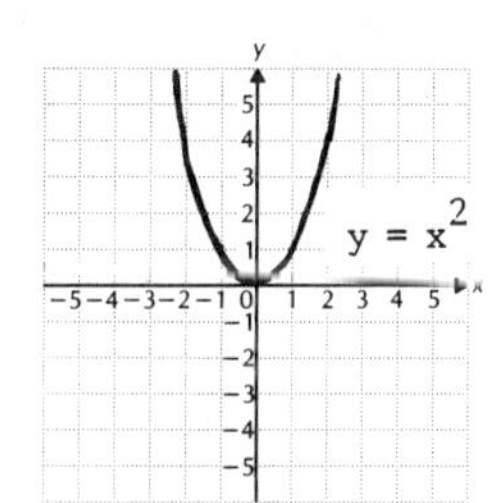

3. The graph of $y = -1\cdot x^2$ opens downward since the coefficient of x^2, which is -1, is negative.

To graph $y = -1\cdot x^2$ we choose some numbers for x and then compute the corresponding values of y.

When $x = -2$, $y = -1\cdot(-2)^2 = -1\cdot 4 = -4$.

When $x = -1$, $y = -1\cdot(-1)^2 = -1\cdot 1 = -1$.

When $x = 0$, $y = -1\cdot(0)^2 = -1\cdot 0 = 0$.

When $x = 1$, $y = -1\cdot 1^2 = -1\cdot 1 = -1$.

When $x = 2$, $y = -1\cdot 2^2 = -1\cdot 4 = -4$.

We plot the ordered pairs resulting from the computations and connect the points with a smooth curve.

x	y
-2	-4
-1	-1
0	0
1	-1
2	-4

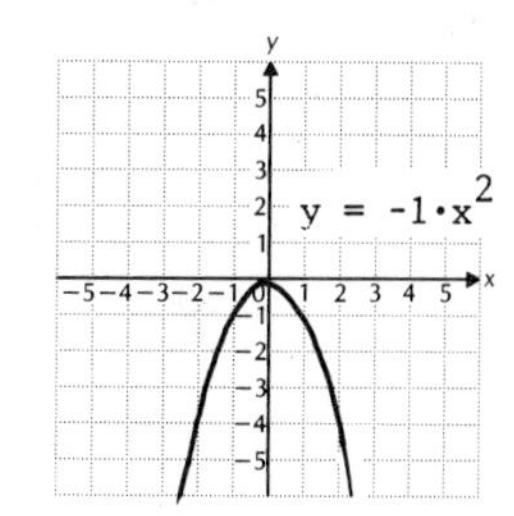

5. The graph of $y = -x^2 + 2x$ opens downward since the coefficient of x^2 is negative.

To graph $y = -x^2 + 2x$ we choose some numbers for x and then compute the corresponding values of y.

When $x = -1$, $y = -(-1)^2 + 2(-1) = -1 - 2 = -3$.

When $x = 0$, $y = -0^2 + 2\cdot 0 = 0 + 0 = 0$.

When $x = 1$, $y = -1^2 + 2\cdot 1 = -1 + 2 = 1$.

When $x = 2$, $y = -2^2 + 2\cdot 2 = -4 + 4 = 0$.

When $x = 3$, $y = -3^2 + 2\cdot 3 = -9 + 6 = -3$.

We plot the ordered pairs resulting from the computations and connect the points with a smooth curve.

x	y
-1	-3
0	0
1	1
2	0
3	-3

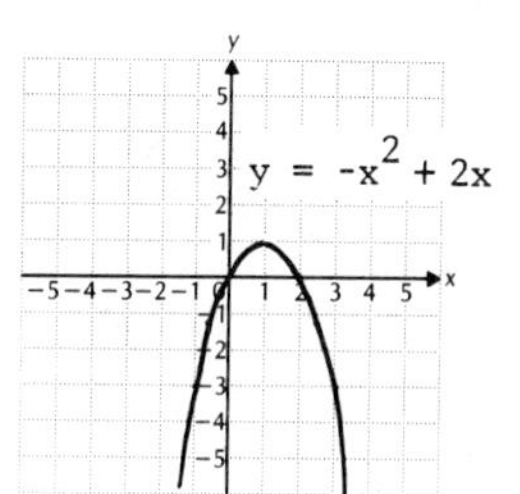

7. The graph of $y = 8 - x - x^2$ opens downward since the coefficient of x^2 is negative.

To graph $y = 8 - x - x^2$ we choose some numbers for x and then compute the corresponding values of y.

When $x = -4$, $y = 8 - (-4)-(-4)^2 = 8 + 4 - 16 = -4$.

When $x = -3$, $y = 8 - (-3) - (-3)^2 = 8 + 3 - 9 = 2$.

When $x = -2$, $y = 8 - (-2) - (-2)^2 = 8 + 2 - 4 = 6$.

When $x = -1$, $y = 8 - (-1) - (-1)^2 = 8 + 1 - 1 = 8$.

When $x = -\frac{1}{2}$, $y = 8 - (-\frac{1}{2}) - (-\frac{1}{2})^2$

$$= 8 + \frac{1}{2} - \frac{1}{4}$$

$$= 8\frac{1}{4}$$

When $x = 0$, $y = 8 - 0 - 0^2 = 8 - 0 - 0 = 8$.

When $x = 1$, $y = 8 - 1 - 1^2 = 8 - 1 - 1 = 6$.

When $x = 2$, $y = 8 - 2 - 2^2 = 8 - 2 - 4 = 2$.

When $x = 3$, $y = 8 - 3 - 3^2 = 8 - 3 - 9 = -4$.

We plot the ordered pairs resulting from the computations and connect the points with a smooth curve.

7. (continued)

x	y
-4	-4
-3	2
-2	6
-1	8
$-\frac{1}{2}$	$8\frac{1}{4}$
0	8
1	6
2	2
3	-4

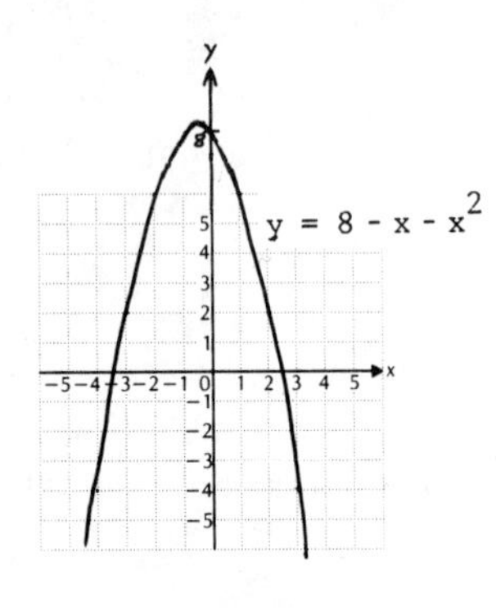

9. The graph of $y = x^2 - 2x + 1$ opens upward since the coefficient of x^2 is positive.

To graph $y = x^2 - 2x + 1$ we choose some values for x and then compute the corresponding values of y.

When $x = -1$, $y = (-1)^2 - 2(-1) + 1 = 1 + 2 + 1 = 4$.

When $x = 0$, $y = 0^2 - 2\cdot0 + 1 = 0 - 0 + 1 = 1$.

When $x = 1$, $y = 1^2 - 2\cdot1 + 1 = 1 - 2 + 1 = 0$.

When $x = 2$, $y = 2^2 - 2\cdot2 + 1 = 4 - 4 + 1 = 1$.

When $x = 3$, $y = 3^2 - 2\cdot3 + 1 = 9 - 6 + 1 = 4$.

We plot the resulting ordered pairs and connect them with a smooth curve.

x	y
-1	4
0	1
1	0
2	1
3	4

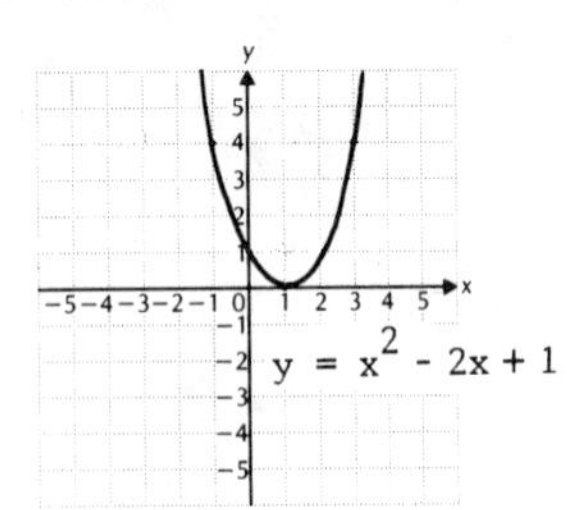

11. The graph of $y = x^2 + 2x - 3$ opens upward since the coefficient of x^2 is positive.

To graph $y = x^2 + 2x - 3$ we choose some values for x and then compute the corresponding values of y.

When $x = -4$, $y = (-4)^2 + 2(-4) - 3 = 5$.

When $x = -3$, $y = (-3)^2 + 2(-3) - 3 = 0$.

When $x = -2$, $y = (-2)^2 + 2(-2) - 3 = -3$.

When $x = -1$, $y = (-1)^2 + 2(-1) - 3 = -4$.

When $x = 0$, $y = 0^2 + 2\cdot0 - 3 = -3$.

When $x = 1$, $y = 1^2 + 2\cdot1 - 3 = 0$.

When $x = 2$, $y = 2^2 + 2\cdot2 - 3 = 5$.

We plot the resulting ordered pairs and connect them with a smooth curve.

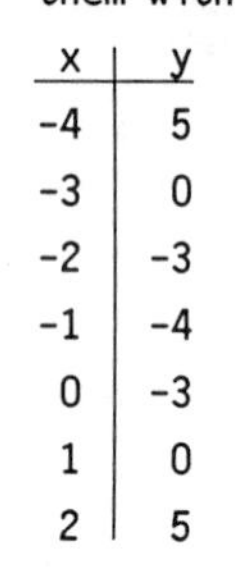

x	y
-4	5
-3	0
-2	-3
-1	-4
0	-3
1	0
2	5

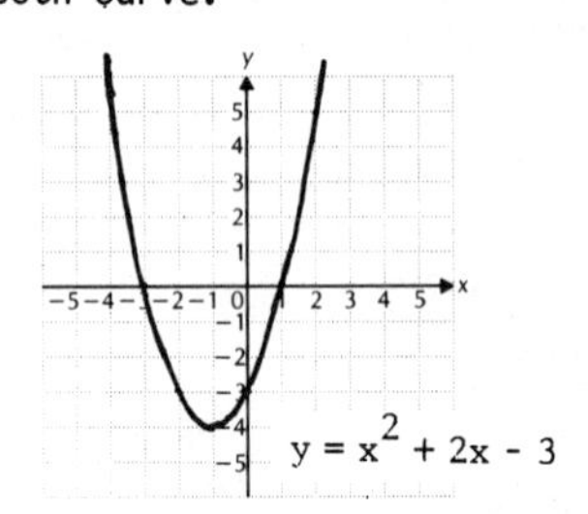

13. The graph of $y = -2x^2 - 4x + 1$ opens downward because the coefficient of x^2 is negative.

To graph $y = -2x^2 - 4x + 1$ we choose some values for x and then compute the corresponding values of y.

When $x = -3$, $y = -2(-3)^2 - 4(-3) + 1$
$= -18 + 12 + 1 = -5$.

When $x = -2$, $y = -2(-2)^2 - 4(-2) + 1$
$= -8 + 8 + 1 = 1$.

When $x = -1$, $y = -2(-1)^2 - 4(-1) + 1$
$= -2 + 4 + 1 = 3$.

When $x = 0$, $y = -2\cdot0^2 - 4\cdot0 + 1 = 0 - 0 + 1 = 1$.

When $x = 1$, $y = -2\cdot1^2 - 4\cdot1 + 1 = -2 - 4 + 1 = -5$

We plot the resulting ordered pairs and connect them with a smooth curve.

x	y
-3	-5
-2	1
-1	3
0	1
1	-5

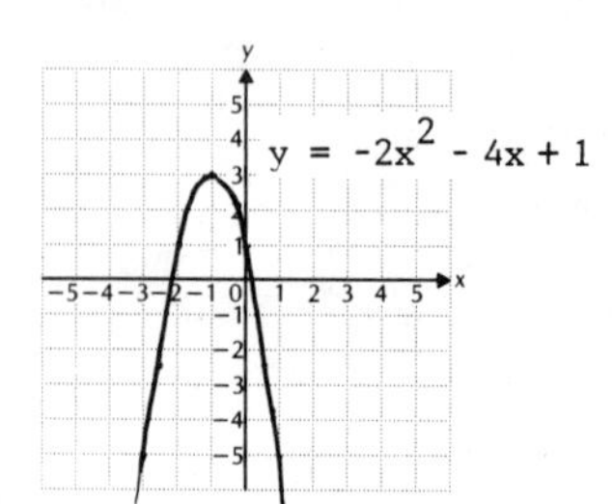

15. Graph $y = x^2 - 5$.

x	y
-3	4
-2	-1
-1	-4
0	-5
1	-4
2	-1
3	4

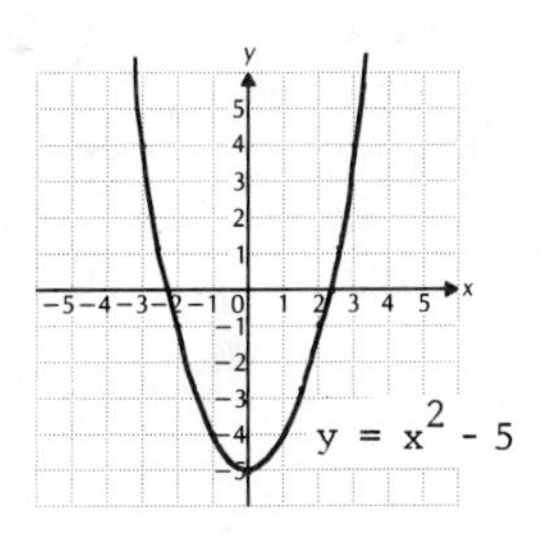

The graph crosses the x-axis at about (-2.2,0) and (2.2,0). These are the x-intercepts. So the solutions of the equation $x^2 - 5 = 0$ are about -2.2 and 2.2.

17. We graphed $y = 8 - x - x^2$ in Exercise 7 above. The graph crosses the x-axis at about (-3.4,0) and (2.4,0). These are the x-intercepts. So the solutions of the equation $8 - x - x^2 = 0$ are about -3.4 and 2.4.

19. Graph $y = x^2 - 8x + 16$.

x	y
2	4
3	1
4	0
5	1
6	4

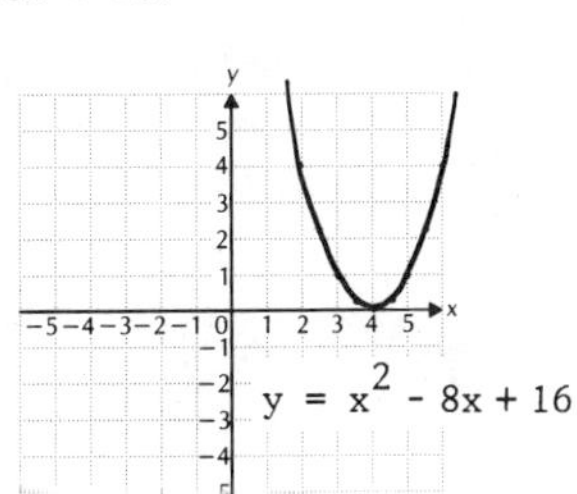

The graph intersects the x-axis at (4,0). The solution of the equation $x^2 - 8x + 16 = 0$ is 4.

21. Graph $y = x^2 + 8$.

x	y
0	8
-1	9
1	9

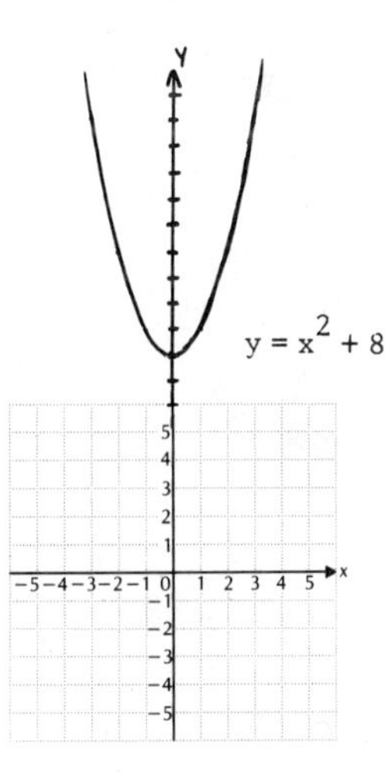

The graph does not cross the x-axis. The equation $x^2 + 8 = 0$ has no solution in the set of real numbers.

23. Graph $y = x^2 + 2x + 3$.

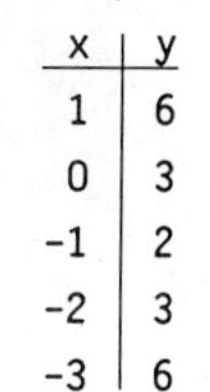

x	y
1	6
0	3
-1	2
-2	3
-3	6

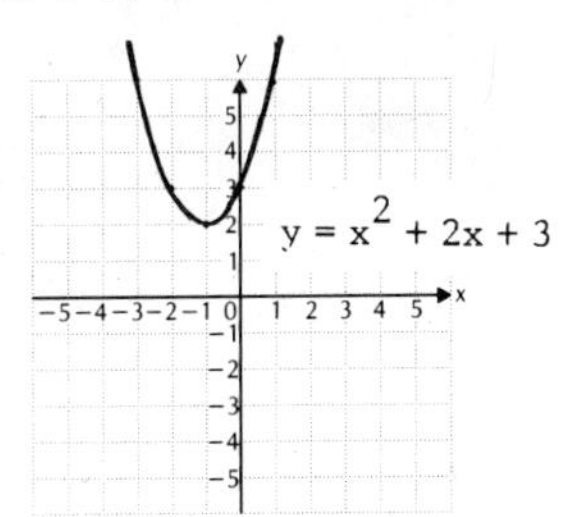

The graph does not cross the x-axis. The equation $x^2 + 2x + 3 = 0$ has no solution in the set of real numbers.

25. $\sqrt{20t^2} = \sqrt{4 \cdot 5 \cdot t^2} = 2t\sqrt{5}$

27. $y = ax^2 + bx + c$

$y = a \cdot 0^2 + b \cdot 0 + c$ (Substituting 0 for x)

$y = 0 + 0 + c$

$y = c$

The y-intercept is (0,c).